Preface

Many options exist for astronomy student lab experiences. Instructors ask students to perform a variety of digital activities, such as examining spectra and getting to know the night sky through an online planetarium. While these experiences can be useful, we believe that there is no substitute for getting students outside to use binoculars, telescopes, and even the naked eye to see the real thing. We feel it is our responsibility as astronomy instructors to instill a sense of wonder in students about the vastness of the universe and the amazing objects that populate it.

This lab manual incorporates hands-on activities so that students can develop critical skills. We emphasize sketching as one of those skills. Don't worry! Your students don't need to be a Van Gogh! Sketching is one of those skills that should help students notice fine detail as they draw what they see through a telescope.

We also have students perform activities that allow them to understand concepts in astrophysics. Examples include using an optical bench to demonstrate how lenses and mirrors work and using spectroscopes to explore the characteristics of light. Students should come away from this lab experience with a broad knowledge of various topics, including how to read star maps, how to interpret a Hertzsprung-Russell diagram, how to classify galaxies, how Kepler's and Hubble's laws work, and how to locate various celestial bodies in the sky.

Background information and procedural steps are clearly written to help students navigate their lab experiences. Data sheets are used in many places to facilitate the collection of data and sketches. Another feature you will find in this lab manual is the inclusion of sidebars to capture students' imaginations. Finally, we selected the beautiful illustrations and images students will encounter from the archives of *Astronomy* magazine, historical collections, the world's best astrophotographers, and observatories both here on Earth and in space.

Keep looking up!

— Mike and Michael

Acknowledgments

First and foremost, the authors would like to thank Holley Y. Bakich for the numerous illustrations she contributed to this work. Her ability to grasp an astronomical concept and transform it into an explanatory diagram never failed to amaze us.

We owe a huge thank you to the astroimagers and sketchers who allowed us to present the results of their hard work collecting light at the telescope and processing it into some of the most beautiful images of the sky ever taken or drawn. Specifically, we owe our gratitude to Anthony Ayiomamitis, Adam Block (for images taken at Kitt Peak National Observatory and for his more recent work at the Mount Lemmon SkyCenter operated by the University of Arizona), John Chumack, David J. Eicher, Bill and Sally Fletcher, Phillip Jones, Damian Peach, Jeremy Perez, Gerard Rhemann, Erika Rix, Chris Schur, Craig and Tammy Temple, and David Tyler.

Thanks to *Astronomy* magazine Editor David J. Eicher for permission to reproduce illustrations from *Astronomy*'s vast collection. Students who use this lab manual will benefit greatly because that artwork made concepts more understandable. Also thanks to telescope manufacturer Celestron for allowing us to use images of any product they make.

We'd like to thank the following reviewers who provided detailed suggestions for improvement:

- Katie J. Berryhill, American Public University System
- Vayujeet Gokhale, Orange Coast College
- Hal Jandorf, Moorpark College and Los Angeles Valley College
- John Salinas, Rogue Community College
- Teresa M. Schulz, Lansing Community College
- Larry Sessions, Metropolitan State University of Denver

We'd like also to thank Mike's former student, Yvonne James, now a master's student at Harvard University, for her comments.

Finally, we want to give a huge thank you to the staff of Morton Publishing. David Ferguson, Marta Martins, Rayna Bailey, Joanne Saliger, and Will Kelley made this entire process as fun and pain-free as we ever could have hoped for.

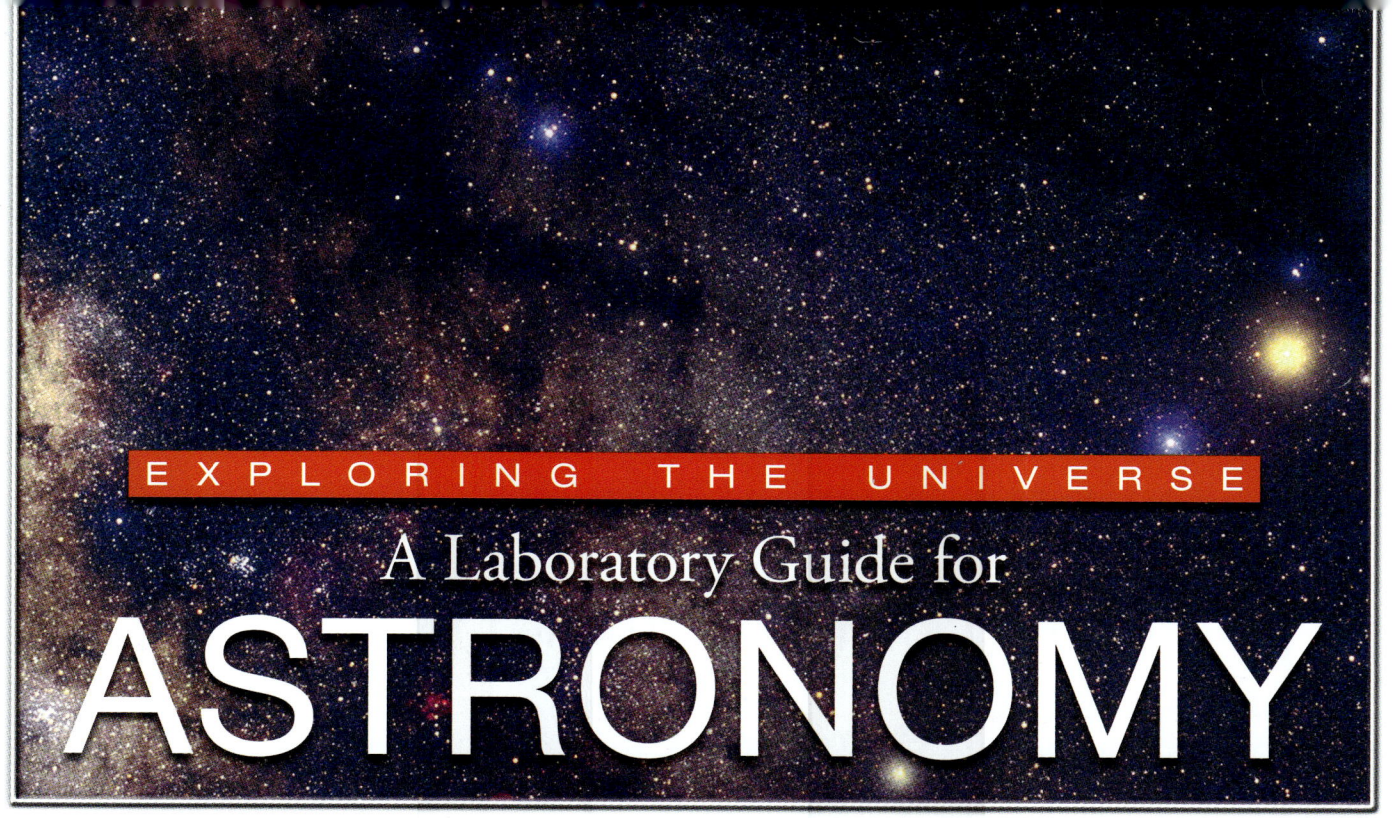

EXPLORING THE UNIVERSE
A Laboratory Guide for ASTRONOMY

Mike D. Reynolds **Michael E. Bakich**

MORTON
PUBLISHING

925 W. Kenyon Ave., Unit 12
Englewood, CO 80110

www.morton-pub.com

Book Team

Douglas N. Morton   Publisher
David M. Ferguson   President
Marta R. Martins   Acquisitions Editor
Rayna Bailey   Project Editor
Sarah D. Thomas   Editorial Assistant
Joanne Saliger   Production Manager
Will Kelley   Production Assistant

Cover images by Dragan Nikin and Bill and Sally Fletcher

Copyright © 2015 by Morton Publishing Company

All rights reserved. No part of this publication may be reproduced, stored in a retrieval system, or transmitted, in any form or by any means, electronic, mechanical, photocopying, recording or otherwise, without the prior written permission of the publisher.

Printed in the United States of America

10   9   8   7   6   5   4   3   2   1

ISBN-10: 1-61731-212-6
ISBN-13: 978-1-61731-212-0

Library of Congress Control Number: 2014947056

About the Authors

Mike D. Reynolds

earned his Ph.D. in science education and astronomy at the University of Florida. He has spent thirty-nine years in astronomy and space sciences in various positions, including high school and university instructor, planetarium and museum director, researcher, and college administrator. Reynolds has written numerous astronomy books and articles, including as an *Astronomy* magazine contributing editor. He has led numerous astronomical expeditions worldwide and has served as an invited speaker internationally at a variety of events, from book signings to lectures on meteorites, the science of science fiction, and general astronomy. Reynolds has also appeared on several Discovery Channel and National Geographic programs.

Reynolds' work in astronomy education has included not only the astronomy classroom and astronomy/STEM outreach for the general public, but also design and implementation of various college-level astronomy courses and astronomy laboratory courses. Additionally, his research has focused on formation of impact craters, both on Earth and throughout the Solar System. Mike has received a number of awards, including NASA's National Teacher-in-Space Finalist Award, Florida State Teacher of the Year Award, and an AstroOscar (an Astronomy Outreach Award).

Mike is Executive Director Emeritus of the Chabot Space and Science Center in Oakland, California, and currently serves as professor of Astronomy and Physics at Florida State College in Jacksonville, Florida.

Michael E. Bakich

has been fascinated with the stars all his life. His astronomical journey began when he was in third grade after his parents bought him a set of constellation flash cards. From that day forward, Michael's goal was to become an astronomer.

Michael realized that goal in 1975 when he graduated with a bachelor's degree in astronomy from The Ohio State University. Michael then attended Michigan State University, where he received a master of arts degree in planetarium education (one of only six such degrees ever awarded) in 1977.

During the past two decades, Michael has worked in seven planetaria and has served as a consultant in the planetarium field. He joined *Astronomy* magazine in February 2003.

Michael has written three books for Cambridge University Press (*The Cambridge Guide to the Constellations*, *The Cambridge Planetary Handbook*, and *The Cambridge Encyclopedia of Amateur Astronomy*) and one for Springer (*1,001 Celestial Wonders to See Before You Die*).

Michael is much sought after as a "tour guide" for eclipses, sky events, and historical astronomy sites. He has conducted tours to the Yucatan Peninsula in Mexico, viewing trips to space shuttle launches, a cruise to see the 1986 appearance of Halley's Comet in Tahiti, and total solar eclipse trips in the United States, Mexico, Peru, Europe, Russia, China, Easter Island, and Australia.

In his spare time, Michael enjoys woodworking, science-fiction movies, and book collecting. Currently, Michael's collection numbers more than 450 individual nineteenth-century first editions—one of the largest private collections anywhere.

Michael has logged thousands of hours observing the sky. He lives in Milwaukee with his wife, Holley.

Contents

UNIT 1 Tools of the Trade

1 Eyes 1
Anatomy of the Eye 2
EXERCISE **1.1** Visual Perception Activities 4
  Afterimages   4
  Blind Spot   5
  Focus   7
  Eye Dominance   8
  Peripheral Vision   8
EXERCISE **1.2** Visual Acuity Tests 10
  Standard Eye Test   10
  Astigmatism Test   11
  Color Vision Test   12
  Depth Perception Test   14

2 Geometrical Optics 15
Optical Components 16
Optics Calculations 18
EXERCISE **2.1** Mathematics of Geometrical Optics 21
EXERCISE **2.2** Lens and Mirror Images 23
EXERCISE **2.3** Optical Bench 24

3 Binoculars 31
Binocular Basics 32
Selecting the Right Equipment 37
EXERCISE **3.1** Comparison of Binoculars 39
EXERCISE **3.2** Estimating Field of View in the Classroom 41

4 Telescopes 43
Refracting Telescopes 44
Reflecting Telescopes 48
Catadioptric Telescopes 51
Telescope Mounts 52
Additional Telescope Accessories 55
EXERCISE **4.1** Getting Familiar with Your Telescope 58
EXERCISE **4.2** Lunar Observations and Impressions 60
EXERCISE **4.3** Constructing a Simple Telescope 60
  DATA SHEET **4.3** Simple Refracting Telescope 61

vii

UNIT 2 Looking Up

5 Star Maps 63
Star Charts.. 64
Star Atlases 70
EXERCISE **5.1** Using a Star Chart 76
EXERCISE **5.2** Using *Atlas of the Stars* Charts 77
    DATA SHEET **5.2** Using *Atlas of the Stars* Charts 85

6 This Season's Sky 89
EXERCISE **6.1** The Autumn Sky.................................... 91
    DATA SHEET **6.1** The Autumn Sky 93
EXERCISE **6.2** The Winter Sky 97
    DATA SHEET **6.2** The Winter Sky................................ 99
EXERCISE **6.3** The Spring Sky 103
    DATA SHEET **6.3** The Spring Sky 105
EXERCISE **6.4** The Summer Sky 109
    DATA SHEET **6.4** The Summer Sky 111

7 Outdoor Sky Observations 115
EXERCISE **7.1** January–February: A Hero for the Ages 116
    DATA SHEET **7.1** January–February 119
EXERCISE **7.2** March–April: Visit the Charioteer 121
    DATA SHEET **7.2** March–April 123
EXERCISE **7.3** May–June: The Big Dipper 125
    DATA SHEET **7.3** May–June 129
EXERCISE **7.4** July–August: A Tail's Tale 131
    DATA SHEET **7.4** July–August 135
EXERCISE **7.5** September–October: Strum the Harp 137
    DATA SHEET **7.5** September–October 141
EXERCISE **7.6** November–December: In the Queen's Court 143
    DATA SHEET **7.6** November–December 147

8 Sketching Techniques 149
Preparation and Materials 150
Techniques 151
EXERCISE **8.1** Sketching Asterisms 156
    DATA SHEET **8.1** Sketching Asterisms 159

9 Magnitude System and Light Pollution 167
The Magnitude System. 168
Light Pollution. 173
EXERCISE 9.1 Star Brightness Comparison 178
EXERCISE 9.2 Limiting Magnitude Evaluation 179
EXERCISE 9.3 Constellation Sketching With Light Pollution 180
DATA SHEET 9.3 Constellation Sketches 181

UNIT 3 Night and Day

10 The Moon 183
The Changing Face of the Moon 184
Features of the Moon 186
Observing the Moon 192
Sketching the Moon 194
EXERCISE 10.1 Lunar Calculations 197
DATA SHEET 10.1 Recording Lunar Data 199
EXERCISE 10.2 Craters of the Moon 203
EXERCISE 10.3 Unaided Viewing of the Moon 205
EXERCISE 10.4 Telescopic Viewing of the Moon 207
DATA SHEET 10.4 Recording Telescopic Observations of the Moon 209

11 The Sun 211
Observing the Sun 212
EXERCISE 11.1 Tracking the Sun Daily 218
DATA SHEET 11.1 Tracking the Sun Daily 219
EXERCISE 11.2 Estimating the Sun's Maximum Altitude 221
EXERCISE 11.3 Determining the Length of the Solar Day 222
DATA SHEET 11.2 Sun's Maximum Altitude Log 223
DATA SHEET 11.3 Length of Solar Days 225
EXERCISE 11.4 Classifying Sunspots 227
DATA SHEET 11.4 Classifying Sunspots 229

UNIT 4 The Solar System

12 Physical Features 231
Our Solar System 232
Classification of Objects in our Solar System 235
EXERCISE 12.1 Comparative Planetology 239
EXERCISE 12.2 Simulating Planetary Atmospheres 248
EXERCISE 12.3 In-Lab Planet Sketching 250
DATA SHEET 12.3 Planet Sketching 251

13 Kepler's Laws of Motion 253
Johannes Kepler's Astronomical Legacy 254
Kepler's Laws 256
EXERCISE **13.1** Applying Kepler's Laws 260
DATA SHEET **13.1** Working with Ellipses 261

14 Observing the Planets 263
Filters. 264
Mercury. 265
Venus 266
Mars: The Red Planet. 268
Jupiter: The King of the Planets 270
Saturn 273
Ice Giants 275
EXERCISE **14.1** Observing Planetary Characteristics 276
DATA SHEET **14.1** Planetary Features through the Telescope. 277
EXERCISE **14.2** Observing Planetary Motions. 279
EXERCISE **14.3** Spotlight on Jupiter. 281
DATA SHEET **14.3** Observing Jupiter 283

15 Minor Bodies 285
Comets 286
Asteroids. 292
EXERCISE **15.1** Making a Comet Model 297
EXERCISE **15.2** Comet Observing and Imaging 298
DATA SHEET **15.2** Observing and Imaging Comets 299
EXERCISE **15.3** Comparing Asteroids 301
EXERCISE **15.4** Asteroid Observing and Imaging. 302
DATA SHEET **15.4** Observing and Imaging Asteroids. 303

16 Meteorites 305
Identification and Classification of Meteorites 306
Starting a Collection 309
EXERCISE **16.1** Meteorite Sample Study. 312
DATA SHEET **16.1** Meteorite Sample Classifications 313
EXERCISE **16.2** Finding the Radiant of the Leonid Meteor Shower 319

17 Transits and Eclipses. 323
Transits 324
Lunar Eclipses 328
Solar Eclipses. 332
EXERCISE **17.1** Calculating Distances Using Transits 338
EXERCISE **17.2** Observing a Lunar Eclipse 340
DATA SHEET **17.2** Lunar Eclipse Observations. 341

UNIT 5 The Deep Sky

18 Stars 343
Double Stars 344
Variable Stars 353
EXERCISE 18.1 Observing Double Stars 360
EXERCISE 18.2 Observing Variable Stars 360
DATA SHEET 18.1 Double Star Observations 361
DATA SHEET 18.2 Variable Star Observations 365
EXERCISE 18.3 Creating a Light Curve for Algol 369

19 Non-Stellar Objects 371
Galaxies 372
Planetary Nebulae 378
EXERCISE 19.1 Classifying Galaxies 382
DATA SHEET 19.1 Galaxy Classification 383
EXERCISE 19.2 Classifying Planetary Nebulae 385
DATA SHEET 19.2 Planetary Nebulae Classification 387

UNIT 6 Classifying Celestial Objects

20 Spectroscopy 389
Historical Discoveries 390
Recent Discoveries 393
A Closer Look at Spectra 395
Wave Characteristics of Light 397
EXERCISE 20.1 Exploring Spectra 400
EXERCISE 20.2 Stellar Spectra Classification 403

21 The H-R Diagram 405
Luminosity and Temperature 406
Reactions in a Main Sequence Star 408
EXERCISE 21.1 Plotting Stars 411
DATA SHEET 21.1 Graph for Plotting Stars 413

22 Radioactivity 415
Radioactive Decay 416
Radioactivity Calculations 419
EXERCISE 22.1 Measuring Radioactivity 422
DATA SHEET 22.1 Ionization Events Count 423

23 Hubble's Law 425
The Relationship between Distance and Velocity 426
EXERCISE 23.1 The Expanding Universe 429
EXERCISE 23.2 Determining Galactic Distances 432

| Appendix | Locations of Major Telescopes | 441 |

Photo Credits **445**

Index **447**

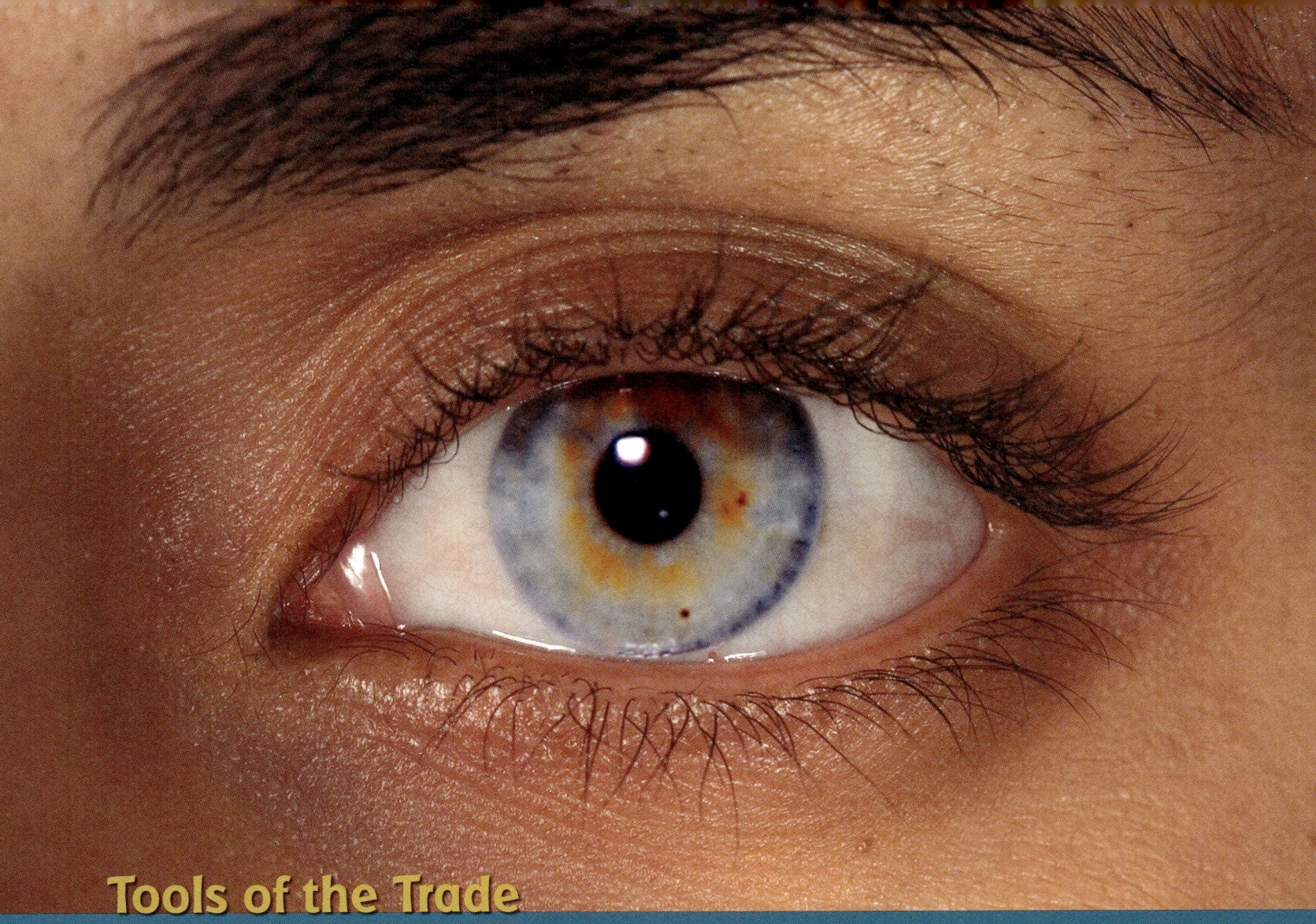

Tools of the Trade

Eyes

1

LEARNING OBJECTIVES

Upon completion of this chapter, you should be able to:

1. Identify the major parts of the human eye.
2. Compare the human eye to a telescope.
3. Explain the process that forms afterimages.
4. Describe why the human eye has a blind spot, and locate it.
5. Identify the prime factor responsible for "aging" of the human eye.
6. Determine if one eye is dominant, and identify which eye that is.
7. Use a standard eye chart to test vision, and describe what the number result means.
8. Conduct simple experiments to test for astigmatism, color blindness, and depth perception.
9. Define all glossary terms.

1

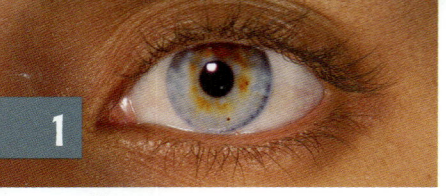

Human eye Courtesy of John Crawley

The eye is one of the organs through which we acquire knowledge of our environment. Our eyes collect light reflected from, or emitted by, objects around us. And although other types of radiation (infrared, ultraviolet, etc.) can enter the eye, the human brain can process information only from visible light.

Information provided by the eyes plays a dominant role in helping us interpret our environment. The power to recognize the contours, colors, and relations of objects with other objects indicates that vision does not merely consist of "seeing," but also of "perceiving and understanding" through the central nervous system. The eye is fundamentally an organizer, actively building a world of objects.

Before the late nineteenth century, all astronomy was visual. Both professional and amateur astronomers used only their eyes to examine the universe around them. And although the eye is a tiny light-gatherer, it is incredibly sensitive to both brightness and color. It also is adaptable. The human eye can function in bright sunlight and faint starlight—an intensity range of more than 10 million.

This lab will demonstrate how the human eye works. You will learn the main parts of the eye and some optical defects that can affect your views through a telescope. Also included are some simple eye tests for you to conduct.

Anatomy of the Eye

The eye consists of the **cornea**, which is the initial structure through which light passes; the **lens**, which focuses the light on the retina; the **iris**, which controls how much light enters; the **pupil**, which is the round hole in the center of the iris; and the **retina**, which senses the light and sends a signal to the brain. The structure of the eye is shown in Figure 1.1.

The retina has two types of specialized nerve cells known as photoreceptors, called rods and cones because of their shapes. **Rods** function in dim light and perceive shades of gray. Each eye contains about 120 million rods, located at the front of the retina. **Cones** function in bright light and provide sharp color images. Human eyes contain about six million to seven million cones concentrated around the center of each retina. There are three different types of cones—red, green, and blue—each sensitive to the color for which they are named. Different types of cones function together to interpret colors other than red, green, and blue. For example, in order for the eye to see yellow light, red and green cones must work together.

GLOSSARY TERMS

Cornea thick, transparent, nearly circular structure covering the lens; "front window" of the eye that is an important part of the eye's focusing system

Lens transparent structure located behind the iris; focuses light rays entering through the pupil to form an image on the retina

Iris pigmented (its color varies from pale blue to dark brown) membrane of the eye located between the cornea and the lens; controls the amount of light entering the eye

Pupil round black hole in the center of the iris; the pupil's size changes automatically to control the amount of light entering the eye

Retina thin, multilayered membrane lining the inside back two-thirds of the eye; contains millions of visual cells and connects to the brain by the optic nerve; receives light and sends electrical impulses to the brain that result in sight

Rod type of nerve cell called a photoreceptor that functions in low-light situations; does not perceive color but produces indistinct images in shades of gray

Cone type of nerve cell called a photoreceptor; cones produce sharp color images and function when your surroundings are well lit

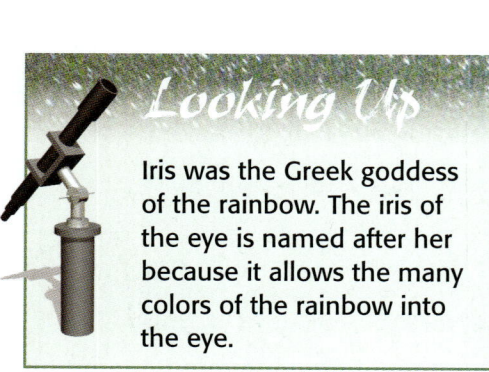

Looking Up

Iris was the Greek goddess of the rainbow. The iris of the eye is named after her because it allows the many colors of the rainbow into the eye.

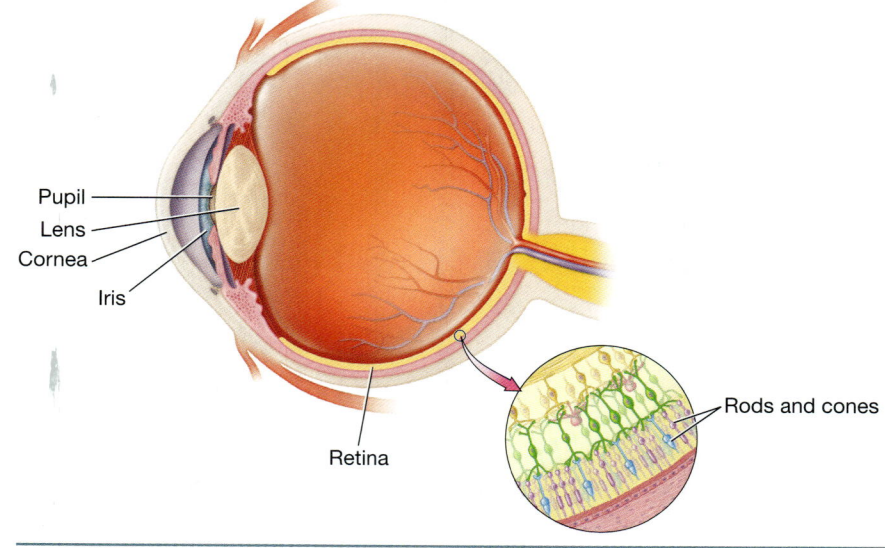

1.1 Anatomy of the human eye.

Check Your Understanding

1.1 What part of the eye lets light enter?

1.2 On what part of the eye does the lens focus images?

1.3 We see some objects because they produce light. Why do we see things that emit no light?

1.4 What is the main link between the eyes and the brain?

1.5. Figure 1.2 shows simple illustrations of anatomical and mechanical lenses. List some similarities among them.

1.6 List some differences among the three illustrations in Figure 1.2.

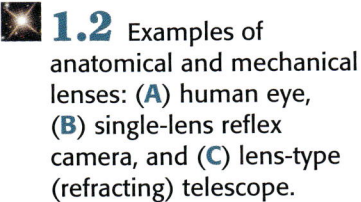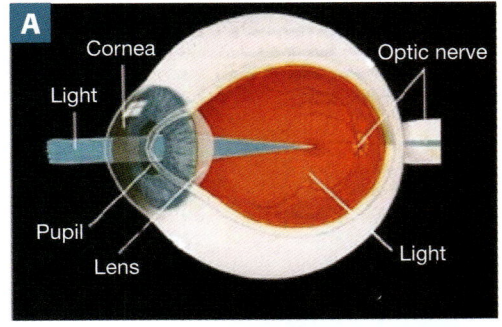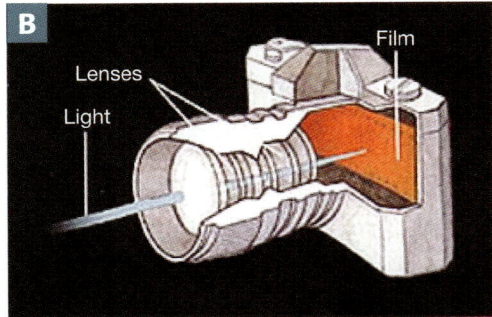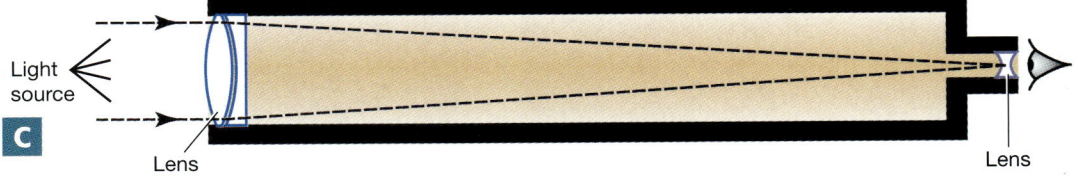

1.2 Examples of anatomical and mechanical lenses: (**A**) human eye, (**B**) single-lens reflex camera, and (**C**) lens-type (refracting) telescope.

Eyes CHAPTER 1 3

Exercise 1.1 — Visual Perception Activities

Afterimages

Sometimes, when you stare intently at an object and then look away, the brain retains an afterimage of what you were viewing that soon disappears. This happens because when you observe an object closely for a long time, the light stimulating the retina eventually fatigues a high percentage of the cones in that area, desensitizing that part of the retina. When you look elsewhere, the less-stimulated (non-fatigued) cones in the affected area still function. The fatigued cones continue to produce the afterimage for up to 30 seconds, but because stimulation (a new image) also is coming from the less-fatigued cones, you perceive the afterimage as a negative image.

In astronomy, afterimages can affect the way an observer sees color in a celestial object. For example, an astronomer may be studying a double star (two stars that appear close through a telescope's eyepiece). If one of the stars is bright blue, staring at that star for even a few seconds will desensitize your eye to blue light. The result is that the second star will appear a bit redder than it would otherwise appear.

 Procedure

Materials
- ❏ Pieces of transparent red and green vinyl
- ❏ Deck of illusion cards

1 Before beginning this procedure, answer the following question: Why do afterimages form?

2 With one eye closed, look at a room light through a piece of red transparent vinyl. After 30 seconds remove the red vinyl and shift your gaze to a piece of blank white paper. Record what afterimage color(s) you see.

3 Repeat step 2 with the other eye using the green transparent vinyl. Record what afterimage color(s) you see.

4 With both eyes open, focus on a king card from the deck of illusion cards (Fig. 1.3). After 60 seconds, shift your gaze from the king card to a piece of blank white paper. Record what afterimage color(s) you see.

1.3 King cards from a deck of illusion cards.

Blind Spot

The human eye contains no rods or cones where the optic nerve penetrates the retina, so no vision is possible in that location, resulting in a **blind spot**. We usually don't notice it because we are looking at things that are large and well lit. But when using a telescope to observe a small, faint object, if your eye's blind spot falls at the object's location, it will be difficult for you to see it. To solve this problem, observers generally move slightly the point where their eye is looking so the object's light doesn't fall on the blind spot. Astronomers call this technique **averted vision**.

In this procedure, you will determine the blind spot locations of both your eyes using a test diagram depicting a circle and cross. Note that the average blind spot in each eye is not centered; rather, it is off-center by 12° to 15° in the direction of the ear closest to that eye and about 1.5° below the horizontal.

> **GLOSSARY TERMS**
>
> **Blind spot** location (sometimes called the optic nerve head) where the optic nerve and major blood vessels exit the eye; the absence of rods or cones here causes a break in the visual field
>
> **Averted vision** a technique observers use to re-aim the eye's blind spot so it doesn't fall at the same location as the (usually faint) celestial object they are viewing; if the object is large or bright, there is no need to use averted vision

Procedure

Materials
- ❏ Blind spot test diagram

Part 1

1. Before beginning this procedure, answer the following questions:
   a. What causes the blind spot in the human eye?

   b. Where is the eye's blind spot located?

2. Either hold the diagram (Fig. 1.4) at arm's length with the circle to the right and the cross in front of your right eye or have your lab partner hold it for you.

3. Cover your left eye, and, while staring directly at the cross with your right eye, slowly move the diagram closer toward your face, continuing to stare at the cross until the circle seems to disappear. When the circle is no longer visible, stop and hold the diagram steady.

4. Have your lab partner measure and record the distance between your face and the diagram.

5. Turn the diagram over so that the cross is to the right and repeat steps 3 and 4 with your left eye open and your right eye closed.

   a. Left eye blind spot is _____ millimeters
   b. Right eye blind spot is _____ millimeters

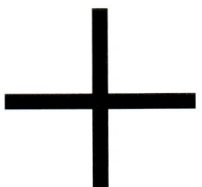

1.4 Blind spot test diagram.

Part 2

Using your blind spot measurements, you can determine the approximate size of the blind spot in each eye. To do this quickly, we make three assumptions that allow us to use simple mathematics:

1. The back of the eye is flat. (It is not.)
2. The distance from the eye lens to the retina is 17 mm. (This distance varies from person to person.)
3. Ignore the distance from the cornea to the lens.

Because of a mathematical relationship called the geometry of similar triangles, you can compare the triangle **ABC** directly to the triangle **CDE** (Fig. 1.5). When two lines cross, the opposing angles are congruent. Using the similar triangle rule, angles x and y in Figure 1.5 have the same measurement, therefore the ratio of line **DE** to line **AB** equals the ratio of line **CD** to line **BC**.

1 Perform the following measurements:

  a Measure the distance from **A** to **B**. This is the distance between the center of the cross and the circle in Figure 1.4:

   AB = _____ mm.

  b The distance from **A** to **C** (or **B** to **C**) is the distance you measured when the circle disappeared:

   AC = _____ mm.

  c Assume the distance from **C** to **D** is 17 mm, so **CD** = 17 mm.

2 Find the size of the blind spot, which is the distance from **D** to **E**.

3 Remember that the similar triangle rule states that the ratio of **DE** to **AB** equals the ratio of **CD** to **BC**. Write this in equation form:

$$\frac{DE}{AB} = \frac{CD}{BC}$$

This also can be written as $DE = \frac{(CD \times AB)}{BC}$.

4 Begin with your right eye. Using the formula in step 3, replace **CD**, **AB**, and **BC** with the numbers arrived at in step 1, and use that information to calculate **DE**.

The blind spot in your right eye is _____ mm across.

5 Repeat steps 1 through 4 for your left eye.

The blind spot in your left eye is _____ mm across.

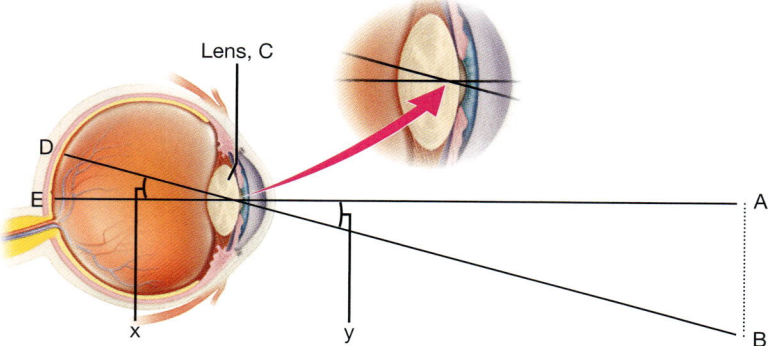

1.5 Geometry of similar triangles.

Courtesy of Holley Y. Bakich

Focus

The eye focuses on objects at different distances by changing the shape of its lens. To **focus** on a distant object, muscles in the eye flatten the lens. To focus on a closer object, the lens becomes more rounded. Therefore, the elasticity of the lens determines how well the eye can focus. Elasticity declines with age—a younger lens can focus more easily on closer objects. The following procedure will determine the "age" of your eyes' lenses.

When we use an astronomical instrument, we want the object we are looking at to be as sharp as possible. Telescopes and binoculars focus a bit differently than the human eye. Their lenses and mirror cannot change shape, so manufacturers build into them standard parts called focusers—mechanical devices that allow you to move the eyepiece(s) closer to the main optics.

> **GLOSSARY TERM**
>
> **Focus** point that lies at the focal length of a lens or mirror; in this chapter, focus specifically refers to the lens in the eye; also called the focal point

Procedure

Materials
- ❏ Meterstick with centimeter scale

1 Before beginning this procedure, answer the following questions:

   a How does the eye focus?

   b What factor changes the eye's elasticity?

If you use no corrective eyewear, complete steps 2 through 5 one time only. If you wear contact lenses, complete steps 2 through 5 one time only with your contacts in. If you wear eyeglasses, complete the steps once with your glasses on, and then repeat them with your glasses off.

2 Place the "0" end of a meterstick on your cheek near the outer edge of your left eye, facing forward away from your nose. Cover your right eye with a blank card.

3 Have your partner hold a pencil so that the sharp end is pointed up, and then slowly slide the pencil along the outer edge of the meterstick toward your left eye.

4 When you can no longer focus on the pencil, immediately have your partner stop moving the pencil and record the number on the meterstick where the pencil point stopped. This is the focus distance.

Near point focus distance for left eye = _____ cm.

5 Repeat steps 2 through 4 with your right eye.

Near point focus distance for right eye = _____ cm.

6 The nearest distance at which the pencil point can be clearly seen (i.e., where it first blurs) is called the near point. Use Table 1.1 to determine the age of your left and right eyes.

   a How "old" is your right eye?

   b How "old" is your left eye?

TABLE 1.1 Age of Eye Determined by Near Point

Near Point (cm)	Age of Eye (years)
9	10
10	20
13	30
18	40
50	50
83	60

Do you wear eyeglasses? If so:

    c How "old" is your right eye without glasses?

    d How "old" is your left eye without glasses?

Eye Dominance

You are familiar with preferences for using a particular hand for tasks such as writing or throwing. Similarly, human eyes also exhibit right-left dominance. Astronomical observers tend to prefer using their dominant eye when looking through a telescope.

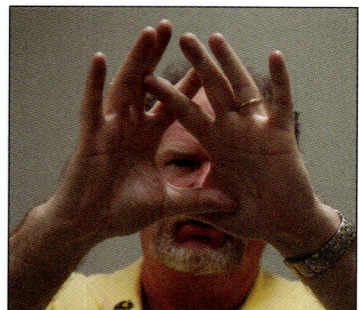

1.6 Checking for your dominant eye.

Procedure

1 Hold your hands at arm's length in front of you and overlap your fingertips and thumbs to form a triangle-shaped opening between your hands as shown in Figure 1.6.

2 With both eyes open, center an object located 3 to 5 meters away in the opening created by your overlapping hands, and focus your eyes on that object.

3 Without moving your hands, close one eye and then the other.

    a With which eye is the object still visible through the opening?

    b With which eye is the object obscured?

    c The eye to which the object is visible is your dominant eye. If the object remained visible for both eyes, you have central dominance. Do you have right, left, or central dominance?

4 Also try this: Have your lab partner look straight at you through your hand triangle to see which eye he or she can see. That's your dominant eye!

Peripheral Vision

Peripheral vision is the ability to see things that fall outside of your direct line of vision. Rods, which are more numerous at the periphery (edge) of the retina, are responsible for this aspect of vision. Peripheral vision is better at detecting motion than forming a clear detailed image, and it tends to be stronger in the dark.

    In astronomy, peripheral vision is more important when you are not using optical equipment like binoculars or a telescope. The greater your peripheral vision, the better chance you will have of detecting an event, for example, a meteor, at the edge of your eyes' field of view.

Procedure

Materials
- ❏ Peripheral vision disk
- ❏ Sight card

1. Before beginning this procedure, answer the following question: Is peripheral vision better during the daytime or at night?

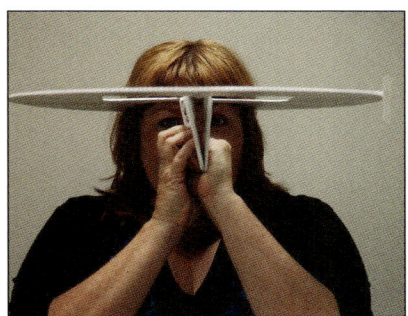

2. Sit at a table or desk and place the peripheral vision disk against your forehead, holding it as shown in Figure 1.7.

3. Brace your elbows on a table or desk top and stare straight ahead at the hole in the disk's focus marker.

4. Have your lab partner move the arm of the vision disk around until it is behind your line of sight. Your lab partner will now clip a sight card to the vision disk arm, and then slowly move the sight card forward from beyond your vision on your left side toward the point where you can see the card.

1.7 Using the peripheral vision disk.

5. Continue looking straight ahead at the hole in the vision disk's focus marker during this exercise.

6. Immediately tell your lab partner when you first see the sight card. Have your lab partner stop moving the vision disk arm and record in the "Detect" column in Table 1.2 the number on the vision disk triangle where the arm is stopped. Note that for this exercise, "detect" means that you can see the sight card at the edge of your vision, but you cannot see it well enough to "read" what it says.

7. Continue focusing straight ahead while your partner resumes slowly moving the vision disk arm around until you can read the information on the sight card. As soon as you can read the card, your lab partner should immediately stop moving the vision disk arm and record in the "Read" column in Table 1.2 the number on the vision disk triangle where the arm is stopped.

8. Repeat steps 4 through 6 for your right eye.

9. Trade places with your lab partner, and again complete steps 2 through 8. Record your partner's data for each eye in Table 1.2.

TABLE 1.2 Peripheral Vision Data

	You		Your Lab Partner	
	Left Eye	Right Eye	Left Eye	Right Eye
Detect				
Read				

Exercise 1.2 Visual Acuity Tests

Standard Eye Test

Perhaps you have heard someone talk about 20/20 vision. These numbers refer to the standard method of determining visual acuity. In 1862, Dutch ophthalmologist Herman Snellen (1834–1908) developed the first eye chart (Fig. 1.8A). On the Snellen chart, the only variable is the size of the letters. In 1976, two Australian ophthalmologists, Ian Bailey and Jan Lovie-Kitchin, published a new variation of the Snellen eye chart, called the logMAR chart, in the shape of an inverted triangle (Fig. 1.8B).

Acuity on the Snellen chart is represented as a fraction with the top number in the fraction being the distance in feet (20) the chart is from the person taking the eye test. The bottom number is the size of the letters in millimeters. Vision experts classify "20/20" as normal vision; thus, when a person is 20 feet from the eye chart and able to read letters 20 millimeters in size he or she has normal vision. Someone who cannot read the largest letter has 20/200 vision (20 feet from the chart, letters 200 millimeters in size) and is considered legally blind if his or her vision is not correctable with eyeglasses or contact lenses. A logMAR chart works similarly but at a distance of six feet.

Procedure

Materials
- ❏ Snellen or logMAR eye chart

1. Before beginning this procedure, answer the following question in the space provided: Based on a Snellen or logMAR eye chart, what numbers represent normal human vision?

2. Using a standard Snellen or logMAR eye chart, determine your visual acuity with both eyes, and then with each eye individually. If you wear eyeglasses, take them off. If you wear contacts, leave them in.

   a. Position the eye chart on a wall or door. Stand exactly 20 feet from the chart if you are using the Snellen chart and exactly 6 feet if you are using the logMAR eye chart. Have your lab partner determine the lowest line you can read clearly with both eyes.

1.8 Examples of (**A**) Snellen chart, and (**B**) logMAR chart.

b Cover your left eye with a card and repeat step a.

  c Cover your right eye with a card and repeat step a.

3 Record the data for your eyes in Table 1.3.

4 Trade places with your partner and complete steps 2a–c. Record the data for your partner's eyes in Table 1.3.

TABLE **1.3** Visual Acuity Data

You			Your Lab Partner		
Left	Right	Both	Left	Right	Both
20/_____	20/_____	20/_____	20/_____	20/_____	20/_____

Astigmatism Test

Astigmatism is a condition in which either the cornea or the lens has an irregular shape. This causes incoming light rays to bend improperly, so they do not focus at a specific point on the retina, resulting in blurred or distorted images. Many people have some degree of astigmatism, whether or not they realize it.

But human eyes aren't the only things prone to optical problems. Take the Hubble Space Telescope (HST), which went into orbit April 24, 1990. Shortly thereafter, engineers discovered that the telescope would not focus properly because of a flaw in its optics (Fig. 1.9A). Both of HST's primary cameras showed the same distortion, which originated in the telescope's main mirror. Astronauts aboard the space shuttle Endeavour installed two corrective devices in HST's optical path in December 1993, fixing the problem (Fig. 1.9B).

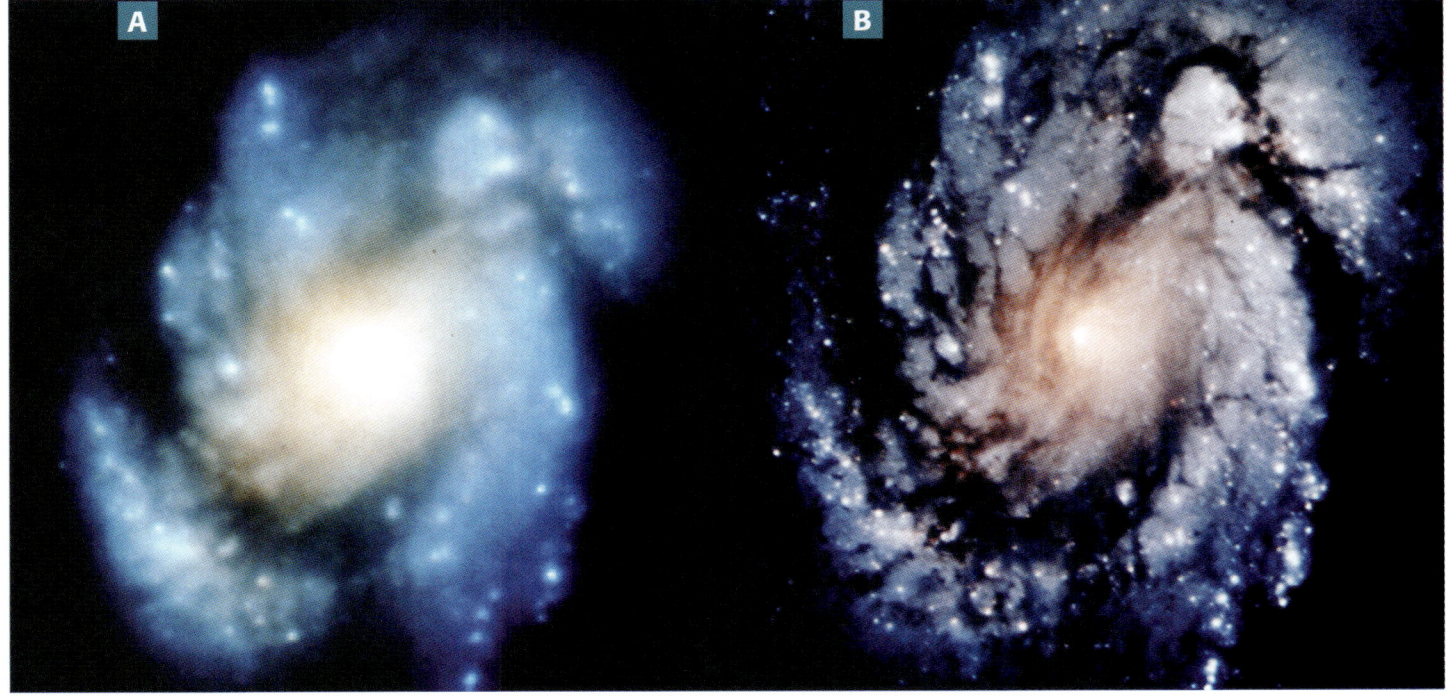

1.9 Spiral galaxy M100 taken by the Hubble Space Telescope: (**A**) before, and (**B**) after astronauts installed a corrective lens in the system.

Photo courtesy of NASA/STScI

 **Procedure**

Materials
❑ Figure 1.10

1 Using an astigmatism chart (Fig. 1.10), note whether you see all the lines equally clear or distinct, with each eye individually and then with both eyes.

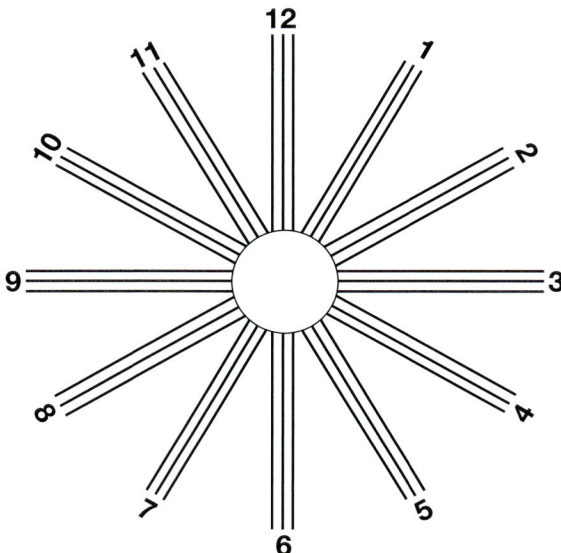

1.10 Astigmatism chart.

2 Record the data for each lab team member in Table 1.4.

TABLE **1.4** Astigmatism Data

You			Your Lab Partner		
Left	Right	Both	Left	Right	Both

Color Vision Test

In astronomy, if you cannot perceive colors correctly, your observations will not agree with those of others. Knowing the degree to which their eyes mistake colors allows astronomers to correct their reports so other researchers can use them.

If any of the three types of cones in your eyes—red, green, or blue—do not function properly, color perception deficiency (often called color blindness) occurs. Total color blindness is rare, but partial states frequently occur, especially among men. The most common type is red-green color blindness, in which an individual cannot distinguish between red and green light.

One of the first—and still most commonly used—tests for red-green color blindness is the Ishihara Color Test, named after its designer, Dr. Shinobu Ishihara

Looking Up

Another way of testing for color blindness is with colored yarns, known as the Holmgren Yarn or Wool Test. Swedish physiologist Alarik F. Holmgren (1831–1897) developed this test in 1874. People with some degree of color blindness are unable to distinguish between specific colored yarns.

(1879–1963), a professor at the University of Tokyo who published his results in 1917. This test consists of 38 colored plates. Another test is the Pseudoisochromatic Plate Ishihara Compatible (PIPIC) Color Vision Test, consisting of 24 plates. The PIPIC Color Vision Test can determine all color deficiencies.

Procedure

Materials
- ❑ Figure 1.11

1. Before beginning this procedure, answer the following questions:

    a Which nerve cells in the retina respond to color?

    b Which colors of cones (red, green, blue) must combine to allow us to see yellow light?

    c Which sex (male or female) suffers most from color blindness?

2. Write down the number you see within each square in Figure 1.11.

    a
    b
    c

3. Does the test indicate that you have any form of color blindness? If yes, describe it in the space provided below.

Note
This simple test is just a demonstration. For a diagnosis, you should visit a vision care professional.

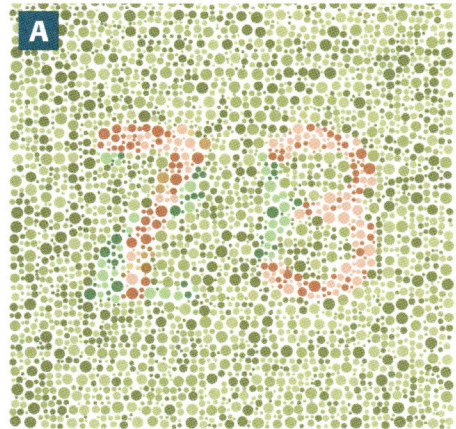  

 1.11 Color blindness tests.

Courtesy of Good-Lite Co.

Depth Perception Test

Depth perception is the visual ability to perceive the world in three dimensions. Although any animal capable of moving around its environment must be able to sense the distance of objects in that environment, the term *perception* is reserved for humans, who are the only animals able to tell each other about their experiences of distances.

In astronomy, depth perception is most important when you are using just your eyes or viewing large objects through binoculars. Depth perception can help you resolve shapes. It is less important when you observe through a telescope. The (usually) high magnification tends to give a flat view because the objects generally lie so far away.

Procedure

Materials
- ❑ Carolina Biological® depth perception tester
- ❑ Ruler

The depth perception tester comes with two background cards, white and black, that fit into the plastic nubs at the end of the tool. You also can use the tool without a card as a background. Do the following procedure with the black card in place.

1. Line up the arrows and see where they appear equal when looking at the top as shown in Figure 1.12.
2. From the side, use the cords to move the arrows out of alignment and then realign them.
3. Use a ruler to be sure the distance from your eyes to the depth perception tool is the same for all three tests.
4. Repeat steps 1 through 3 with the white card in place.
5. Repeat steps 1 through 3 with no card in the tool.
6. Record your data in Table 1.5.
7. Repeat steps 1 through 5 for your lab partner.
8. Record your partner's data in Table 1.5.
9. When finished with the test, wrap the cords carefully around the tool before storing.

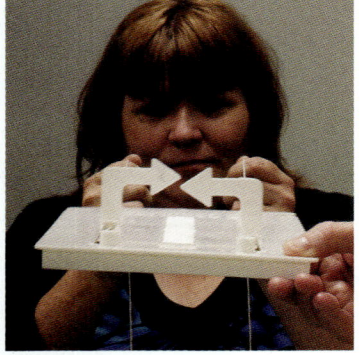

1.12 Using the Carolina Biological® depth perception tester.

TABLE **1.5** Depth Perception Data

You			Your Lab Partner		
Black	White	None	Black	White	None

Tools of the Trade

Geometrical Optics

LEARNING OBJECTIVES

Upon completion of this chapter, you should be able to:

1. Identify the most common types of mirrors, lenses, and lens combinations.
2. Compare and contrast how lenses and mirrors affect light passing through or reflecting off of them.
3. Explain how light-gathering power affects the brightness of an image.
4. Describe the differences between first- and second-surface mirrors.
5. Determine the focal length, object distance, or image distance in an optical system when two of the quantities are known.
6. Determine various characteristics of optics using an optical bench.
7. Define all glossary terms.

2 Prism refracting light
Courtesy of Rapho Agence/Science Source

GLOSSARY TERMS

Electromagnetic wave energy generated by variations of electric and magnetic fields; visible light is the best-known type of electromagnetic wave

Optical bench platform used to support equipment for optical experiments

Lens piece of transparent material (such as glass, quartz, or plastic) that has two opposite surfaces, either both curved or one curved and the other plane (flat), which is used either singly or combined in an optical instrument for forming an image by focusing rays of light

Concave hollowed or curving inward; in optics this refers to the shape of a mirror or lens

Convex bowed or curving outward; in optics this refers to the shape of a mirror or lens

Aperture diameter of a lens or mirror

Optics are all around us. As children, you may have played with magnifying glasses, small telescopes, or toy microscopes. Most of us use some sort of optical system or components numerous times each day. Many of us wear eyeglasses or contact lenses. We use mirrors in various everyday applications, from rearview mirrors in automobiles to bathroom and cosmetic mirrors. Cameras are not only stand-alone units but we also now find them in nearly every cell phone and many computers, tablets, and similar devices.

In astronomy and physics, geometrical optics describes how light interacts with lenses and mirrors. Light is described as an **electromagnetic wave**, which is a wave with predictable electric and magnetic properties. How light interacts with lenses and mirrors can be described by a concept called a light ray. A light ray represents light as it travels in a straight line until it strikes matter. At this point, whatever the light ray hits can absorb, reflect, or refract the light. A light ray is shown on paper by a thin line or thin parallel lines. The line is usually indicated by an arrow that shows the direction of the incoming light.

This lab will demonstrate how lenses and mirrors work. You will calculate focal length, object distance, and image distance. You also will use an **optical bench**.

Optical Components

Lenses

The exercises in this chapter deal with thin lenses, which have a small central thickness. To help your understanding, you also can represent lenses on paper as a cross section. Imagine taking a **lens** out of a pair of eyeglasses and cutting the lens in half so you can see its curve, or cross section. Lenses come in several shapes, but are primarily three main types: flat, **concave** (the surfaces curve inward), or **convex** (the surfaces curve outward).

Some lenses are combinations of the three main types (Fig. 2.1). They include double concave (also called biconcave), double convex (also called biconvex), convexo-concave (also called negative meniscus), plano-concave, plano-convex, and concavo-convex (also called positive meniscus, in which the convex part of the lens is more curved than the concave side of the lens).

To describe the diameter of a lens or mirror, scientists and engineers often use the term **aperture**. Just remember that "aperture" equals "size."

Looking Up

How old are lenses?

- The oldest known lens is the Nimrud lens of ancient Assyria, perhaps used as a magnifying glass or to start fires.
- Egyptian hieroglyphs dating back to the eighth century BCE depict simple lenses.
- In 424 BCE, the Greek writer Aristophanes, in his play *The Clouds*, mentions a lens used to focus sunlight to start fires.
- Pliny, who lived from 23 to 79 CE, recorded the use of lenses in the Roman Empire.
- Reading stones, used in the eleventh to thirteenth centuries, were made from glass spheres cut in half.
- Spectacles, better known today as eyeglasses, or simply glasses, probably made their first appearance in Italy in the late thirteenth century.

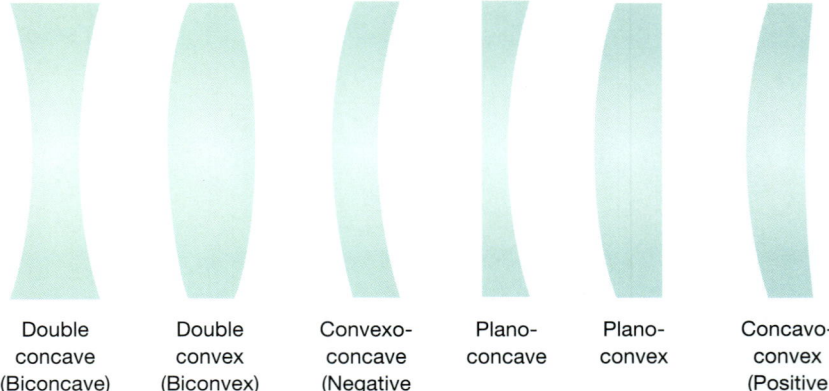

2.1 Configurations of simple lenses.
Courtesy of Holley Y. Bakich

Mirrors

Mirrors also come in the three basic shapes: flat, concave, and convex (Fig. 2.2). However, they function differently than lenses. Concave mirrors reflect and converge light, whereas convex mirrors reflect and diverge it. Flat mirrors cause light to simply be reflected.

Where the manufacturer applies a mirror's reflective coating determines how it interacts with incoming light. A **first-surface** mirror (Fig. 2.3A) has a coating on its exposed surface, so the light interacts with the coating, reflecting light off its front surface. In **second-surface** mirrors (Fig. 2.3B), the coating is on the back of the glass surface, so the light first passes through the glass, then reflects off the back surface, and again passes through the glass. Many everyday items—for example, bathroom mirrors and rearview mirrors in cars—are second-surface mirrors. Fine optical instruments, such as reflecting telescopes, employ only first-surface mirrors, which eliminates the extra passage (and bending) through the glass.

How an optic interacts with light depends on its shape. Concave lenses of all types (plano-concave, double concave, or negative meniscus) cause light rays to converge, or pinch inward. Convex lenses (plano-convex, double convex, or positive meniscus) cause light to diverge, or spread out. The lens bends the light as it passes through it.

> **GLOSSARY TERMS**
>
> **Mirror** optical component, generally covered on its figured end with a very thin coat of aluminum; designed to reflect light
>
> **First-surface** mirror with the reflective surface above the glass backing
>
> **Second-surface** mirror with the reflective surface below the glass backing

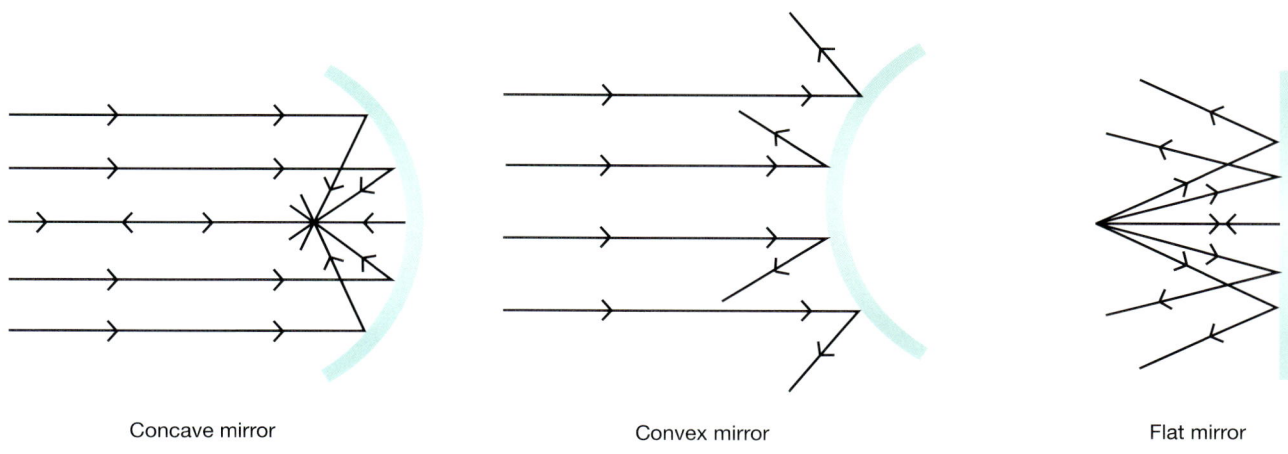

2.2 Three types of mirrors.

Courtesy of Holley Y. Bakich

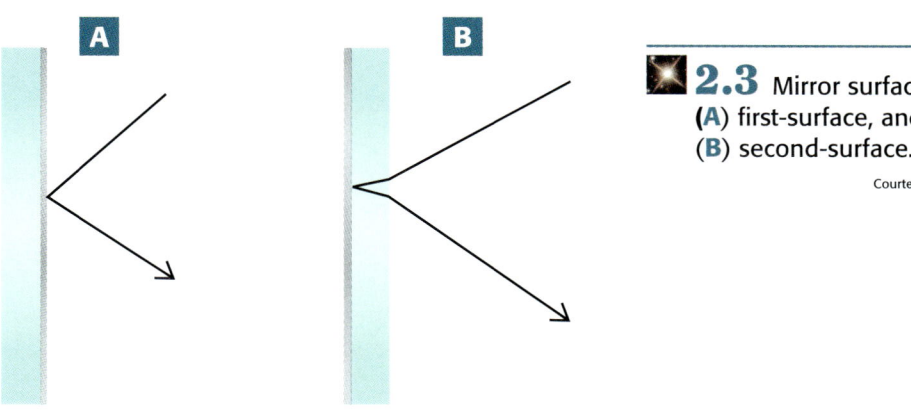

2.3 Mirror surface types: **(A)** first-surface, and **(B)** second-surface.

Courtesy of Holley Y. Bakich

Check Your Understanding

1.1 List three examples of optical items you use on a regular basis, and how you use each.

1.2 What do scientists use to describe how light interacts with optics?

1.3 How far back can the use of lenses be traced?

1.4 List six types of lenses.

1.5 Which type of mirror, first-surface or second-surface, is best for astronomy? Explain your answer.

Optics Calculations

Focal Point

> **GLOSSARY TERM**
> **Focus** point that lies at the focal length of a lens or mirror; also called the focal point

When working with converging lenses and concave mirrors, you can determine the **focus** of the converging lens (or concave mirror) and how the image forms, that is, how you see it. Where the lens or mirror comes to focus is the focal point, represented as x (Fig. 2.4). You can visualize the source as being either relatively close or at what appears as an infinite distance (such as with astronomical objects).

In geometrical optics, a real image is one in which the outgoing rays from points on the object pass through a single point behind the lens. A virtual image is an image in which the outgoing rays from points on the object diverge. The virtual image will appear to converge in the lens or mirror. With both real and virtual images, a light ray, or series of light rays, usually represent the route through which light actually passes.

Light reflected by a concave mirror comes to a focal point (x). For spherical concave mirrors, the focal point is equal to one-half the radius of the curvature of the mirror. As

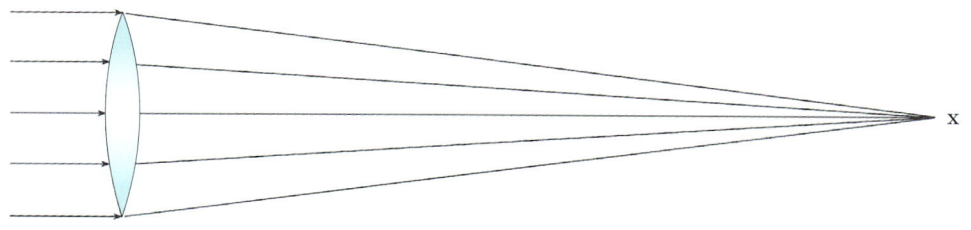

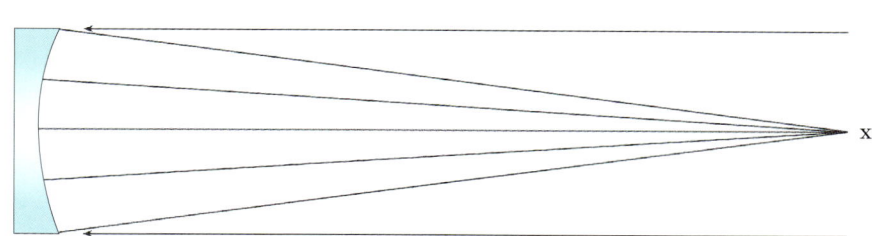

2.4 Examples of focal point.
Courtesy of Holley Y. Bakich

As previously noted, convex mirrors, like concave lenses, cause light to diverge. For spherical convex mirrors, the focal point is also precisely equal to one-half the radius of the curvature of the mirror, but it is negative.

The **focal length** (f) of a lens or mirror is related to the object's distance (d_o) and the image's distance (d_i) by the equation:

$$\frac{1}{f} = \frac{1}{d_o} + \frac{1}{d_i}$$

GLOSSARY TERM

Focal length distance from a lens or mirror to the focus in an optical system

Consider a distant object, such as a star, a galaxy, or some other astronomical object. If $\frac{1}{d_o}$ is a large number, then $\frac{1}{d_i}$ becomes very small:

$$\frac{1}{f} = \frac{1}{d_o} + \frac{1}{d_i} = \frac{1}{d_i} \rightarrow f = d_i$$

This means that at astronomical distances (usually meaning any distance beyond our solar system), the image distance (d_i) equals the optic's focal length (f).

Magnification

The **magnification** (M) of the lens (Fig. 2.5) is calculated by:

$$M = -\frac{d_i}{d_o} = \frac{f}{f - d_o}$$

GLOSSARY TERM

Magnification process of enlarging the image of an object and not the object itself

When calculating magnification, if $M > 1$, the image produced is larger than the object. For astronomical objects, d_o is many, many times greater than d_i. Since the denominator is many, many times larger than the numerator, you can say that the magnification (M) is zero.

The ratio of the image height, represented by h_i, to the object height, represented by h_o, is given as:

$$M = \frac{h_i}{h_o}$$

Note the relationship between the two equations:

$$M = \frac{h_i}{h_o} = \frac{d_i}{d_o}$$

This relationship does not imply that $h_i = d_i$ or $h_o = d_o$. It does, however, show there is a proportional relationship between them.

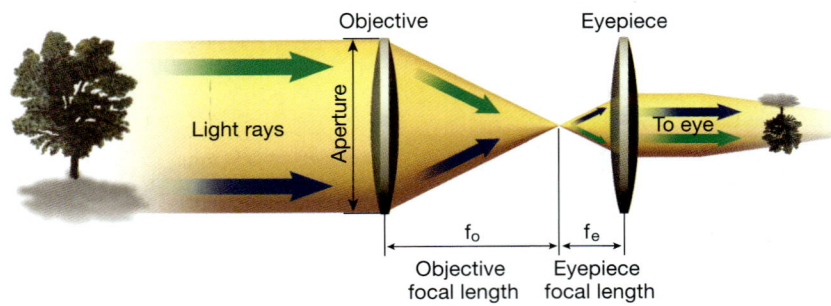

GLOSSARY TERM

Light-gathering power measure of a lens' or mirror's ability to gather light; this is solely dependent on the clear aperture of the optic

 2.5 Example of lens magnification.

Courtesy of Roen Kelly, *Astronomy* magazine

2.6 A 6-inch lens has a light-gathering power four times that of a 3-inch lens.

Courtesy of Roen Kelly, *Astronomy* magazine

Light-gathering Power

Light-gathering power (**LGP**) is an important characteristic of optical systems, especially telescopes. LGP is a measurement of the amount of light collected. Light-gathering power is directly proportional to the area of the lens or mirror, that is, the light-collecting surface. The light-collecting surface has an area proportional to the square of the diameter of a circular lens or mirror:

$$LGP \propto A \propto D^2$$

This can be represented as a set of ratios:

$$\frac{LGP_x}{LGP_y} = \frac{A_x}{A_y} = \frac{\pi r^2_x}{\pi r^2_y} = \left(\frac{D_x}{D_y}\right)^2$$

This means that a lens or mirror twice as large as another captures four times as much light (Fig. 2.6).

 # Check Your Understanding

2.1 Define focal point.

2.2 Which type of lens and mirror has a positive focal length?

2.3 What does it mean if a magnification calculation is greater than 1?

2.4 If a lens or mirror has four times the diameter of a second lens or mirror, how much more light does the first lens or mirror capture?

Unit 1 Tools of the Trade

Mathematics of Geometrical Optics

Exercise 2.1

In this exercise, you will calculate an unknown characteristic of a lens or mirror by using the known characteristics in the equations that determine focal ratio or magnification. Calculating these quantities will help you to better understand how lenses and mirrors work in astronomical equipment.

Procedure

1 Determine the focal length (f) of a concave mirror if the object's distance (d_o) is 400 mm, and the image's distance (d_i) is 100 mm.

2 Determine an object's distance (d_o) if the focal length (f) of a lens is 100 mm, and the image's distance (d_i) is 50 mm.

3 Determine an image's distance (d_i) if the object's distance (d_o) is 20,000 mm, and the focal length (f) of a lens is 500 mm.

4 Determine the magnification of the mirror in problem 1.

5 Determine the magnification of the lens in problem 2.

6 Determine the height of an image (h_i) if the magnification (**M**) of a lens is 2.0, and the height of the object (h_o) is 50 mm.

7 Consider two lenses, one 10 cm in diameter and the second 20 cm in diameter. Simple examination will show that the 20 cm diameter lens collects more light (light-gathering power) than the 10 cm diameter lens. Approximately how much more light is collected by the 20 cm diameter lens than by the 10 cm diameter lens?

Lens and Mirror Images

Exercise 2.2

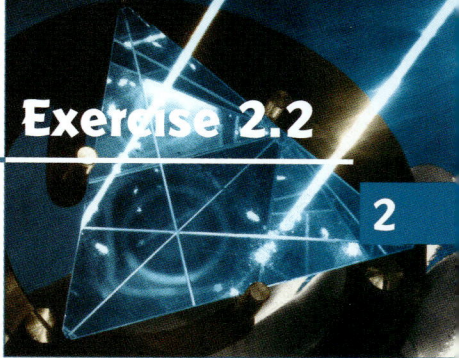

In this exercise, you will describe the images you see through various types of lenses and when reflected by various types of mirrors. Comparing them will form a basis that will help you understand their similarities and differences.

Procedure

Materials
- ❏ Simple convex and concave lenses
- ❏ Simple convex and concave mirrors
- ❏ First- and second-surface flat mirrors

1 One at a time, hold a concave and convex lens up to your eye, and describe the image you see when you look through each lens.

Concave lens: _____

_____

Convex lens: _____

_____

2 Repeat step 1 with a concave and convex mirror. Describe the image you see by looking at the reflection from each mirror.

Concave mirror: _____

_____

Convex mirror: _____

_____

3 If possible, look at the Moon or stars through the concave and convex lenses. Describe the image you see through each lens.

Concave lens: _____

_____

Convex lens: _____

_____

4 Look at the first- and second-surface flat mirrors. Place the tip of a pencil or pen directly on the surface of each. What do you see? Can you see more than one reflected image in either the first- or second-surface mirror?

_____

_____

CAUTION

NEVER look at the Sun through a lens or reflected off a mirror. Also, don't look at a laser or any other intense light.

Exercise 2.3

Optical Bench

In this exercise, you will use various optics attached to a device known as an optical table, or optical bench (Fig. 2.7). Such a platform allows you to precisely align components (lenses, mirrors, a light source, and other optical components), and it also serves as a stable base to reduce vibrations.

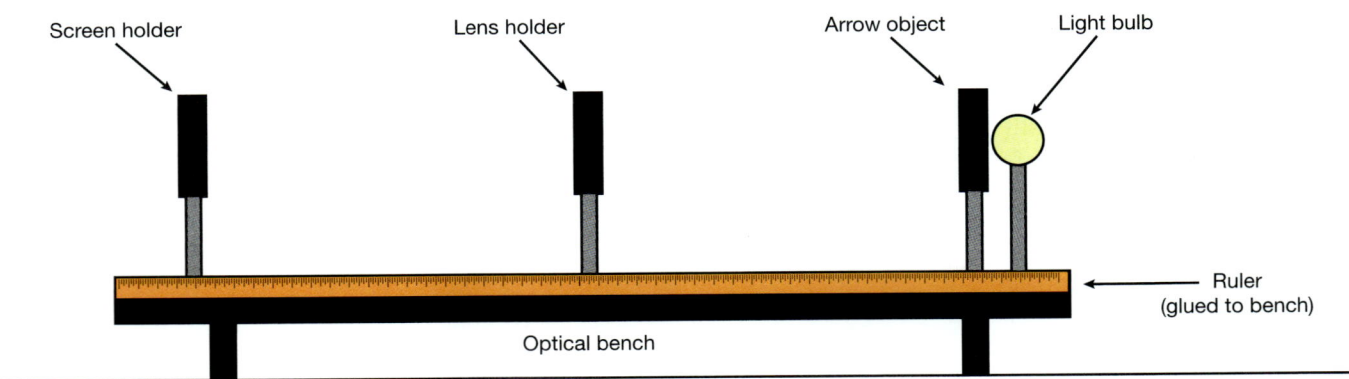

2.7 Optical bench.

Courtesy of Holley Y. Bakich

Procedure

1. Arrange the optical bench provided in the lab so that you can conveniently read the long scale. Place the object and the light source at one end of the optical bench near an electrical outlet. Make certain the light source's wire won't cause someone to trip or create a personal hazard.

2. Place a screen on the optical bench at the end opposite of the object and light source. Place one of the convex lenses in a lens clamp or holder, and attach it to the optical bench. Adjust the height of the lens, object, light source, and screen so that their centers are the same heights above the optical bench. Finally, position the screen 1,000 mm (1 m) from the object and the light source.

3. Determine the two points where the image is in focus on the screen by sliding the lens back and forth between the object, light source, and screen. Determine the object distance (d_o) and the image distance (d_i) where the focus occurs. Enter your data in Table 2.1.

Materials

- Optical bench
- Meterstick
- Two to three convex lenses with different focal lengths and diameters of approximately 50mm
- Lens clamps or holders
- Screen
- Lens masks with 5, 10, and 20mm diameter apertures

TABLE 2.1 Object and Image Distance Data

Object distance 1, d_o	mm	Object distance 2, d_o	mm
Image distance 1, d_i	mm	Image distance 2, d_i	mm

4 With the above data, use the formula $\frac{1}{f} = \frac{1}{d_o} + \frac{1}{d_i}$ to determine the focal length of the lens. Remember that you will be able to determine the focal length of the lens using the object and image distances 1 and 2. Showing your work below, determine the focal length using both sets of data and then fill in Table 2.2.

TABLE 2.2 Lens Focal Length Data

Lens focal length (f) for distance 1	mm
Lens focal length (f) for distance 2	mm

5 Place the screen a distance of $3f$ from the object/light source. Sketch your optical configuration in the space provided.

6 The lens should not come to focus at any point; verify this experimentally. Place the lens half way between the screen and the object/light source. Move the screen away from it and the lens by a couple of centimeters, and once again center the lens between the screen and object/light source. Repeat this step until the image is in focus. This step should determine the *minimum* distance between the object/light source and the focused image. Record these points below, noting the similarities between the two distances.

Minimum object distance, d_o = _____ mm

Minimum image distance, d_i = _____ mm

7 From step 6, record in the space below what occurs when the object distance (d_o) and image distance (d_i) are equal to $2f$. Verify that the minimum distances determined in step 6 are indeed equal to $4f$. And verify experimentally that the focus cannot be achieved when the object and object/light source are closer than $4f$. Record your results in the space below.

8 With the screen and object/light source $4f$ apart, move the lens so that the image on the screen is slightly out of focus. Sketch your optical configuration in the space provided.

9 Measure the lens, and record its diameter in Table 2.3. Reduce the diameter of the lens by holding a lens mask with a hole centered in front of the lens. Record the changes in the image on the screen in Table 2.3. Measure the diameter of the lens mask opening to the nearest millimeter, and record it in Table 2.3. In the space below, calculate the ratio of light-gathering power for the lens unmasked and masked:

$$\frac{LPG_x}{LPG_y} = \frac{A_x}{A_y} = \frac{\pi r^2_x}{\pi r^2_y} = \left(\frac{D_x}{D_y}\right)^2$$

TABLE **2.3** Lens and Hole Diameter Data

Aperture Mask 1	Lens diameter	mm
	Hole diameter, measured	mm
	Changes to screen image	
Aperture Mask 2	Lens diameter	mm
	Hole diameter, measured	mm
	Changes to screen image	
Aperture Mask 3	Lens diameter	mm
	Hole diameter, measured	mm
	Changes to screen image	

10 What visual and light-gathering power effects did the mask have on the image being produced? Show calculations and response below.

11 Repeat step 9 with two other aperture masks and record your data in Table 2.3.

12 Move around the aperture masks. Note what happens when the aperture masks are not centered (off-axis).

  a What effect does reducing the effective diameter of the lens have on the range over which an image appears to focus?

Geometrical Optics CHAPTER **2** **27**

b Is the image brighter with smaller or larger diameter aperture masks?

c Is the image easier to bring into focus with smaller or larger diameter aperture masks?

13 Based on these results, what do you think are the main differences between the view through a small-diameter and a large-diameter lens-type telescope? Which do you think would be preferred for astronomical research?

14 Position the converging lens a few centimeters from the object/light source and look through the lens toward the object/light source at a distance of about 6 cm from the converging lens. Sketch your optical configuration in the space provided.

15 Observe the magnified object/light source as you slide the lens away from the object, keeping your eye about 6 cm from the lens. Repeat this observation several times. Describe the image (if one exists) in Table 2.4 when the object distance is: $d_o < f$; $d_o = f$; and $d_o > f$.

TABLE **2.4** Image Descriptions at Varying Lengths

$d_o < f$	
$d_o = f$	
$d_o > f$	

16 a With your eye approximately 6 cm from the lens, what is the greatest distance between the lens and the object at which the virtual image is still in focus?

_____

b The distance should be close to, but less than, the focal length (f) of the lens. How does the object distance (d_o) compare to the focal length determined in steps 1 and 2?

_____

_____

c Calculate the percentage error, using the focal length you determined in the previous section as the actual focal length value and record your values in Table 2.5.

TABLE 2.5 Focal Length Values and Percentage Error

Maximum object distance, d_o	mm
Object distance, d_o	mm
Percentage error	%

17 Place a lens 40 cm from the object/light source. Compute the image distance for the setup, using an object distance (d_o) = 40 cm and the focal length (f) previously determined for the lens. Place the screen at this location. Make minor adjustments to achieve focus, and record the distance of the screen from the center of the lens. The object distance (d_o) is _____ mm.

18 Compute the magnification of this arrangement using:

$$M = -\frac{d_i}{d_o}$$

The calculated magnification is _____ ×.

19 Measure the size of the image and the size of the object (object/light source) to the nearest millimeter.

a Image size, h_i = _____ mm

b Object size, h_o = _____ mm

20 Calculate the ratio of these measurements to determine magnification:

$$M = \frac{h_i}{h_o}$$

21 Compare these two values for magnification by computing the percentage difference between them to determine your percentage error. What is your percentage error?

Tools of the Trade: Binoculars

LEARNING OBJECTIVES

Upon completion of this chapter, you should be able to:

1. Discuss why an observer would want to use binoculars.
2. Identify the major parts of binoculars.
3. Explain how binoculars keep images "correct."
4. Describe the proper application of low and high magnifications.
5. Identify advantages to using binoculars.
6. Determine the field of view of binoculars in the lab.
7. Observe various objects using binoculars.
8. Conduct experiments to determine which binoculars are best for astronomy.
9. Define all glossary terms.

The Hercules Cluster (M13)
Courtesy of NASA/ESA/
Hubble Heritage Team (STScI/AURA)

3

GLOSSARY TERMS

Resolution angular measure that indicates the minimum distance between two objects (usually stars) binoculars will separate

Eye relief distance (in millimeters) your eye must be from the eyepiece for you to see a focused view; the longer the eye relief, the easier binoculars are to use

Prism optical part that rotates or flips the image so it appears the same as it would without using the binoculars

Eyepiece either of the two back lenses (and their holders) of binoculars

Light loss difference between the amount of light that enters the binoculars and the amount of light that passes through the eyepieces; light loss occurs because each optical surface reflects some light that hits it; the glass in lenses and prisms also absorbs a tiny percentage of the light

Sometimes, two eyes are better than one. This lab introduces the main types of binoculars and their components, as well as what magnification, field of view, and exit pupil mean. Also discussed are the celestial objects best viewed through these devices and how to choose suitable binoculars.

Binocular Basics

The word *binocular* comes from the Latin word for "two eyes." This type of observing is more comfortable, and there are real advantages to using both eyes. Using both eyes means up to 40 percent more light reaches the brain, which translates to the ability to observe fainter objects. Another important reason for using both eyes is that **resolution**—the ability to separate two close objects—increases. Image contrast—the ability to differentiate between fine details—also increases. Finally, using both eyes increases the ability to detect color for many observers; and, for most people, observing with two eyes open rather than one seems more natural and comfortable.

Some amateur astronomers consider binoculars an accessory, but most regard them as a necessity. In some ways, binoculars may be a better choice for observing than a telescope, especially for those starting out in astronomy.

Binoculars have a wide field of view and provide right-side up images, making objects easy to find. They require no effort or expertise to set up—just sling them around your neck, step outside, and you are ready to go. That portability also makes binoculars ideal for those clear nights in the middle of the week when you don't have the time or inclination to set up a telescope.

Additionally, binoculars often are superior in one respect than many telescope/eyepiece combinations: they offer more **eye relief**. Eye relief is a term optical designers use to describe how far (almost always measured in millimeters) your eye must be from the eyepiece for you to see any object (earthly or celestial) in focus. Some eyepieces used in telescopes have such short eye relief that your eye nearly touches the lens, and your eyelashes certainly do. Most binocular manufacturers, however, build in an adequate amount of eye relief. The result is that you can position your eyes slightly away from the lenses, which makes for more comfortable observing.

Binoculars also offer a more affordable way to tour the sky than a telescope. Even if the appeal of stargazing wanes, binoculars can be used for other pursuits. In common usage, you may hear someone refer to binoculars as a "pair of binoculars." That is not correct; the unit is simply "binoculars."

Some binoculars go by other names. "Opera glasses" are small, low-power binoculars that got their name when people started buying them and taking them to opera performances. "Field glasses" are binoculars that received that name from nature-buffs who took them on field trips.

Prisms

Binoculars use a series of **prisms** (Fig 3.1) to bend incoming light several times before it reaches your eyes through the **eyepieces**. Manufacturers must position all optical components correctly so that there is no **light loss**. Prisms also keep the image being observed upright and correct left-to-right. Telescopes, due to their optical design, invert an object (that is, turn it upside down and with some types of telescopes reverse left and right). Adding the optics required to correct this decreases the incoming light, which is especially important when viewing fainter objects. But inverted images simply would not be acceptable for binoculars, especially for everyday uses, such as sports or birding.

Binocular prisms come in two basic designs: roof prisms and Porro prisms (capitalized because they are named for nineteenth-century Italian inventor, Ignazio Porro). Roof prisms are lighter and smaller, but they don't work well.

Porro prisms are better and are made of the highest quality glass available. Most high-quality binoculars are multicoated on all optical surfaces. You will see this referred to as "fully multicoated."

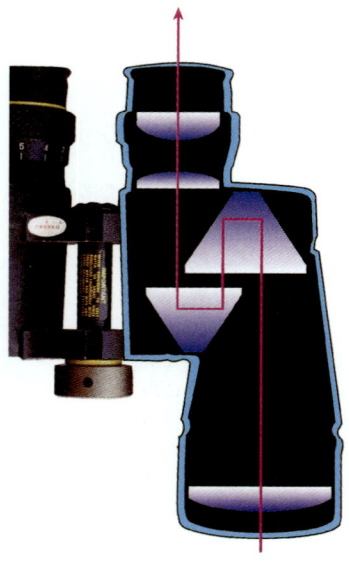

3.1 Placement of prisms in binoculars to bend incoming light, so images appear right-side up and correct left-to-right.
Courtesy of Holley Y. Bakich

32 Unit 1 Tools of the Trade

Magnification

Every binocular has a two-number designation, such as 7×50. The first number (in this case, 7) is the **magnification**, or power. A binocular's magnification is a direct (and practical) outcome for the values you calculated in the Geometrical Optics lab (Chapter 2). The second number (50) is the diameter in millimeters of each **objective** lens.

A magnification of 7 is in the "medium" range, just high enough to bring out some detail in large astronomical objects (like the Moon). Figures 3.2A and 3.2B are examples of two of Celestron's binoculars in the medium range. Magnification that is too high, above 10, will highlight involuntary shaking of your hands, which will cause viewed objects to move around. A high magnification also limits the field of view, making objects more difficult for beginners to find. Binoculars with magnifications below 5 are considered low-power units, and they have little use for astronomy.

Figures 3.2A and 3.2C show binoculars in which the **aperture** (size) of each front lens is 50 millimeters. The larger this number, the more light the binoculars collect, and that makes the target brighter. Binoculars with 50-millimeter front lenses collect more than twice as much light as those measuring 35 millimeters across. The disadvantages to larger front lenses are that the binoculars will be larger, heavier, and more expensive.

Like binoculars with large-aperture front lenses, those with higher magnifications also tend to require support, because every wiggle and vibration will be more pronounced in higher magnification binoculars. Most observers like to tripod-mount binoculars above 10–12 power. Giant binoculars, for example, would need to be tripod mounted.

Giant binoculars (Fig. 3.3) have front lenses that are 4 inches (102 mm) or larger. Classifying giant binoculars by size is a bit of a gray area, however. Some manufacturers label the largest binoculars they sell "giant," which may be true for them.

> **GLOSSARY TERMS**
>
> **Magnification** how much larger any object appears through binoculars; also called "power"
>
> **Objective** either of the two front lenses of binoculars
>
> **Aperture** size of each front lens of binoculars, given in millimeters

3.2 Celestron binoculars: (**A**) Oceana 7×50, (**B**) UpClose 7×35, and (**C**) UpClose 10×50.

Courtesy of Celestron

3.3 Giant binoculars, such as Celestron's Skymaster 25×100, must be used with a tripod mount. They measure 12.5 inches (32 centimeters) long and weigh 8.8 pounds (4 kilograms).

Courtesy of Celestron

As with smaller models, magnification in giant binoculars varies. Some giant binoculars even allow you to change eyepieces to increase or decrease the magnification. No giant binoculars can be hand-held; they are simply too heavy to hold.

In addition to magnification and size, people gauge how binoculars feel in their hands. Comfort while holding even relatively lightweight binoculars, especially for an extended period of time, is important. Many manufacturers have taken this into account for birders, hunters, and nature observers. "Feel" considers the shape of the binoculars, the outside covering, and how well they fit in your hands.

Focus

> **GLOSSARY TERM**
>
> **Focuser** device or devices that focus binoculars

A major mechanical feature of binoculars is its **focuser**. Binoculars focus either through two individual eyepiece systems or a central system, which is referred to as center focus (Fig. 3.4). Less expensive binoculars, especially some of the mini-giants and giants, usually employ individual eyepiece focus systems. There is nothing wrong with this as long as the focus operates smoothly.

Center focusing can be accomplished through several means. The most common include a knurled focusing knob; a rocker-like bar or plate; or a lever, with the eyepieces joined through what is called the eyepiece bridge. The lever system is not generally preferred for astronomy; it is best for users who must continually change the focus as they track a moving object (such as a bird or race car). In astronomy, once you focus on a celestial object, you should be set for the evening, perhaps needing only occasional fine adjustments (unless someone else uses your binoculars and needs to adjust the focus for their eyes).

With the vast majority of center-focus binoculars, you will be able to fine-focus one eyepiece. This allows you to correct for focusing differences in each eye. Focus using the main control while looking through the set eyepiece first (usually the left eye). Most people close their right eye while doing this. Then, once you have focused the left eye, close it and adjust the focus for the right eye. Some binoculars have numbers on the right eyepiece to help people return to a previous setting.

Field of View

> **GLOSSARY TERM**
>
> **Field of view** angle (in degrees) you can see through any binoculars; also refers to the actual area in view when you look through the eyepieces

One of the specifications you will find noted with nearly every binoculars is the **field of view**. This number refers to how much of the sky or an area you can see. It is measured in degrees, but sometimes a manufacturer lists the field of view's width (in feet) at a distance of 1,000 yards from the observer. Remember that from horizon to horizon is an angular

3.4 Examples of center focusers.

Courtesy of Celestron

distance of 180°. Your eyes can take in about 60° of the sky. But individual sky objects are small. Even the full Moon's disk spans only about 0.5°.

An example of a listed specification might be, "Field of view is 350 feet at 1,000 yards." This tells you that you can see something 350 feet across that lies 1,000 yards (3,000 feet) away. For astronomy, listing the field of view in degrees is more useful. Also, in general, the bigger the field of view, the better it is. If the field of view measures less than 4°, it becomes a little difficult to orient yourself; if the field of view is less than 2°, you may need a low-power finder scope, like one that you would use with a telescope.

Exit Pupil

One of the most important terms when dealing with binoculars is **exit pupil** (Fig. 3.5). This is the diameter of the shaft of light coming from each side of the binoculars to your eyes. If you point the front of the binoculars at a bright surface, light, or the sky, you will see two small disks of light exiting at the eyepieces. The diameter of either of these disks is the size of the exit pupil.

> **GLOSSARY TERM**
> **Exit pupil** diameter of the shaft of light exiting each eyepiece

3.5 Binocular eyepieces showing exit pupils.

The diameter of the exit pupil (in millimeters) equals the binocular aperture divided by the magnification. So for 7×50 binoculars, the exit pupil diameter would be 50 divided by 7, or roughly 7 mm. For astronomy, you want to maximize this number, because the pupils in our eyes dilate (expand) in darkness. The wider the shaft of light, the brighter the image will be, because light is hitting more of the eye's retina. This rule, however, is only true up to a point. If a binoculars' exit pupil is too large to fit into your eye, you will lose some of the incoming light.

For most observers, binoculars with exit pupils measuring 7 mm are just right. Some people have dark-adapted pupils measuring nearly 9 mm in diameter, others have small pupils less than 5 mm across. Our pupils are largest when we are young. From age 30 on, they begin to contract, but the process slows in our later decades. Women tend to have larger pupils than men of the same age, on average. Unfortunately, there is no hard and fast rule that correlates pupil size with age.

Check Your Understanding

3

1.1 A binoculars' front lenses measure 50 mm across and the eyepiece lenses measure 10 mm across. What is the aperture of these binoculars?

1.2 List some possible reasons the amount of light coming out of the binoculars' eyepieces is less than the amount of light entering the objective lenses.

1.3 Give two advantages of using both eyes for observing.

1.4 Why are Porro prisms better than roof prisms?

1.5 What do the two main numbers marked on all binoculars stand for?

1.6 What is the smallest aperture of a binocular labeled "giant"?

1.7 How many full Moons could fit side by side in the view of binoculars with a field of view of 7°?

1.8 Assume that at night the pupils in your eyes dilate (expand) to 7 mm. Why should you not use binoculars that have an exit pupil of 9 mm?

Selecting the Right Equipment

Binoculars

How should you choose your first binoculars? First, pick up the binoculars and shake them gently. Next, twist them gently. Then move the focusing mechanisms several times. Finally, move the barrels together, then apart. What you are assessing is quality of workmanship. If you hear loose parts or if there is any play when you twist or move the binoculars, select another one. Another thing to consider at this point is the weight of the binoculars. If you are going to be hand-holding them, try to imagine what they will feel like at the end of a long observing session.

Look into the front of the binoculars and check for dirt or other contaminants. You can ignore a small amount of dust on the outside of the lenses. The inside of the binoculars, however, should be immaculate. If not, select another one.

Hold the binoculars in front of you with the eyepieces toward you. Point them at a bright area. You will see the exit pupils—disks of light formed by the eyepieces. They should be round. If they are not round, the optical alignment of the binoculars is bad and the prisms are not imaging all the light.

Of course, you must look through the binoculars. Try to do this outdoors and at night. Nothing will reveal flaws in the design of binoculars more than star images. If it is impossible to test the unit at night, or even outdoors, look through a door or window at distant objects. How well do the binoculars focus? Are objects clear? If there is any sign of a double image, the two barrels are not aligned; select another one.

If you are wearing glasses—and if you plan to observe with them on—can you get your eyes close enough to the binoculars to see the entire field of view? Move the binoculars visually across a straight line such as a phone wire or the horizon. Does the line look straight? A tiny amount of bending near the edges of the field of view is not a big problem.

The Astronomical League awards its Binocular Messier Club certificate to those who have observed with binoculars 50 or more deep-sky objects out of the 109 on French comet hunter Charles Messier's famous list.

Binocular Mounts

For the steadiest images possible, nothing beats mounting your binoculars to a tripod or custom binocular mount. Smaller, well-mounted binoculars with less magnification will, after a few minutes of continuous use, beat hand-held binoculars of larger aperture and magnification.

The simplest binocular mount (Fig. 3.6A) is a metal "L" bracket (a piece of metal bent into an L-shape that contains at least two holes). This bracket is attached to the mounting hole on the binoculars' center post, and the other end of the L attaches to a camera tripod. This setup generally is adequate if the objects you are observing aren't too high in the sky. For objects near the zenith (the overhead point), tripod-mounted binoculars are uncomfortable—and in some cases impossible—to use.

A parallelogram binocular mount (Fig. 3.6B) allows people of all heights to view the same object in the sky. This type of mount has a swinging arm that can be moved up or down to the observer's eye level. A movable counterweight lets you use binoculars of different weights.

Another option is to purchase or build a custom-made binocular mount. Plans for binocular mounts are readily available, but most users purchase commercially made

3.6 Mounting binoculars: **(A)** Celestron's Binocular Tripod Adapter (approximately 6 inches [15 centimeters] long); and **(B)** a homemade parallelogram binocular mount (stands approximately 6 feet 6 inches [2 meters] high at maximum).

(A) Courtesy of Celestron

mounts. Such units employ a design based on an arrangement that keeps the binoculars pointed at an object over a wide range of motion, allowing people of varying height to use them. This setup is ideal for observing sessions in which both adults and children will be viewing the same objects.

When selecting a binocular mount, choose one that is sturdier than you require. That will allow you to easily upgrade your binoculars to a larger (heavier) model in the future. A binocular mount is rugged if a few seconds after you have located a celestial object the image settles down and shows no vibration (unless a strong wind is blowing).

The other necessity for high-power binocular observing is the tripod for the mount. Most camera tripods are inadequate for use with high-power binoculars. They simply are not robust enough to handle the weight of the binoculars plus the weight of the mount. With camera tripods balance also could be a problem. If you have a tripod, by all means try it. You will know immediately if it is up to the task.

Check Your Understanding

2.1 List several factors that will let you know you are holding high-quality binoculars.

2.2 Why are the highest-power binoculars not always the best for skywatching?

2.3 Why might you want a binocular mount rated for larger binoculars than you own?

Comparison of Binoculars

Exercise 3.1

Procedure

Materials
- ❏ Binoculars of various types, six or more if available

This procedure uses different types of binoculars to observe a specific object. The observations through each binoculars selected should be of the same object (sign, tree, window, the Moon, etc.) so you are able to make the required comparisons.

1. Choose six binoculars (or as many different types available) in no specific order.
2. Record the manufacturer, model, focus type, and other information about each binoculars in the correct place in Table 3.1 on the next page.
3. Observe your selected object through each binoculars, and record the information in the correct place in Table 3.1.

> For practice with binoculars,
> try one of the exercises in Chapter 7.

TABLE **3.1** Binocular Observations

Binoculars	Focus Type	Manufacturer/ Model (if any)	Sketch Sketch the object, which represents the binoculars' field of view	Binocular Characteristics	Additional Notes
1	___ × ___		○		
2	___ × ___		○		
3	___ × ___		○		
4	___ × ___		○		
5	___ × ___		○		
6	___ × ___		○		

40 *Unit 1* Tools of the Trade

Estimating Field of View in the Classroom
Exercise 3.2

Procedure

Materials
- ❏ Binoculars
- ❏ Tape
- ❏ Tape measure
- ❏ Calculator with trigonometric functions

1. Select one of the available binoculars.
2. Stand with your back against one wall.
3. View the opposite wall through the binoculars, focusing as best you can.
4. Have your partner mark (with tape) the left and right extent of your field of view.
5. Measure the distance between the walls. Convert the measurement to inches. Record this number as A.

    A = _____ inches

6. Measure the distance between the pieces of tape. Convert the measurement to inches and divide by 2. Record this number as B.

    B = _____ inches

7. Now use trigonometry to figure out the angle that equals half of the field of view. Divide A into B and record the answer as X.

    X = _____ inches

8. Using a calculator, enter X and press tan⁻¹ (that is the key for the arctangent function) and record the answer.

    X (tan⁻¹) = _____

9. Multiply the answer to problem 8 by 2. This is the approximate field of view of the binoculars in degrees.

    Binoculars field of view = _____ °

10. Record the field of view in the correct column in Table 3.2.
11. Repeat steps 1 through 10 for each of the remaining binoculars.

TABLE 3.2 Field of View Comparison Data

Binocular #	Field of View (1 = smallest field of view to 6 = largest field of view)	Binoculars Rating
1		
2		
3		
4		
5		
6		

12 After recording your field-of-view data for each binoculars in Table 3.2, answer the following questions.

    a Compare the field of view for each of the binoculars, and record your ratings in Table 3.2.

    b What binocular characteristics are factors in helping to set the field of view?

    c Would you rather use center-focus or individual-focus binoculars? Give several reasons for your answer.

    d Which binoculars would you prefer for general purpose observing? Why?

    e How would the binoculars you chose in the previous question work for astronomical viewing? Why do you think this is?

    f Based on this procedure and the binoculars you used, which binoculars would you prefer for astronomy? Explain your answer.

Tools of the Trade

Telescopes

LEARNING OBJECTIVES

Upon completion of this chapter, you should be able to:

1 Describe the three main types of telescopes and identify important names and dates associated with their invention.

2 Describe the function of a telescope mount and its importance in the whole system.

3 Identify the most important telescope accessories and describe the use of each.

4 Calculate the magnification of any telescope/eyepiece combination.

5 Examine telescopes and determine characteristics such as optical type, mount, focal length, and focal ratio.

6 Identify different types of eyepieces by size and focal length.

7 Determine how an object moves through the field of view when a telescope is moved in various directions.

8 Define all glossary terms.

Hubble telescope orbiting Earth
Courtesy of NASA

4

Dutch spectacle maker Hans Lipperhey (1570–1619) constructed the first telescope. This groundbreaking invention consisted of a tube with a convex lens (the surfaces curve outward) in the front and a concave lens (the surfaces curve inward) in the rear. The device magnified objects approximately three times.

Shortly after, in 1609, Italian scientist Galileo Galilei (1564–1642) began to build his own telescopes. He was the first to use the new device to study celestial objects, and what he saw revolutionized astronomy. Galileo (Fig. 4.1) saw the imperfections of our Moon, four moons circling Jupiter, and Venus going through phases. The great scientist's observations helped overthrow the prevailing scientific beliefs of his time and helped place the Sun in its proper position at the center of the solar system.

Galileo Galilei

4.1 Early Galileo telescope.

Galileo did it all with a telescope 1 inch in diameter. He fashioned his telescope from glass lenses, which refracted the light passing through them to form images. Such refracting telescopes were all the rage among observers until 1668, when Isaac Newton built the first reflecting telescope (like Galileo's, it was 1 inch in aperture) that used a mirror to gather and focus light. Since then, telescope design has become more complex, involving additional lenses and mirrors in the construction of compound telescopes.

In this lab, you will learn how to construct a telescope and use a telescope to observe actual objects (celestial or terrestrial) outdoors or, in the case of inclement weather, terrestrial objects through the classroom's windows. Also discussed in this lab is the importance of focusing, judging atmospheric conditions, and determining a telescope's field of view.

Refracting Telescopes

Refraction is the bending of light as a result of it passing from one medium (such as air) to another (such as glass). A **refracting telescope** (Fig. 4.2) makes use of this property by using a lens with curved surfaces. As light passes from air to glass and then back to air, its path (Fig. 4.3) bends toward the lens' optical axis (the imaginary line that passes through the center of the lens). If the lens surfaces are shaped properly, the light is brought to a focus. Table 4.1 lists advantages and disadvantages of using refracting telescopes.

> **GLOSSARY TERM**
>
> **Refracting telescope** design of telescope that uses lenses to focus the light from distant celestial objects

4.2 Refracting telescope.
Courtesy of Celestron

44 Unit 1 *Tools of the Trade*

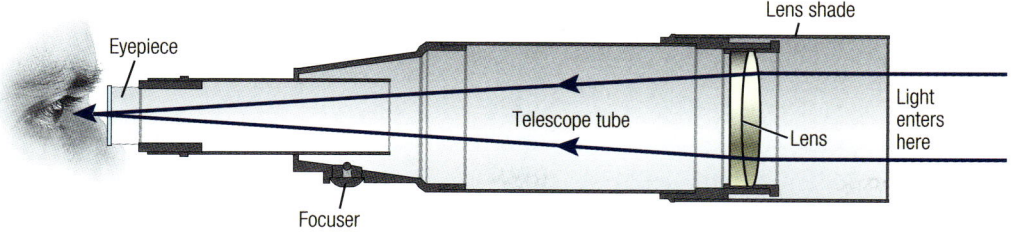

4.3 Pathway of light in a refracting telescope.

Courtesy of Roen Kelly, *Astronomy* magazine

TABLE **4.1** Advantages and Disadvantages of Refracting Telescopes

Advantages	
Image Contrast	Refractors have a totally clear aperture. This means no central obstruction causes light to scatter from brighter to darker areas, so image contrast is better through refractors.
Low Maintenance	Unlike mirrors, lenses do not require recoating. In addition, a refractor's optics generally do not require adjustment (a process called collimation). The lens is fixed into the tube and, without some major trauma, usually does not become misaligned.
Disadvantages	
Cooldown Time	Because a refractor is a closed-tube assembly, it can require a long amount of time to cool to outside temperature; a tube made from thin aluminum reduces this time significantly.
Chromatic Aberration	Chromatic aberration commonly manifests itself as faint fringes of color around bright objects like the Moon or Jupiter.
Expense	An apochromatic triplet lens has six surfaces that a manufacturer must grind to a perfect figure. The cost ratio between a 6-inch apochromatic lens and a high-quality 6-inch mirror (which has only one figured surface) is at least 10 to 1.
Size Limit	The largest telescope lens ever built measures 1.016 m across. Larger lenses are impractical because the glass (which is a liquid) would distort under its own weight

> **GLOSSARY TERM**
>
> **Focal ratio** ratio between the focal length and the diameter of a mirror or lens; for example, a 100mm diameter lens with a 1,500mm focal length would be *f*/15

The earliest refracting telescopes had poor optical quality, and the lenses had many aberrations (irregularities). Telescope makers, such as Dutch astronomer Christiaan Huygens (1629–1695) and Polish astronomer Johannes Hevelius (1611–1687), were resourceful, however. They found that if they made telescopes with lenses that had a large **focal ratio** they could minimize the optical defects. This is because lenses with large focal ratios focus far from the lens, so the light doesn't have to bend very much. A lens with a short focal ratio focuses close to the lens, so the light has to bend a lot as it passes through the lens. Such drastic changes in the direction of the light introduce optical defects and degrade the images viewed through such lenses.

Achromatic Lenses

One of the problems with early lenses was an optical defect known as chromatic aberration (Fig. 4.4). White light is a combination of all colors of light. Unfortunately, the colors, when passed through a simple lens, do

> *Looking Up*
> - Huygens was the first to correctly identify the nature of Saturn's rings.
> - Hevelius' instruments were incredibly long, massive units held together by wooden frames, coupled by pulleys, and moved about with the help of a team of assistants. Hevelius' telescopes were difficult to point accurately and useless if the wind was blowing.

GLOSSARY TERM

Achromatic lens two-element lens corrected to bring two wavelengths (usually of red and blue light) to a common focal point to reduce the defect found in single lenses known as chromatic aberration; achromatic literally means "without color," but such lenses do exhibit slight chromatic aberration

not focus at the same point. Blue light is refracted more than red light. Chromatic aberration shows up as a reddish or bluish fringe around bright objects like Venus.

In 1729 Chester Moore Hall (1703–1771) devised a lens with two elements (Fig. 4.5), one made of crown glass and the other of flint glass. This is an **achromatic lens,** meaning "not color dependent." Hall's lens produced a much better image.

The nineteenth century saw a tremendous increase in the technical quality of glass and achromatic lenses. In America Alvin Clark (1804–1887) and his sons began making high-quality refracting telescopes. Their crowning achievement was the largest refractor ever built (Fig. 4.6). This telescope began operation in 1897 for Yerkes Observatory of the University of Chicago. Its lens measures 1.016 m in diameter.

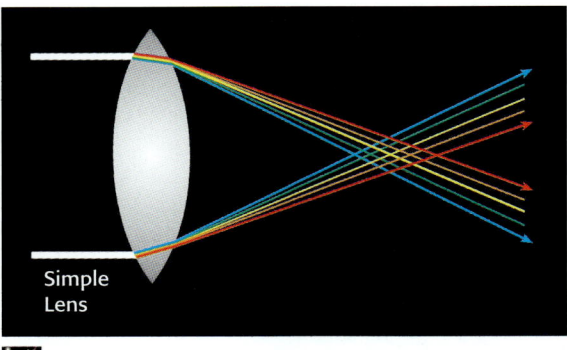

4.4 Chromatic aberration. *Courtesy of Holley Y. Bakich*

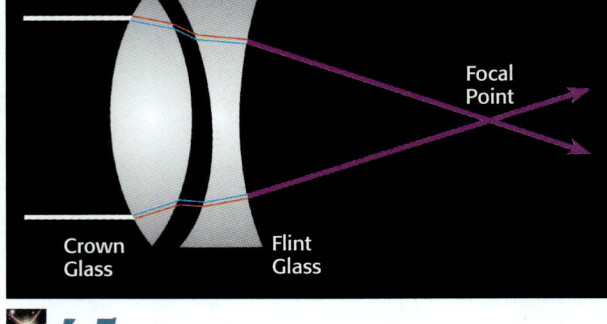

4.5 Lens with two elements. *Courtesy of Holley Y. Bakich*

4.6 Yerkes Observatory in Williams Bay, Wisconsin, home of the world's largest refracting telescope.

Courtesy of Yerkes Observatory, Richard Dreiser, Photographer

Apochromatic Lenses

The first refractor lens to be labeled "color free" was a triple-lens system offered by Astro-Physics, Inc., in 1981. This was the beginning of the new age of refractors with **apochromatic lenses** (Fig. 4.7). Although the word *apochromat* means "without color," different wavelengths of light do not come to exactly the same focus, although much closer than in achromats. Apochromatic objectives now have two to four lens elements. At least one element is made with either fluorite or ED (extra-low dispersion) glass, which gives superb color correction.

> **GLOSSARY TERM**
> **Apochromatic lens** optical design, usually incorporating three elements, that is virtually free of residual color

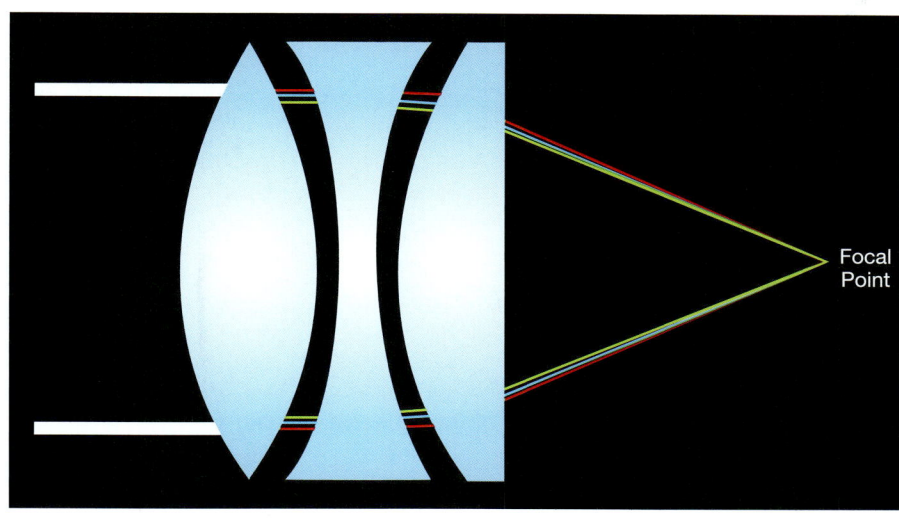

 4.7 Apochromatic lenses.
Courtesy of Holley Y. Bakich

 ## Check Your Understanding

1.1 Who invented the refracting telescope?

1.2 If a lens has chromatic aberration, what do you see?

1.3 How many elements does an achromatic lens have?

1.4 Where is the world's largest refractor?

1.5 Why are no refractors larger than the Yerkes Refractor?

4.8 Reflecting telescope.
Courtesy of Celestron

Reflecting Telescopes

Reflection occurs when light strikes a mirror and changes its direction. A rule regarding reflected light is that the light reflects from a mirror at the same angle as it strikes it, although not always in the same direction. A **reflecting telescope** (Fig. 4.8) makes use of this property by incorporating carefully manufactured mirrors to bring light to a focus. Different types of reflecting telescopes use mirrors with different shapes. Table 4.2 lists the advantages and disadvantages of reflecting telescopes.

TABLE **4.2** Advantages and Disadvantages of Reflecting Telescopes

	Advantages
Color Free	Because the light is reflected, not refracted through a lens, reflecting telescopes suffer no chromatic aberration.
Easy to Produce	Mirrors have only one optical surface. An apochromatic lens has between four and eight. Amateur telescopes with apertures larger than 200 mm are all reflectors or catadioptrics.
Cost	Reflectors offer the best aperture-to-dollar ratio, by far.
No Size Limit	Because mirrors in large telescopes are supported from the bottom, no theoretical size limit exists.
	Disadvantages
Central Obstruction	A reflector's secondary mirror creates an obstruction that scatters light from bright objects, reducing image contrast.
Coma	All Newtonian reflectors suffer from **coma**, a defect that causes stars to flare, making them look like tiny comets.
Maintenance	Mirrors eventually require recoating.
Frequent Collimation	Optics in reflectors are sensitive to bumps or jostling. A reflector should be collimated before each observing session.
Cooldown Time	Large reflectors contain thick primary mirrors. The thicker they are, the more time they take to cool.
Limited Access	Large Newtonian reflectors require the observer to use a ladder to view objects near the zenith.

The Newtonian Reflector

In 1663 Scottish mathematician James Gregory (1638–1675) published a description of what would come to be known as the reflecting telescope, but he never actually made the telescope. It is British mathematician Isaac Newton (1643–1727) who constructed the first working reflecting telescope in 1668. It contained a spherical mirror with a 25 mm **aperture** (diameter) and a tube length of 150 mm. Not satisfied with his first effort, Newton completed an improved reflector with an aperture of nearly 50 mm. The first "Newtonian" reflector was presented to the Royal Astronomical Society in 1671, at which time Newton was made a full member.

Early reflectors had mirrors made of speculum, an alloy roughly 80 percent copper and 20 percent tin. Once figured and polished, this shiny metal would begin to corrode after only a few weeks. It then would need to be repolished. During each repolishing, care had to be taken to keep the same curve on the mirror. The **Newtonian telescope** uses a **primary mirror** ground into the shape of a parabola (an approximately U-shaped curve) and a flat **secondary mirror**. Light reflects through a hole in the primary mirror, which allows an eyepiece or camera to be mounted at the back end of the tube (Fig. 4.9). Although the secondary mirror in all reflecting telescopes blocks some of the light from hitting the primary mirror, it does not create a hole in the center of the image because each part of the telescope's primary mirror is reflecting the entire image to the focal point.

GLOSSARY TERMS

Reflecting telescope telescope that uses mirrors to focus the light

Aperture diameter of the main lens or mirror in an optical telescope, or the antenna size in a radio telescope

Newtonian telescope first practical reflecting optical telescope, invented by Isaac Newton, consisting of a concave, paraboloidal primary mirror, and flat secondary mirror that diverts light out the side of the tube

Primary mirror main light-gathering mirror of a reflecting telescope

Secondary mirror flat mirror used in Newtonian telescopes to reflect the image formed by the primary mirror into the eyepiece; when used in this way, the secondary mirror is often called the diagonal mirror

Coma inherent property of telescopes that use parabolic mirrors, like Newtonian reflectors; incoming light rays that strike the mirror can be parallel to the telescope tube or at a small angle to it; parallel rays cause no problems; when incoming light rays strike the mirror at an angle, however, individual rays do not reflect to the same point, resulting in an image that looks wedge-shaped, or like a tiny comet; the more the angle differs from parallel, the worse this effect is

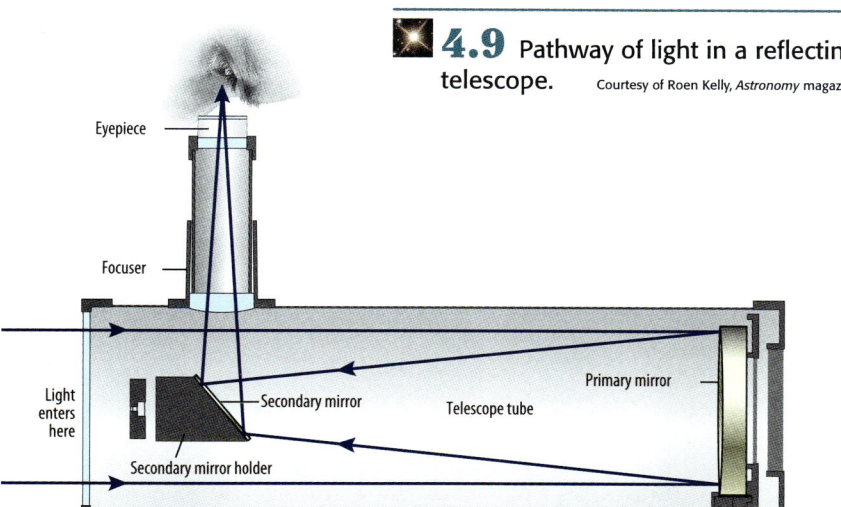

4.9 Pathway of light in a reflecting telescope. *Courtesy of Roen Kelly, Astronomy magazine*

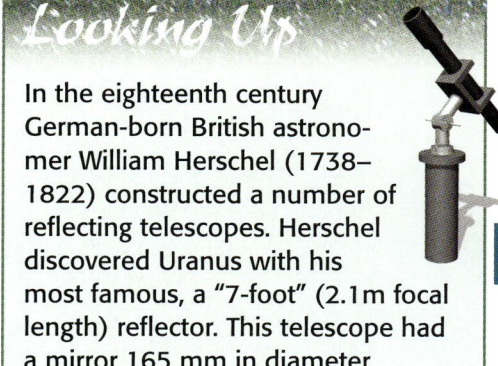

In the eighteenth century German-born British astronomer William Herschel (1738–1822) constructed a number of reflecting telescopes. Herschel discovered Uranus with his most famous, a "7-foot" (2.1m focal length) reflector. This telescope had a mirror 165 mm in diameter.

In 1835 German chemist Justus von Leibig (1803–1873) developed a process for depositing a thick layer of silver on glass. This was a major step forward, because when the silver tarnished it could be chemically removed and a new layer redeposited without altering the mirror's curve.

Apart from tarnishing, silver is not the ideal reflective surface for a telescope mirror. Aluminum, for example, reflects 50 percent more light. John Donovan Strong (1905–1992), a physicist at the California Institute of Technology, was one of the first to coat a mirror with aluminum. The first mirror he aluminized, in 1932, is the earliest known example of a telescope mirror coated by this technique. Now all reflectors have aluminized mirrors.

Construction of Large Reflecting Telescopes

Small telescopes can do only so much. Bigger telescopes collect more light and reveal finer detail. The technology needed to build larger telescopes advanced slowly, however. American astronomer George Ellery Hale (1868–1938) oversaw construction of the first 100-inch (2.5 meters) telescope in 1917. That telescope still stands on Mount Wilson in California, overlooking Los Angeles.

The next record came in 1948 with another Hale project, the 200-inch (5 meters) Hale Telescope on California's Palomar Mountain (Fig. 4.10). By the late twentieth and early twenty-first centuries, big telescopes were popping up around the world, mostly on isolated mountaintops where the skies are clear and steady.

Larger reflecting telescopes are being planned and built. The Giant Magellan Telescope (Fig. 4.11A) is well on its way from concept to completion. The University of Arizona's Steward Observatory Mirror Lab has cast two of the telescope's seven 8.4-meter mirrors and nearly finished polishing the first. Construction workers expect to break ground for the project at Las Campanas Observatory in Chile in 2020.

Other telescopes currently in the planning stage are the Thirty Meter Telescope and the European Extremely Large Telescope (Figs. 4.11B and C). The Thirty Meter Telescope

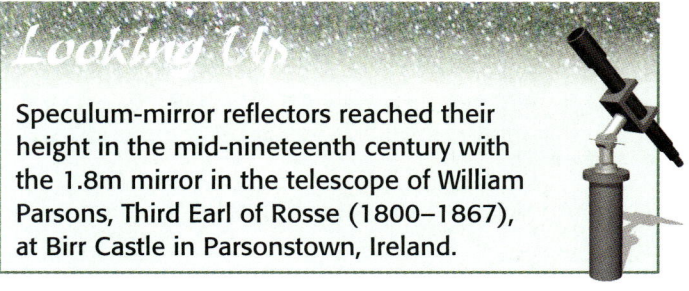

Speculum-mirror reflectors reached their height in the mid-nineteenth century with the 1.8m mirror in the telescope of William Parsons, Third Earl of Rosse (1800–1867), at Birr Castle in Parsonstown, Ireland.

4.10 Palomar Observatory 200-inch Hale reflector, the world's largest telescope from 1948 to 1976. *Courtesy of Palomar Observatory/California Institute of Technology*

will sit atop Hawaii's Mauna Kea when completed around 2020. Its primary mirror will be composed of 492 hexagonal segments 1.44 meters across. In the early 2020s, the largest telescope in the world may well be the 39.3-meter European Extremely Large Telescope. It will view the cosmos from Chile's Cerro Armazones.

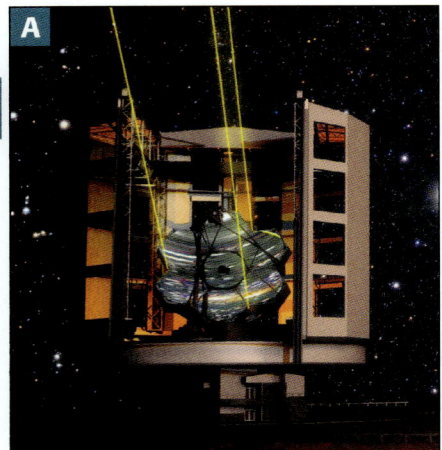  

4.11 (**A**) Giant Magellan Telescope, (**B**) Thirty Meter Telescope (TMT), and (**C**) European Extremely Large Telescope (E-ELT).

Courtesy of (A) Giant Magellan Telescope—GMTO Corporation, (B) TMT Observatory Corporation, (C) ESO

Check Your Understanding

2.1 Who invented the reflecting telescope?

2.2 Who built the first working reflector?

2.3 What do all telescope makers coat their mirrors with now?

2.4 What do you think is the main advantage of a reflecting telescope?

2.5 Where does the light go after it reflects off the secondary mirror of a Newtonian reflector?

Catadioptric Telescopes

Catadioptric means pertaining or due to both reflection and refraction of light. **Catadioptric telescopes** also are known as compound telescopes (Fig. 4.12) and have a mix of refractor and reflector elements in their design.

The first catadioptric (compound) telescope was made by the German astronomer Bernhard Schmidt (1879–1935) in 1930. The Schmidt telescope had a spherical primary mirror at the back of the telescope and a glass corrector plate in the front of the telescope to remove spherical aberration. This defect occurs when only a spherical mirror is used to focus light. The different light rays striking the mirror do not meet at a single focus. The farther the rays are from the center of the mirror, the closer to the mirror they focus. The telescope becomes a Schmidt camera used for photography by placing photographic film at the prime focus.

Schmidt's design is the precursor of today's most popular telescope, the **Schmidt-Cassegrain telescope (SCT)**. A combination of the Cassegrainian design with Schmidt's corrector plate occurred in the 1960s. Like the Cassegrain reflector, a secondary mirror bounces light through a hole in the primary mirror to the eyepiece or a detector (Fig. 4.13).

> **GLOSSARY TERMS**
>
> **Catadioptric telescope** compound telescope that employs elements of both refracting and reflecting telescopes; uses a spherical primary mirror with the addition of a correcting plate
>
> **Schmidt–Cassegrain telescope (SCT)** telescope design that includes a spherical primary mirror, a full-aperture corrector plate, and a negative secondary in a Cassegrain configuration

4.12 Catadioptric, or compound, telescope.
Courtesy of Celestron

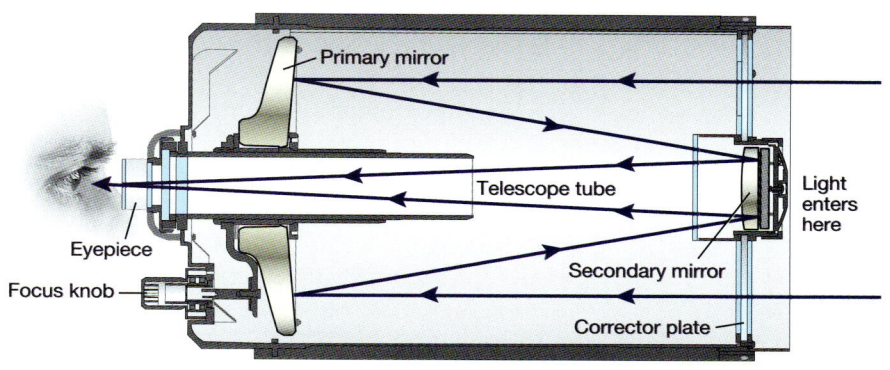

4.13 Pathway of light in a catadioptric (compound) telescope.
Courtesy of Roen Kelly, Astronomy magazine

Check Your Understanding

3.1 Give another name for a catadioptric telescope.

3.2 Who invented the catadioptric telescope?

3.3 In what year did the Schmidt-Cassegrain telescope appear?

3.4 Look at Figures 4.3, 4.9, and 4.13. Explain one advantage the catadioptric telescope has over the other two designs.

3.5 A Schmidt-Cassegrain telescope has a _____ in its primary mirror.

GLOSSARY TERM

Alt-azimuth (alt-az) mount type of telescope mount in which one axis points to the zenith and allows rotation along the horizon and the other allows changes in altitude, or distance above the horizon; used for most small telescopes; a Dobsonian mount is an alt-az system; also used for larger telescopes with the addition of computer-controlled drives for both axes

Telescope Mounts

Astronomers call their instruments "telescopes," but half of any telescope is the mount. An unstable mount will render the finest apochromatic refractor unable to deliver quality images. If the mount is not large enough, wind—the bane of most large telescopes—will not be the only problem; you also will experience bouncing images even when focusing.

The best possible mount for a telescope would be one affixed to a permanent pier sunk into concrete in an observatory. Most observers, however, travel and want their telescopes to be portable. The degree of portability is a trade-off, to be sure.

The two main types of mounts are alt-azimuth mounts and equatorial mounts. In the simplest (nonmotorized) versions, an observer using an alt-azimuth mount must move it in two directions to follow an object across the sky, but an observer using an equatorial mount need only make a single correction. Because alt-azimuth mounts are easier to make, they cost less than their equatorial counterparts.

Alt-azimuth Mounts

An **alt-azimuth mount** (Fig. 4.14) is the simplest type of telescope mount, and also the least expensive. It moves in two directions: left to right, or up and down. The word *alt-azimuth* is a combination of altitude and azimuth.

Altitude is the distance in degrees an object lies from the horizon to the zenith (the overhead point). An object on the horizon has an altitude of 0°. If an object is at the zenith, it has an altitude of 90° (Fig. 4.15).

Azimuth is a measurement, also in degrees, beginning at the north point of the horizon and proceeding through east. The azimuth of any object you observe in the sky ranges from 0° to 360°. If a celestial object lies due north, its azimuth is 0° (Fig. 4.15).

Likewise, an object with an azimuth of 90° lies in the east. An azimuth of 180° places it due south, and if the object lies directly west, its azimuth is 270°.

Dobsonian Mounts

Chinese-born American amateur astronomer John Dobson (1914–) built his first telescope in 1956. The alt-azimuth mount he invented (Fig. 4.16A) revolutionized amateur astronomy.

A **Dobsonian mount** is a simple, dual-pivot mount almost always combined with a Newtonian optical tube assembly (Fig. 4.16B). All an observer needs to do is gently move the tube up or down to change the altitude and move it left or right (it will spin in a full circle) to change the azimuth.

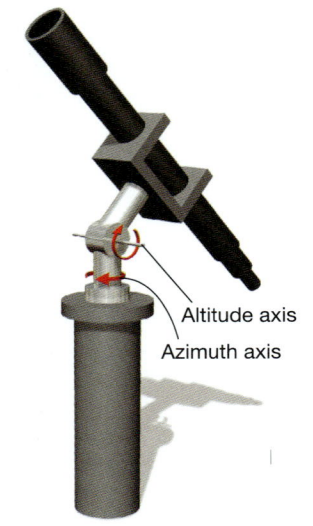

4.14 Example of an altitude-azimuth (alt-azimuth) mount. *Courtesy of Roen Kelly, Astronomy magazine*

GLOSSARY TERM

Dobsonian mount simple alt-azimuth mount invented by John Dobson

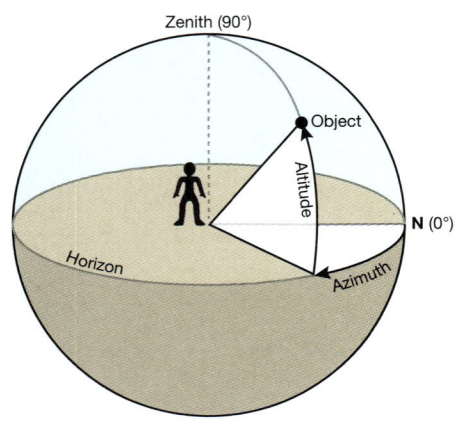

4.15 Examples of altitude, and azimuth. *Courtesy of Holley Y. Bakich*

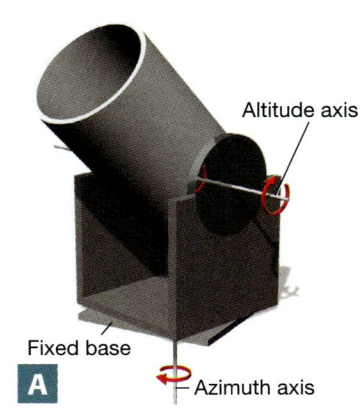

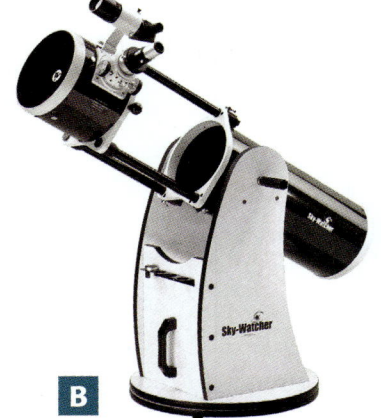

4.16 **(A)** Dobsonian mount, and **(B)** Newtonian reflector in a Dobsonian mount. *Courtesy of (A) Roen Kelly, Astronomy magazine, (B) Celestron*

Driven Alt-azimuth Mounts

A recent development in alt-azimuth mounts is the *driven* alt-azimuth mount. With motors attached to the mount's altitude and azimuth axes, the telescope either can 1) track an object across the sky once the object is placed in the **field of view,** or 2) interface with a computer to both find and track an object. Once found, an object can be followed without the observer moving the telescope.

> **GLOSSARY TERM**
> **Field of view** area visible through the lens of an optical instrument; measured in terms of angular diameter (e.g., 20 minutes of arc)

Equatorial Mounts

If Earth did not rotate on its axis once every 24 hours, an alt-azimuth mount is all any of us would ever need. But Earth does spin, and that must be dealt with when observing with a telescope. The equatorial mount (Fig. 4.17) tracks the apparent motion of the stars. It does this by aligning one of its axes parallel to Earth's axis. Two types of equatorial mounts exist: manual and motorized.

The technique of properly setting up an equatorial mount is called polar alignment (Fig. 4.18). All professional telescopes are polar aligned to high accuracy. Amateur astronomers must position small telescopes so an imaginary line extends from the mount's polar axis to the north celestial pole. For visual observing, rather than imaging, the axis can be aligned with Polaris, the North Star. Polaris is not at the exact north celestial pole, so the telescope will be a little misaligned. Celestial objects, however, will remain in the eyepiece's field of view for quite a while. If, on the other hand, the telescope will acquire images, alignment must be more precise.

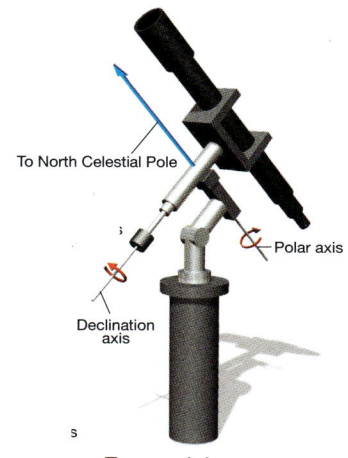

4.17 Example of an equatorial mount.
Courtesy of Roen Kelly, *Astronomy* magazine

Manual Equatorial Mounts

A manual equatorial mount is one that does not have an attached motor. Once the user has polar aligned such a setup, he or she must move the mount to follow objects across the sky. Almost all manual equatorial mounts incorporate a slow-motion control, which is simply a knob that slowly moves the telescope.

Motorized Equatorial Mounts

Motorized equatorial mounts, also known as go-to systems, combine a digital coordinate system with a dual-axis motor driver. The go-to drives in these mounts require the user to center one or more stars for initial alignment.

Most go-to drives contain many objects in their databases. In fact, some small telescopes equipped with go-to contain objects in their databases that the telescopes have no hope of revealing. Note that although manufacturers tout "go-to telescopes," technically only the drive is go-to.

> **Note**
> More about how to polar align a telescope can be found online. For example, telescope maker Celestron offers detailed instruction on its website: http://www.celestron.com/c3/support3/index.php?_m=knowledgebase&_a=viewarticle&kbarticleid=1732

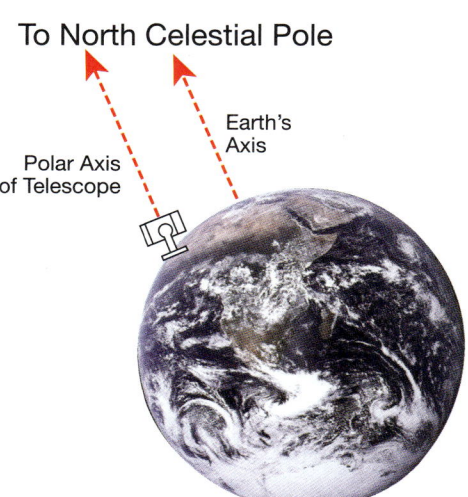

4.18 Polar alignment.
Courtesy of Holley Y. Bakich

Telescopes **CHAPTER 4** 53

Recently, manufacturers have paired telescope go-to drives with global positioning system (GPS) electronics and electronic compasses. The result is an almost hands-free setup. The computer figures out the position (longitude, latitude, and elevation) of the telescope. Next, it finds north, then true north, which is offset from the magnetic north a compass would show. Finally, the go-to system compensates for a non-level setup.

Many amateur astronomers believe a person's first telescope should not have a go-to drive, and their opinions have nothing to do with cost. They contend that all beginners should learn the sky, and star-hopping (locating a celestial object by moving from one identifiable star to the next) is part of that basic education.

On the other hand, most celestial objects are faint, and locating them can be difficult, especially for a beginner. Failure to learn how to find "things in the sky" is a common reason why people quit amateur astronomy. Because of this, we favor go-to mounts, and not just for beginners. Experienced observers who know the sky also can benefit from go-to mounts. Many times, experienced amateur astronomers have said, "Since I got my go-to drive, I spend much more time observing!"

Check Your Understanding

4.1 On a certain date, the planet Jupiter sits halfway up in the southeastern sky.

   a What is its altitude? _____

   b What is its azimuth? _____

4.2 What are the motions of an alt-azimuth mount?

4.3 What is the purpose of an equatorial mount?

4.4 When you polar align a mount, what do you align it to?

Additional Telescope Accessories

After you have made a careful selection of a telescope, mount, and starter eyepieces, you may want to add additional accessories. Think of it like buying a car. After some shopping around at various dealerships, you choose the make, model, and color. However, when you turn on the radio, you may realize your vehicle is incomplete and want to add a high-quality sound system. So it is with telescopes.

Eyepieces

Observers often choose **eyepieces** (Fig. 4.19) based on financial considerations. The best eyepieces are not cheap. For some observers, it is difficult to justify spending as much on eyepieces as on their telescopes. The investment, however, has to be looked at over the long term. If you upgrade your telescope (a flaw in the human spirit sometimes called "aperture fever"), you can use the same eyepieces. Even if your new telescope uses 2-inch eyepieces and your old one used 1.25 inch, many manufacturers produce inexpensive adapters.

An observer thinking about adding an eyepiece should, if possible, try it first in his or her own telescope. Borrow one from somebody at a star party, or perhaps a vendor may let you try out an eyepiece before you buy it.

Remember, too many personal variables exist to take someone else's word about the best eyepiece for you. Are you nearsighted? Farsighted? Wear eyeglasses when observing? The list goes on. Consider an eyepiece as a long-term investment for your future observing.

What an eyepiece shows to your eye is the true field of view, usually less than 2°. Manufacturers, however, often list an eyepiece's apparent field of view, which is the angular size of the light cone able to enter the eyepiece. Eyepiece apparent fields range from 25° to 84°.

When using eyepieces, notice the contrast between the object you are observing and the dark sky. Observing a bright planet or the lunar terminator (the dividing line between the Moon's light and dark areas) is a good test. How much of the light scatters into the dark area?

Edge sharpness provides another test for eyepiece quality. For this test, nothing beats star images moved to the edge of the field of view.

You can calculate the magnification of any eyepiece (Fig. 4.20) by using the following formula:

$$M = \frac{f_t}{f_e}$$

In this instance, **M** equals magnification, f_t equals telescope **focal length**, and f_e equals eyepiece focal length. For example, a telescope's aperture is 300mm and its focal length is 3,048 mm. A 22mm eyepiece gives a magnification of 3,048/22, which equals 138.54, or approximately 139×. Note that aperture has nothing to do with magnification.

> **GLOSSARY TERM**
> **Eyepiece** lens (or combination of lenses) at the eye end of an optical instrument through which an image is viewed

4.19 Example of an eyepiece. *Courtesy of Celestron*

> **GLOSSARY TERM**
> **Focal length** distance from the objective (lens or mirror) to the focal point in an optical system

4.20 Finding magnification. *Courtesy of Roen Kelly, Astronomy magazine*

Star Diagonals

Newtonian reflectors have a built-in 90° bend in the light path; refractors and SCTs do not. Because the best views of celestial objects come when they are high in the sky (when the amount of air between the object and the observer is minimized), observing with a "straight-through" arrangement often proves uncomfortable. The solution is a star diagonal (Fig. 4.21), a machined piece with one side that fits into the focuser and the other side accepts an eyepiece.

4.21 Star diagonals that bend light (**A**) 45°, (**B**) 90°.
Courtesy of Celestron

Barlow Lenses

A **Barlow lens** (Fig. 4.22) increases the effective focal length of an objective lens. Barlows are rated by magnification factors. One might magnify $2\times$, another $3\times$, etc. A Barlow lens can effectively double the number of eyepieces in your set. For example, in a telescope with a 1,500 mm focal length, a 15mm eyepiece will magnify $100\times$. If you insert a $2\times$ Barlow, the magnification will be $200\times$. So, the *effective* focal length of the objective has changed from 1,500 mm to 3,000 mm.

Finder Scopes

The world's best telescope is useless if you can't find anything with it. A quality **finder scope** (Fig. 4.23) will help point you in the right direction. Finder scopes come in two optical configurations: straight-through and right angle.

The straight-through finder inverts (flips) the image, but it allows you to site down the telescope tube, which is more intuitive for beginners than using a right-angle finder. For some observers, looking "at the telescope" while trying to find objects is difficult.

Use a finder that has a front lens at least 50 mm across. Such a finder will let in enough light so you can locate fainter objects. Magnifications between $7\times$ and $9\times$ work best.

Be sure to align your finder scope with your telescope. Check your telescope's owner's manual to learn how to do this simple procedure. If you travel to observe, align the finder prior to each session.

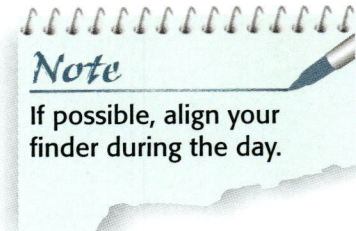

Note
If possible, align your finder during the day.

> **GLOSSARY TERM**
>
> **Barlow lens** negative lens that increases the effective focal length of an objective lens or mirror; rated by magnification factor $2\times$, $3\times$, etc.
>
> **Finder scope** low (or unity) power telescope attached to a larger telescope whose optical axis is aligned to the larger telescope; usually has a larger field of view than the telescopes to which they are attached

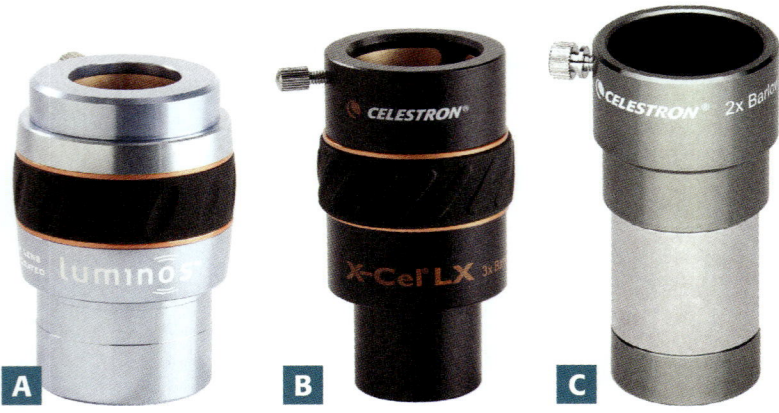

4.22 Barlow lenses: (**A**) Luminos, (**B**) X-Cel, (**C**) Omni.
Courtesy of Celestron

4.23 Finder scope.
Courtesy of Celestron

Focusers

A chain is only as good as its weakest link. For some telescope systems, the weakest link is the focuser. Quality focusers are not cheap. A focuser (Fig. 4.24) should hold your eyepieces securely, regardless of the telescope's position. Select a focuser that will accommodate both 2-inch and 1.25-inch eyepieces (with a slide-in adapter).

Storage Cases

Quality optical equipment must be treated with care. Not doing so invites dirt, scratches, and dents. Most observers acquire some type of foam-padded storage case.

A good storage case should be sturdy with quality hardware, and will survive the occasional minor ding, drop, or fall with minimum damage to the case itself and no damage to the contents. Such cases usually are sealed to keep out dust, are water tight, and foam filled. Most manufacturers use foam in the form of precut (though not quite all the way through) cubes. To use, simply figure out what size "hole" your equipment needs and remove just enough of the cubes.

Note
Removing one fewer cube than necessary provides a tighter fit. Whether or not your case has foam, always keep the plastic covers on your eyepieces for maximum protection against dust.

4.24 Examples of telescope focusers.

Courtesy of Celestron

 ## Check Your Understanding

5.1 A 6-inch (150mm) telescope has a focal length of 1,000 mm. What focal length eyepiece would you have to insert to get a magnification of 100×?

5.2 By how much does a star diagonal bend the light's path?

5.3 What is a good finder scope magnification?

Exercise 4.1 — Getting Familiar with Your Telescope

Procedure

Materials
- ☐ Telescopes
- ☐ Assorted eyepieces
- ☐ Finder scopes

1 Record the general procedures used to set-up the telescope.

    a What telescope are you using?

    b What is the diameter of your telescope?

    c What is the focal ratio of your telescope?

2 What do you see looking through the focuser *without* an eyepiece?

3 How does the telescope mount work? Move the telescope around, noting such things as how the telescope moves, any challenges in pointing the telescope, etc.

4 View through the telescope's finder scope, and point at an object (terrestrial or astronomical). How well does the finder scope work? Was the object easy to locate in the finder scope? What object did you aim at through the finder scope?

5 Place an eyepiece in the focuser and focus as directed. List each eyepiece and its magnification in Table 4.3. Select an object for observation, and view it with the naked eye. Describe what you see in the correct column in Table 4.3. View the selected object through the first eyepiece, and describe it in the correct column in Table 4.3. Repeat for each eyepiece listed. Use the same object for each observation in this step.

TABLE **4.3** Eyepiece vs. Naked Eye Observations

Eyepiece	Magnification	Object Viewed with Naked Eye	Object Viewed Through Eyepiece

6 Using a low-power eyepiece, move the telescope as listed in Table 4.4, and note what direction the object appears to move.

TABLE **4.4** Direction of Object Movement through the Telescope

Direction	How Does the Object Move through the Field of View?
Eyepiece end down	
Eyepiece end up	
Eyepiece end to the right	
Eyepiece end to the left	

Exercise 4.2 Lunar Observations and Impressions

Procedure

Materials
- ❑ Telescopes
- ❑ Eyepieces

1. Note the observing conditions at your site.

   a Observing site _____

   b Sky conditions _____

   _____

   c Viewing time: Start _____ End _____

2. Note the specifications of your telescope.

   a Telescope _____

   b Eyepiece _____

   c Magnification _____

3. Make your lunar observations. Describe how the Moon "appears" to you.

   _____
   _____
   _____
   _____

Exercise 4.3 Constructing a Simple Telescope

Procedure

Materials
- ❑ Simple Refracting Telescope kit

1. Construct the simple refracting telescope according to the instructions.

2. Put the simple refracting telescope to use, looking through the foam eyepiece end. To focus, push-pull the sliding cardboard tube assembly.

3. Use the telescope on a bright, easy-to-identify object in the lab. Take your time, and practice focusing the telescope.

4. Answer the questions in the data sheet, pages 61–62.

5. Upon completion of this procedure, carefully disassemble all of the telescope parts. Place the parts back into the baggie. Return the parts baggie and sliding telescope tubes to the lab table.

CAUTION Do not look at the Sun under any circumstances.

Name _____   Section _____   Date _____

Simple Refracting Telescope

1 Take the eyepiece assembly off of the telescope. Look through the end where the eyepiece is supposed to be placed, not the objective end. What do you see through the telescope without the eyepiece?

_____

_____

2 Put the eyepiece back onto the telescope. Once the telescope is focused, describe the image you see when looking through the telescope. Is it right-side up? (Note that many students have better success first using the telescope outdoors.)

_____

_____

3 Take the telescope outside, and look at four different objects located some distance away. In Table 4.5, list the object, what you can see of the object without the telescope, and what you can see with the telescope. **Have some patience as you focus the telescope; ask for help if you are having any difficulties.**

_____

_____

TABLE 4.5 Object Comparison

Object	How the Object Appears to the Naked Eye	How the Object Appears through the Telescope
1		
2		
3		
4		

4 One of the characteristics of the telescope is that it magnifies objects. Estimate the magnification, or power, of this simple refracting telescope:

_____ ×

5 How did you arrive at this estimate?

_____

_____

_____

> **Note**
> Hints for "guesstimating" magnification: 1) compare sizes of what you see through the telescope vs. the unaided eye, and 2) keep both eyes open as you look through the telescope.

Exercise 4.3 Data Sheet

Exercise 4.3 Data Sheet

6 Focus on a nearby object, noting where the two sliding tubes come together. Now focus on an object farther away (you might need to be outside), again noting where the two sliding tubes are positioned. How do these two positions compare?

7 Look at an object through the telescope. As you move the telescope's objective end to the right, what does the object appear to do?

8 What happens when you move the objective end to the left?

9 Again look at the same object, this time moving the telescope's objective end up. What does the object appear to do?

10 What happens when you move the objective end down?

11 After using this simple refracting telescope, what do you think would be some of the challenges using this telescope for:

  a Terrestrial (Earth-based) observing?

  b Astronomical observing?

Unit 1 Tools of the Trade

Looking Up

Star Maps

LEARNING OBJECTIVES

Upon completion of this chapter, you should be able to:

1. Explain the function of a star atlas.
2. Compare and contrast a variety of star atlases to select the best atlas for a particular level of observer.
3. Locate specific objects on several star charts.
4. Explain the role latitude plays when selecting a star chart.
5. Determine the scale on a star chart.
6. Describe how to use right ascension and declination to locate celestial features.
7. Identify stars, star clusters, nebulae, galaxies, and coordinates on a variety of star charts.
8. Define all glossary terms.

Constellations of the Northern Sky
Courtesy of Larry Landolfi/Science Source

GLOSSARY TERMS

Star chart map showing a particular part of the sky, or a map of the sky above a certain location at a certain time

Constellation one of 88 officially recognized areas of the sky according to the International Astronomical Union; also may refer to one of the 88 officially recognized groups of visible stars within those areas

Double star any two stars that appear close to one another in the sky; may be physically linked by gravity, or they may be chance alignments along our line of sight

Open cluster mostly young groups of between a few hundred and a few thousand stars; few open clusters are old, because a cluster's stars disperse over millions of years due to gravitational interactions within the cluster

Globular cluster spherical collection of old stars found within the halos of spiral galaxies and, more commonly, around elliptical galaxies; globular clusters usually are as old as their galactic hosts

Nebula Latin for "cloud"; a cloud of interstellar dust and gas; nebulae can be dark (blotting out the light from stars behind them) or bright (glowing because the gas absorbs energy from nearby stars and re-emits it as light)

Galaxy collection of up to thousands or billions of stars, dust, and gas, held together by gravity

Milky Way name of our galaxy; also, the band of diffuse starlight that stretches across the sky, best seen in the Northern Hemisphere during evenings in summer and winter

Messier, Charles French astronomer and comet hunter whose most famous work was a list of 109 objects he published that were not comets; astronomers designate each object with an M and the object number, such as M1, M2 . . . M109

Celestial sphere apparent background of the stars, assumed to be of infinite extent in all directions; the sky

5

This lab will introduce you to various types of star charts, also known as star maps, including what role your latitude plays when selecting a star chart. You will learn about chart designations and the coordinates right ascension and declination, and how to use a simple star chart.

Star Charts

Our ancestors watched the Sun, Moon, and planets move against the background of stars. They learned to tell time and predict the changing of the seasons by observing the heavens. Modern astronomical observers have a variety of tools to understand the motions of objects in the night sky. The simplest of the devices used to locate objects is a **star chart**, also known as a star map.

Star charts are designed for specific latitudes, but they work well in a range of 10° north and 10° south of the listed latitude. Star charts can be either circular or rectangular, and most have the directions printed along their circumferences.

When you use a star chart, you will immediately notice several things:

- Charts show star brightness by the size of the dots representing the stars
- Larger charts show constellation boundaries and the **constellations**
- Some of the brightest stars are labeled
- Symbols (see the star chart's key) may also indicate the positions of deep-sky objects, such as **double stars, open clusters, globular clusters, nebulae,** and **galaxies**
- Some charts show the outlines of the **Milky Way** as it appears in the sky

A notable aspect of star charts is that many of the stars are labeled with Greek letters. The process of labeling the brightest stars in a constellation with lowercase Greek letters began in 1603 when German uranographer (celestial cartographer) Johann Bayer (1572–1625) published his celestial atlas, *Uranometria*. In most cases, Bayer labeled a constellation's brightest star with the first letter in the Greek alphabet, Alpha (α). To the second-brightest star he assigned letter number two, Beta (β), and so on. Large, bright constellations often have many Greek letters, whereas small, faint star patterns may contain only an Alpha.

In the eighteenth century French comet-hunter **Charles Messier** compiled a catalog of interstellar objects. As he searched for comets through his small telescope, Messier occasionally encountered objects that looked fuzzy, like comets. But, unlike comets, they didn't move against the background stars. He published a list of 109 objects for the benefit of other comet hunters. Today, amateur astronomers regard the objects on Messier's list as some of the finest telescope targets in the sky.

Messier isn't the only one who cataloged interstellar objects. You might see the letters NGC followed by a number. These designations come from an astronomical object list called the *New General Catalogue*. English astronomer John Louis Emil Dreyer (1852–1926) published this list of 7,840 objects in 1888.

The Celestial Sphere

When orienting yourself with a star chart, it is useful to imagine objects in the sky attached to a sphere with Earth at the center, which is how the ancients thought of it. The sphere is called the **celestial sphere** (Fig. 5.1). Relative to Earth's vantage point, the celestial sphere rotates from east to west. Our planet is really the object in motion, and it moves west to east, so everything in the sky appears to travel in the opposite direction.

If rotation were our planet's only motion, the stars wouldn't change from night to night. All the stars would rise at the same time, no matter what time of year. But Earth's motions include more than rotation. We also orbit the Sun once a year; astronomers call this motion revolution. Because of revolution, we are in a different position in space each night, so our planet has to rotate approximately four additional minutes to realign with the stars. The combination of rotation and revolution, therefore, is the reason that the stars rise four minutes later each night.

The **celestial equator** is a projection of Earth's equator onto the celestial sphere. If you are at Earth's equator, the celestial equator runs from the east point on the horizon through the overhead point back down to the west point on the horizon.

Likewise, the **celestial poles** are an extension of Earth's north and south poles. Our North Star, Polaris (Alpha [α] Ursae Minoris), lies within 1° of the North Celestial Pole (NCP). Unfortunately, for those observing from the Southern Hemisphere, the South Celestial Pole (SCP) has no relatively bright star at or near it.

Finding the Scale

To determine the **scale** of the star chart you are using you must calculate the number of degrees each millimeter represents. Make this calculation for the star chart in Figure 5.2.

For what latitude does the chart show the sky? We call this number **L**. Figure 5.2 shows a sky at 40° north latitude.

One fact about the sky is that the altitude (in degrees) of the North Celestial Pole (NCP) equals the latitude of the observer. For an observer at the North Pole (latitude = 90°), the NCP is directly overhead; that is, it has an altitude of 90°. For an observer at the equator (latitude = 0°), the NCP lies on the horizon at an altitude of 0°.

To find the scale of a star chart, first measure the distance in millimeters between the NCP and the north point on the horizon. We call this distance **M**, and we show it on a simplified star chart in Figure 5.2. To find the number of degrees per millimeter, simply divide **M** by **L**. On the example star chart, note that the NCP lies 55.176 millimeters from the north point on the chart's horizon. That is the distance **M**. To find the scale of the

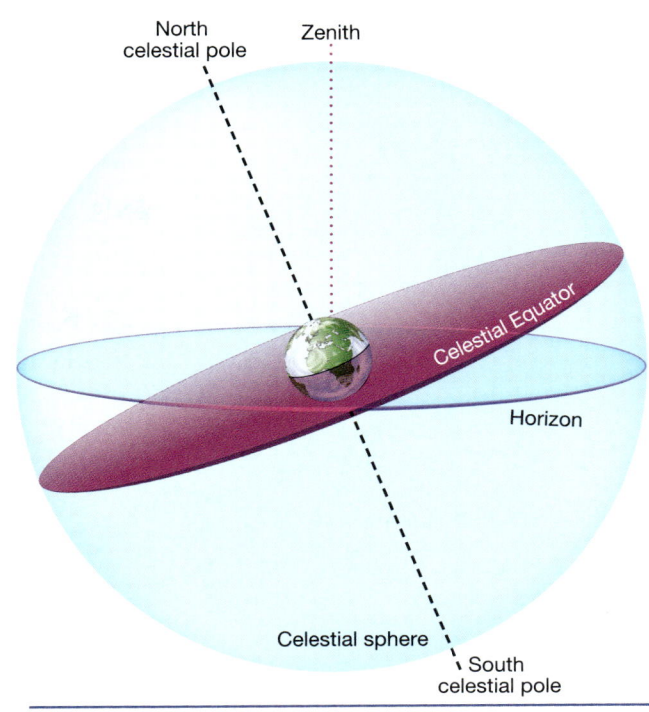

5.1 Celestial sphere as it appears from approximately latitude 30° north.
Courtesy of Holley Y. Bakich

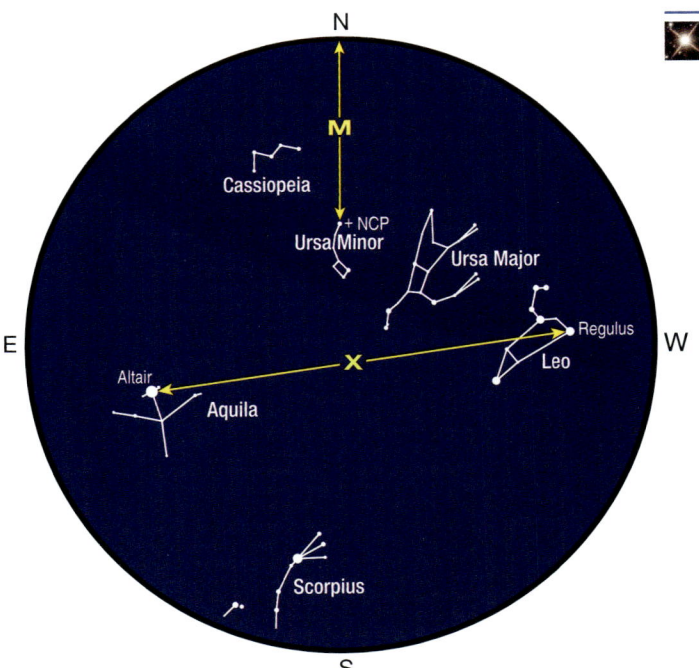

5.2 A 40° north latitude star chart showing determination of scale (M) and degrees of separation between the stars Altair and Regulus (X).
Courtesy of Holley Y. Bakich

GLOSSARY TERMS

Celestial equator intersection of Earth's equatorial plane with the celestial sphere; the projection of Earth's equator onto the sky

Celestial poles two points in the sky located above Earth's North and South Poles

Scale ratio of the distance on a star chart to the angular distance in the sky

Note

To use a star chart, hold it nearly overhead so the direction you are facing is at the bottom. For example, if you are facing north, the north side of the star chart should be at the bottom.

GLOSSARY TERMS

Circumpolar star star whose apparent daily path through the sky is located completely above an observer's horizon; at the equator no star is circumpolar; at the North or South Pole all stars are circumpolar; observers at any other latitude, a star whose declination is greater than 90° minus the observer's latitude will be circumpolar. Example: At latitude 40°, any star with a declination greater than 50° (90° minus the observer's latitude) is circumpolar

Declination angle similar to latitude on Earth; measured from the celestial equator to a point on the celestial sphere; declination is positive if the object is north of the celestial equator and negative if the object is south of the celestial equator

Right ascension angle measured eastward along the celestial equator from the vernal equinox to the intersection of the hour circle passing through an object; usually expressed in hours, minutes, and seconds from 0 hours to 24 hours, where one hour of right ascension equals 15°

Arcminute also known as a minute of arc, this is an angle equal to 1/60 of 1°

Arcsecond also known as a second of arc, this is an angle equal to 1/3600 of 1°, or 1/60 of 1 arcminute

---

chart in degrees per millimeter, simply divide L by M. The scale, therefore, is 40/55.176 mm = 0.725°/mm. Now note the location of the stars Regulus in the constellation Leo and Altair in the constellation Aquila. The distance between the two stars is marked by X. These two stars lie 123.585 millimeters apart. Therefore in the sky they are separated by 123.585 mm × 0.725 °/mm = 170.5°.

Not all stars are visible from specific latitudes, because Earth is in the way and blocks half of the celestial sphere from your view at any time (Fig. 5.3). Stars that never set for observers at any given latitude within the number of degrees that equal that latitude of the NCP are called **circumpolar stars**. They remain in the sky, although it is too bright in the daytime to see them.

Consider the location of the South Celestial Pole (SCP) as seen from latitude 30° north. It is situated 30° below the southern horizon. Any objects within 30° of the SCP will never be seen in this latitude; they are always below the horizon.

Celestial Coordinates

As with earthly longitude and latitude, any point on the celestial sphere with celestial coordinates can be specified. Latitude tells how far north or south you are from Earth's equator. Longitude tells how far east or west you are of the Prime Meridian, an imaginary line, or starting point, that runs through Greenwich, England. On the celestial sphere, **declination** (Dec.) replaces latitude, and **right ascension** (R.A.) replaces longitude.

Declination (Fig. 5.4) is the measure how far north or south of the celestial equator an object is in the sky. The celestial equator lies at declination 0°. As with Earth's latitude, declination is measured in degrees, minutes, and seconds. There are 60 **arcminutes** (designated ') in 1°, and 60 **arcseconds** (designated ")

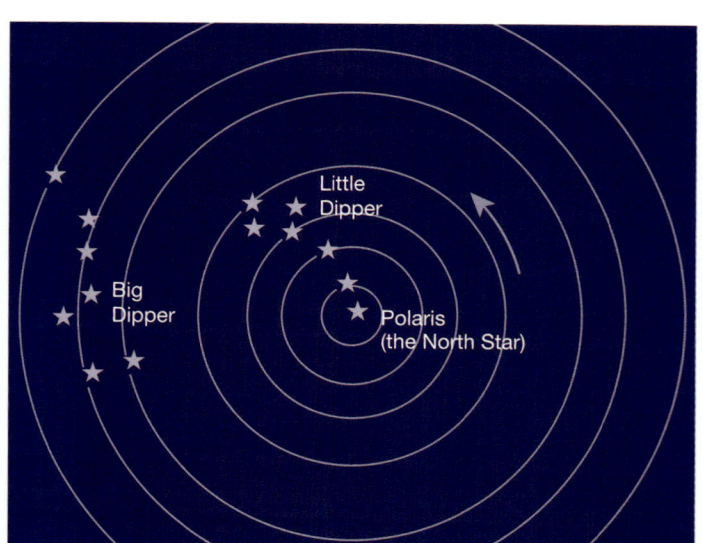

5.3 As the Earth rotates, all of the stars of the Little Dipper (and one star in the Big Dipper) remain in view. These are called circumpolar stars. The rest of the stars shown, however, set at some point.
Courtesy of Holley Y. Bakich

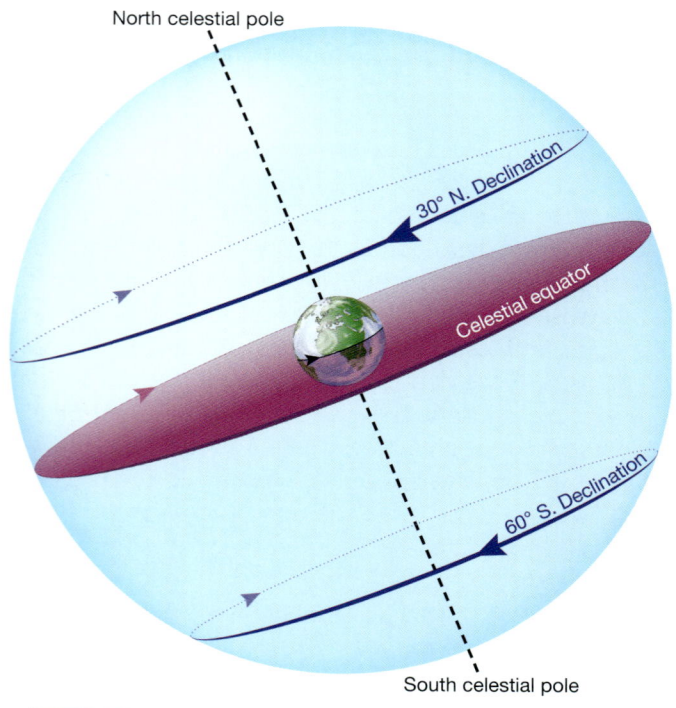

5.4 Examples of different declinations.
Courtesy of Holley Y. Bakich

in 1'. All circles of declination are parallel to the celestial equator.

Right ascension (Fig. 5.5) is trickier than declination. Like Earth's longitude, the starting spot for right ascension is arbitrary, but there is a reason behind it. Astronomers chose the Sun's position on March 21 (spring equinox) as the starting point. This is one of the intersections between the celestial equator and the ecliptic. They designated this position 0 hours (written as 0h). R.A. runs from 0h to 24h; each hour equals 60', and each minute equals 60".

Because of Earth's spin, the sky rotates once every 24 hours (approximately) and there are 24 hours of right ascension. Therefore, astronomers have divided the sky into distances called hours, and the whole system represents a big clock. A circle contains 360°, so each hour of R.A. contains 15°.

Figure 5.5 also shows the ecliptic, a circle on the celestial sphere that shows the Sun's apparent path during the year to an observer on Earth. The 12 famed constellations of the zodiac (Aries the Ram, Taurus the Bull, Gemini the Twins, Cancer the Crab, Leo the Lion, Virgo the Maiden, Libra the Scales, Scorpius the Scorpion, Sagittarius the Archer, Capricorn the Goat, Aquarius the Water-Bearer, and Pisces the Fish) all lie on or near the ecliptic.

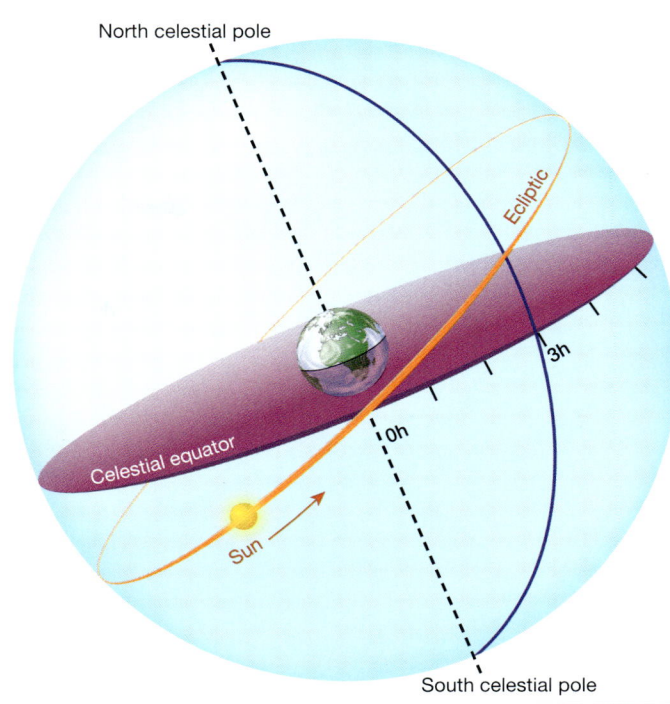

5.5 Illustration showing the line of right ascension equal to three hours, and the ecliptic. Any celestial object on this line, therefore, would have a right ascension of 3h. The ecliptic shows the Sun's apparent path seen from Earth during the year. Courtesy of Holley Y. Bakich

Looking Up

Astrology and the Zodiac

What's Your Sign?

There is less than a one-in-eight chance that the Sun was "in" a constellation given by any astrological horoscope. The reason lies in a long-term motion of Earth called precession. First described by the Greek scientist Hipparchus in the second century BCE, precession involves a continuous westward motion of the vernal equinox along the ecliptic. Both astronomers and astrologers use the vernal equinox, whose date defines the first day of spring in the Northern Hemisphere (and the first day of autumn in the Southern Hemisphere), as the beginning of the zodiac.

One complete precessional cycle takes 25,800 years. So, Sun-sign dates established more than 2,000 years ago are now out of sync by almost one-twelfth the circumference of the sky. Unfortunately, for those who believe in astrology, this means the signs are displaced with respect to their dates by almost one constellation, and it will only continue to get worse as time goes on. The next time the Sun signs and the dates given by astrologers agree won't occur for almost 24,000 years!

Also, note how the Sun's time in each constellation varies greatly. It spends 45 days within the boundaries of Virgo, 18 days in Ophiuchus, and only seven days in Scorpius. Horoscopes you commonly see list between 29 and 32 days as the time the Sun spends in each sign.

The Missing Sign?

Recently, there was a debate about adding Ophiuchus to the list of astrological signs. You might wonder how early skygazers missed Ophiuchus as a Sun sign. Our daytime star spends nearly three times as much time in it as in neighboring Scorpius. The reason is that, 2,000 years ago, like today, the number 13 was considered unlucky. On the other hand, 12 was a "magic" number, because you can divide it evenly by 1, 2, 3, 4, 6, and itself. That made it much easier for numerologists to work with. So, early astrologers simply chose not to refer to Ophiuchus, which lies between the much better known constellations of Scorpius and Sagittarius.

Check Your Understanding

1.1 Calculate M for Figure 5.6. M = _____

5

5.6 January star chart.

Courtesy of Richard Talcott and Roen Kelly, *Astronomy* magazine

1.2 Using Figure 5.6, measure the distance between each pair of objects.

    a Betelgeuse and Sirius: _____

    b Mira and NGC 253: _____

    c Aldebaran and Algol: _____

    d Castor and Pollux: _____

    e M31 and M33: _____

    f Polaris and Capella: _____

1.3 Using your calculated M value and measured distances, calculate the number of degrees that separate each pair of objects in the sky.

    a Betelgeuse and Sirius: _____ °

    b Mira and NGC 253: _____ °

    c Aldebaran and Algol: _____ °

    d Castor and Pollux: _____ °

    e M31 and M33: _____ °

    f Polaris and Capella: _____ °

1.4 What is the starting point for measurements of declination? _____

1.5 What is the maximum declination a celestial object can have? _____

1.6 How many degrees are in 1 hour of right ascension? _____

Star Atlases

As an astronomy lab student, one of your most important accessories is a good star atlas. A star atlas provides a road map of the sky (Figs. 5.7 and 5.8). As such, it can start you on an adventure that will last your entire life. The objects it highlights and the information it contains will guide you on a tour of the sky in a way similar to a map of the Earth.

If you are just beginning to explore amateur astronomy, pick an atlas that displays the stars you can see with just your eyes. These atlases show stars to about **magnitude** 6 (the same as the faintest star most people can see from a dark viewing site). Magnitudes range from small (and even negative) numbers, which represent the brightest objects, to larger numbers. That is because the first person to divide stars into magnitudes, the Greek astronomer Hipparchus, defined 1st-magnitude stars as the brightest, then came stars of the 2nd magnitude, and so on to 6th magnitude; the larger the magnitude, the fainter the

> **GLOSSARY TERM**
>
> **Magnitude** measure of the amount of light (or other radiation) received from a luminous celestial object

5.7 Constellation Scorpius: photograph.

Courtesy of Bill and Sally Fletcher

5.8 Constellation Scorpius: star chart.

Courtesy of Richard Talcott and Roen Kelly, *Astronomy* magazine

object, and vice versa. Since the invention of the telescope, astronomers have found stars and other objects with magnitudes far fainter than Hipparchus saw, so they often use numbers larger than 6. Beginner atlases also show a wide swath of sky on each page and may include constellation outlines.

These atlases also limit the number of plotted **deep-sky objects** to only those visible through a 4-inch telescope. That list includes the 109 **Messier objects**, selected objects from the *New General Catalogue* (NGC), and the brightest and most colorful double stars and **variable stars**. Following are some suggested introductory and advanced star atlases.

Introductory Atlases

Abrams Planetarium *Sky Calendar*

Although the Abrams Planetarium *Sky Calendar* (Fig. 5.9) is not a traditional star atlas, for four decades it has promoted basic skywatching. As its name implies, the sheet for each month takes the form of a calendar. Diagrams track the Moon's motion past the planets, bright stars, and notable groupings of sky objects. The reverse side (Fig. 5.10) shows the month's mid-evening sky for latitude 40° north, which makes it useful anywhere

> **GLOSSARY TERMS**
>
> **Deep-sky object** any celestial object not part of our solar system
>
> **Messier object** one of the objects on French comet-hunter Charles Messier's list of 109 objects that, he thought, astronomers of his time could confuse with comets
>
> **Variable star** star whose brightness changes by any means

© ABRAMS PLANETARIUM

SKY CALENDAR MAY 2013

An aid to enjoying the changing sky

Selected highlights for May: Jupiter (mag. –2.0) begins May as the most prominent evening "star", despite Venus' greater brilliance of mag. –3.9. In early May, Venus is less noticed because it sets in bright twilight, just 0.7 hr after sunset on May 1, compared to Jupiter's 2.9 hrs after sunset. Diagram for May 1, 30 min after sunset, shows Venus very low in WNW, 27° LR of Jupiter. On each successive evening, the gap between the two bodies shrinks by 1°. **On Memorial Day, Mon. May 27, Jupiter will be seen just 1.2° left of Venus**, and the two planets will set together 1¼ hrs after sunset. **On the next evening, Tues. May 28, these two brightest planets form their closest pair, with Jupiter just over 1.0° south (LL) of Venus.** On these nights when the two brightest planets appear closest to each other, **Mercury will appear 2.4°–2.8° above Venus.** Before Venus overtakes Jupiter on May 28, Mercury passes superior conjunction behind Sun's disk on May 11 and within a week emerges to LR of the two brighter planets. Using binoculars and diagrams on this calendar, try for Mercury within 5° LR of Venus on May 18; and within 2.5° LR on May 21. Even though Mercury fades from mag. –1.7 to –0.4 during May 16-31, it gets easier to spot daily, as it sets later in darker twilight. **On Fri. May 24, Mercury passes within 1.4° N of Venus. This is the first of six evenings of a Venus-Jupiter-Mercury trio**, defined by astronomer Jean Meeus as when three planets fit within a field no more than 5° across. *These evenings, May 24-29, will provide spectacular views for binoculars!* On the third evening of the trio, Sunday May 26, Mercury passes 2.4° N of Jupiter, and all three planets fit into a field less than 3° across, as they do again on the following evening. For more about the trio, and on the Moon and planets in May, visit www.pa.msu.edu/abrams/May2013Bonus.pdf

This issue may be reprinted for free distribution.

Planetarium business office: (517) 355-4676
Night Sky Notes on World Wide Web:
http://www.pa.msu.edu/abrams/nightskynotes/

Robert C. Victor
ISSN 0733-6314

Subscription: $11.00 per year, starting anytime, from Sky Calendar, Abrams Planetarium, Michigan State University, 755 Science Rd, East Lansing, MI 48824 or online at www.pa.msu.edu/abrams/SkyCalendar/

Courtesy of Abrams Planetarium, Michigan State University

Abrams Planetarium: *Sky Calendar.*

This chart is drawn for latitude 40° north, but should be useful to stargazers throughout the continental United States. It represents the sky at the following local daylight times:

Late April	11 p.m.
Early May	10 p.m.
Late May	9 p.m.

This map is applicable one hour either side of the above times.

May Evening Skies

© 2013 Abrams Planetarium
Subscription: $11.00 per year, from *Sky Calendar*, Abrams Planetarium, 755 Science Rd, East Lansing, MI 48824 or online at www.pa.msu.edu/abrams/SkyCalendar/

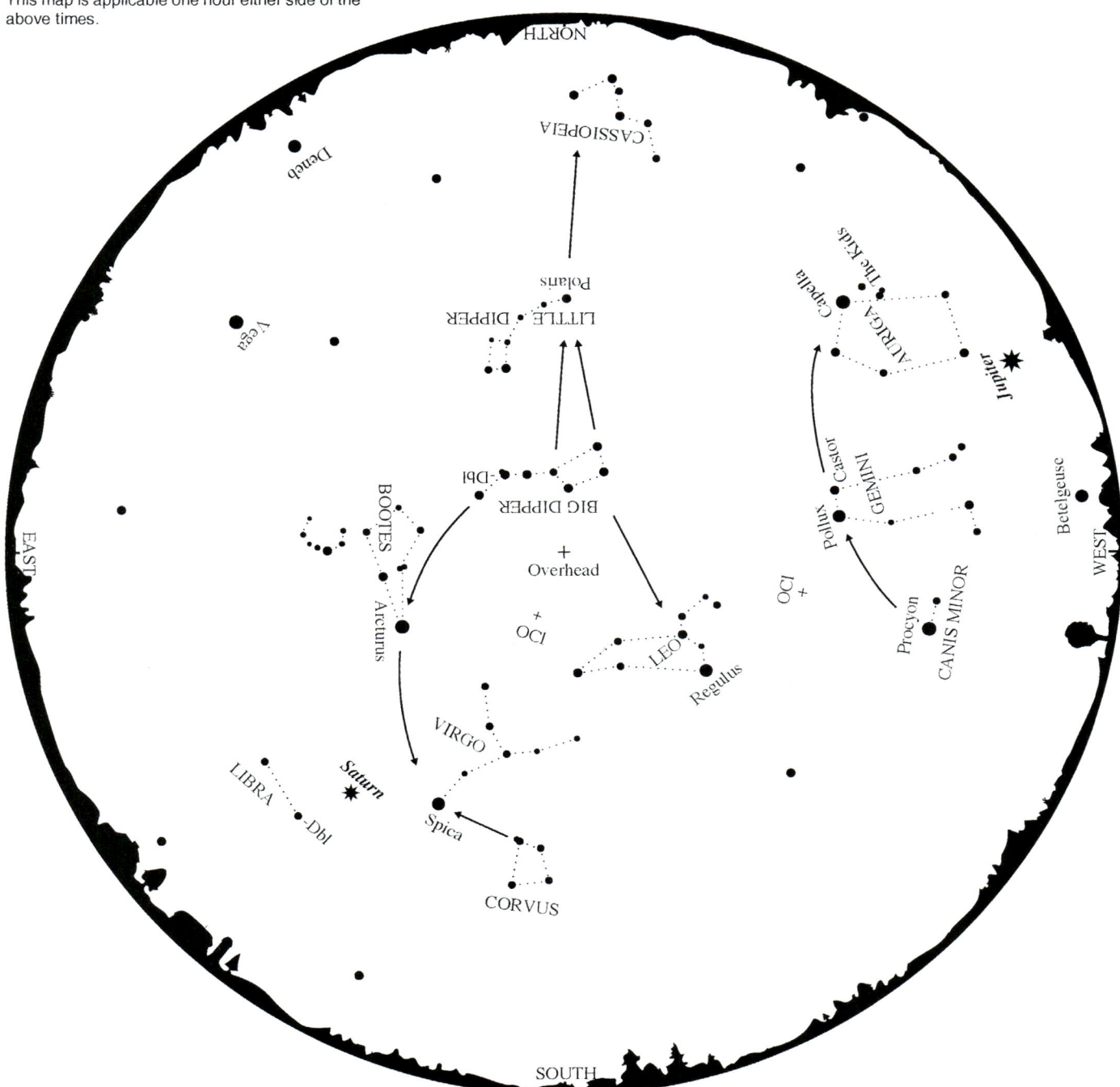

The planets Jupiter and Saturn are plotted for mid-May 2013. Eleven objects of first magnitude or brighter are visible. In order of brightness they are: Jupiter, Arcturus, Vega, Capella, Saturn, Procyon, Betelgeuse, Spica, Pollux, Deneb, and Regulus. In addition to stars, other objects that should be visible to the unaided eye are labeled on the map. The double star (Dbl) at the bend of the handle of the Big Dipper is easily detected. The double in Libra is more challenging. The open or galactic star cluster (OCl) known as the "Beehive" can be located between the Gemini twins and Leo. Coma Berenices, "The Hair of Berenice," is another open cluster (OCl), between Leo and Bootes. Try to observe these objects with unaided eye and binoculars.

—D. David Batch

5.10 Abrams Planetarium: star chart.

Courtesy of Abrams Planetarium, Michigan State University

> **Note**
> Abrams Planetarium's *Sky Calendar* is loose-leaf and mailed quarterly (three months per mailing). Additional information is available on Abrams Planetarium's website, www.pa.msu.edu/abrams/SkyCalendar/index.html.

in the continental United States. And while some star atlases contain thousands of stars, the *Sky Calendar* shows only the brightest 400 or so, depending on the month. That number works well for beginners or for those who observe under moderate light pollution.

Norton's Star Atlas and Reference Handbook

The oldest atlas on this list is *Norton's Star Atlas and Reference Handbook*, which first appeared in 1910. *Norton's* contains two circular charts showing the celestial poles and six rectangular charts, each four hours of right ascension wide. Approximately 8,800 stars to magnitude 6.5 are displayed, as are hundreds of double and variable stars, and more than 600 deep-sky objects. All eight charts have two pages of lists and notes about that particular chart's most interesting objects.

Bright Star Atlas

The *Bright Star Atlas* by Wil Tirion is for users of small backyard telescopes. It shows stars to magnitude 6.5 and includes about 600 deep-sky objects. This atlas divides the sky into four polar and six equatorial zones. The charts include constellation boundaries (the edges of each constellation) but no constellation figures (the lines that connect the stars and form the pattern the constellation represents).

Advanced Star Atlases

If you want to delve more deeply into the night sky, you will need more information than the above atlases can provide. When you scan the sky through large telescopes—those with apertures of 10 inches or more—a 6th-magnitude limit doesn't take you far. For large-telescope use, turn to advanced star atlases that show stars to 9th magnitude and fainter, and plot tens of thousands of deep-sky objects.

Uranometria 2000.0

The two-volume *Uranometria 2000.0* contains 220 double-page atlas charts, which collectively show more than 280,000 stars to magnitude 9.75, and 30,000 deep-sky objects. The atlas also includes a set of 22 "Uranometria Star Maps" at the front of each volume. These wide-scale charts serve as keys to the more detailed charts. *Uranometria* organizes its charts by declination. Within each declination range, the charts display objects from east to west, which allows each chart to flow across the book's binding, creating continuous two-page charts.

Sky Atlas 2000.0

Wil Tirion's *Sky Atlas 2000.0* contains 26 charts that show 81,312 stars down to magnitude 8.5, as well as more than 2,700 deep-sky objects. Unlike other atlases, *Sky Atlas 2000.0* is available in several different versions. The field version features white stars and deep-sky objects against a black background, just like a view through a telescope. The desk version plots black stars and objects on a white background, which works better for daytime reading. Both the field and desk versions are unbound. A spiral-bound deluxe version features black stars and color-coded deep-sky objects on a white background.

Great Atlas of the Sky

The final atlas on the list, the *Great Atlas of the Sky*, Jubilee Edition, is the world's largest printed atlas. It contains 296 charts. This atlas plots 2,430,768 stars to magnitude 12. Atlas charts also show more than 70,000 deep-sky objects. The *Great Atlas of the Sky*'s binding allows for easy removal of individual charts for outdoor use. The atlas also comes with a coordinate grid overlay to help you plot additional objects with known coordinates, like comets.

Check Your Understanding

2.1 How is the Abrams Planetarium *Sky Calendar* different from the other beginning atlases described above?

2.2 Beginning atlases are appropriate for use with approximately what size telescope?

2.3 Which of the atlases listed above displays the most stars?

2.4 Which advanced atlas listed here displays the fewest stars?

2.5 List several reasons an observer would benefit from a close-up chart of a deep-sky object or a region of space.

Looking Up
Alternative Star Charts

The Planisphere

A planisphere is a type of star chart that allows you to display the sky on any date and at any time of night. It does this by using two adjustable disks that rotate around a central pivot point. The upper disk has a cut-out, the bottom edge of which represents the horizon. The bottom disk is the star chart. Because the bottom disk shows the entire sky at all times, the top disk's cutout is necessary to limit the user's view to just the stars visible on a particular date and time.

Manufacturers alter the horizon line on the upper disk and sometimes the position of the star chart on the bottom disk to create a planisphere that shows the sky from a specific latitude, for instance 40° north. Although that is the label the planisphere carries, such a device is useful from latitudes 25° to 55°. Likewise, a 30° planisphere would have an approximate useful latitude range from 15° to 45°.

Star-Charting Apps

Many planetarium software apps have been developed for Apple, Android, and other products. A few examples:

Dark Sky Meter — This app lets you measure how dark your observing location is.

Star Walk — This app pinpoints the exact location of the celestial object you want to see. You then can use the precise location to find the object in your telescope.

Pocket Universe — This app for Apple users provides a virtual tour of the cosmos, including quizzes and event notifications.

Mobile Observatory — This app for Android users comes with updated star maps, has interactive views of the solar system, and "zoomable" views of the sky and planets.

Online Planetarium Software

Also known as star-charting software, you can download one of these packages and create digital or printed charts that show certain areas of the sky with as many details as you wish to include. The main features of each software package are similar. Best of all, each one is free.

- *Carte du Ciel,* www.ap-i.net/skychart/en/start
- *Hallo Northern Sky,* www.hnsky.org/software.htm
- *Stellarium,* www.stellarium.org

Exercise 5.1 Using a Star Chart

Procedure

Materials
- ❏ Star chart
- ❏ Binoculars
- ❏ Telescope (optional)

1 Answer the following questions:
   a What time did your observing session begin?

   b Note the overall sky conditions, including light pollution.

   c Is the Moon visible? If so, what is its phase?

   d Are any planets visible? If so, list them.

2 List five constellations you used your star chart to find.
   1.
   2.
   3.
   4.
   5.

3 With binoculars and/or telescope, examine five stars you used your star chart to locate, then complete Table 5.1.

TABLE 5.1 Star Chart Data

Name	Constellation	Color	Equipment Used

4 What time did your observing session end?

76 Unit 2 Looking Up

Using *Atlas of the Stars* Charts

Exercise 5.2

In 2006 *Astronomy* magazine produced *Atlas of the Stars*, an easy-to-use and cost-effective publication. The 24, two-page, full-color star charts display all of the constellations, more than 5,000 stars, and more than 1,000 deep-sky objects (star clusters, nebulae, and galaxies). Each chart includes a two-page introduction that points observers to the best objects for viewing. There is also a list of the finest double stars for each chart and complete lists of Messier and Caldwell objects.

This lab shows you how to use the *Atlas* and its many features. It also will be a review of some of the fundamentals of finding celestial bodies, the terminologies used, and an introduction to some of the universe's spectacular objects that are visible from our vantage point.

Note
The complete digital version of *Atlas of the Stars* is available for purchase and download at www.astronomy.com.

Procedure

1 Look at the three, two-page charts included in this lab (Figs. 5.11–5.13). In general, how is the atlas laid out?

2 Answer the questions on the data sheets, pages 85–88.

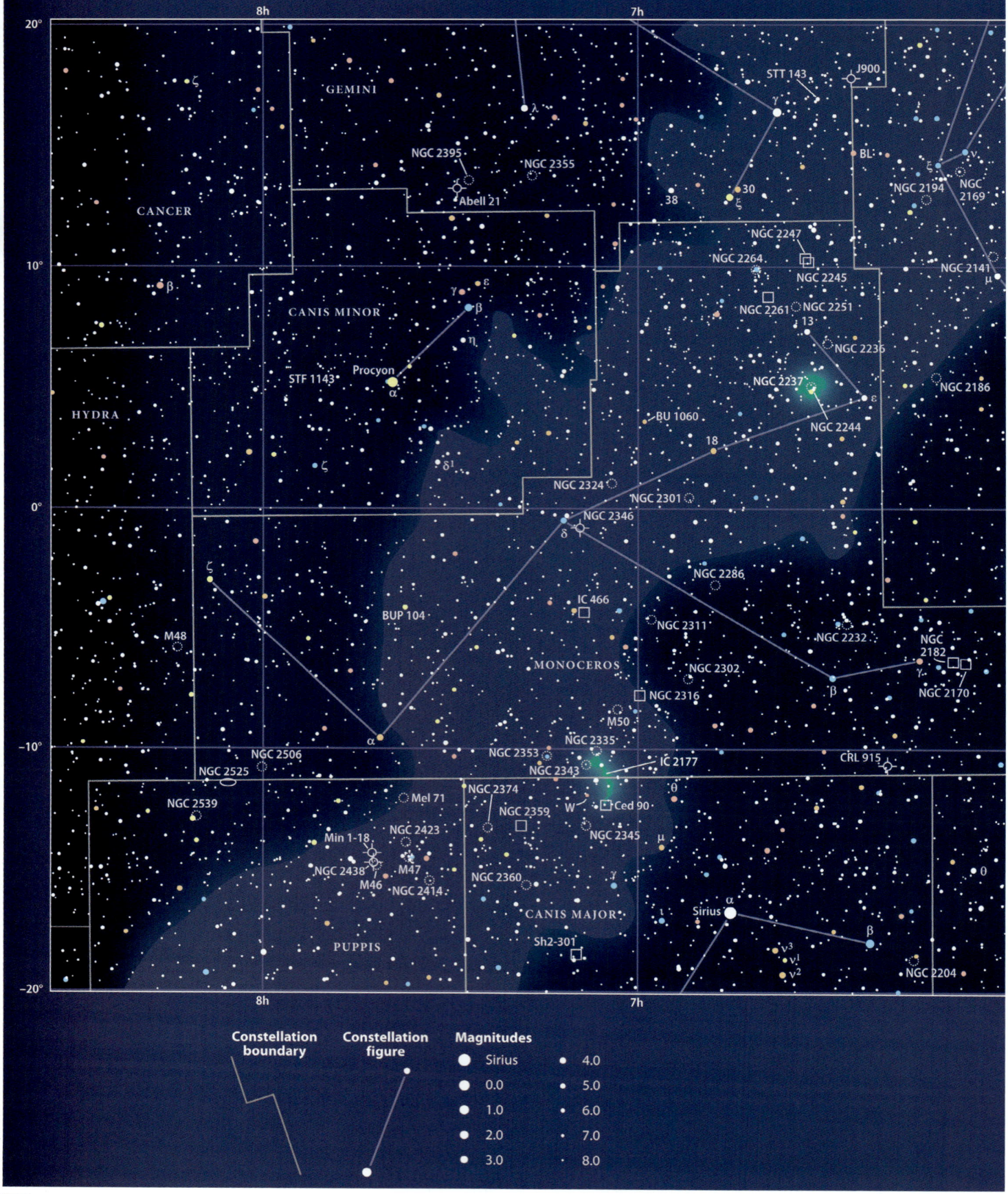

5.11 Star chart featuring the constellation Orion the Hunter and some of the star patterns that border it.

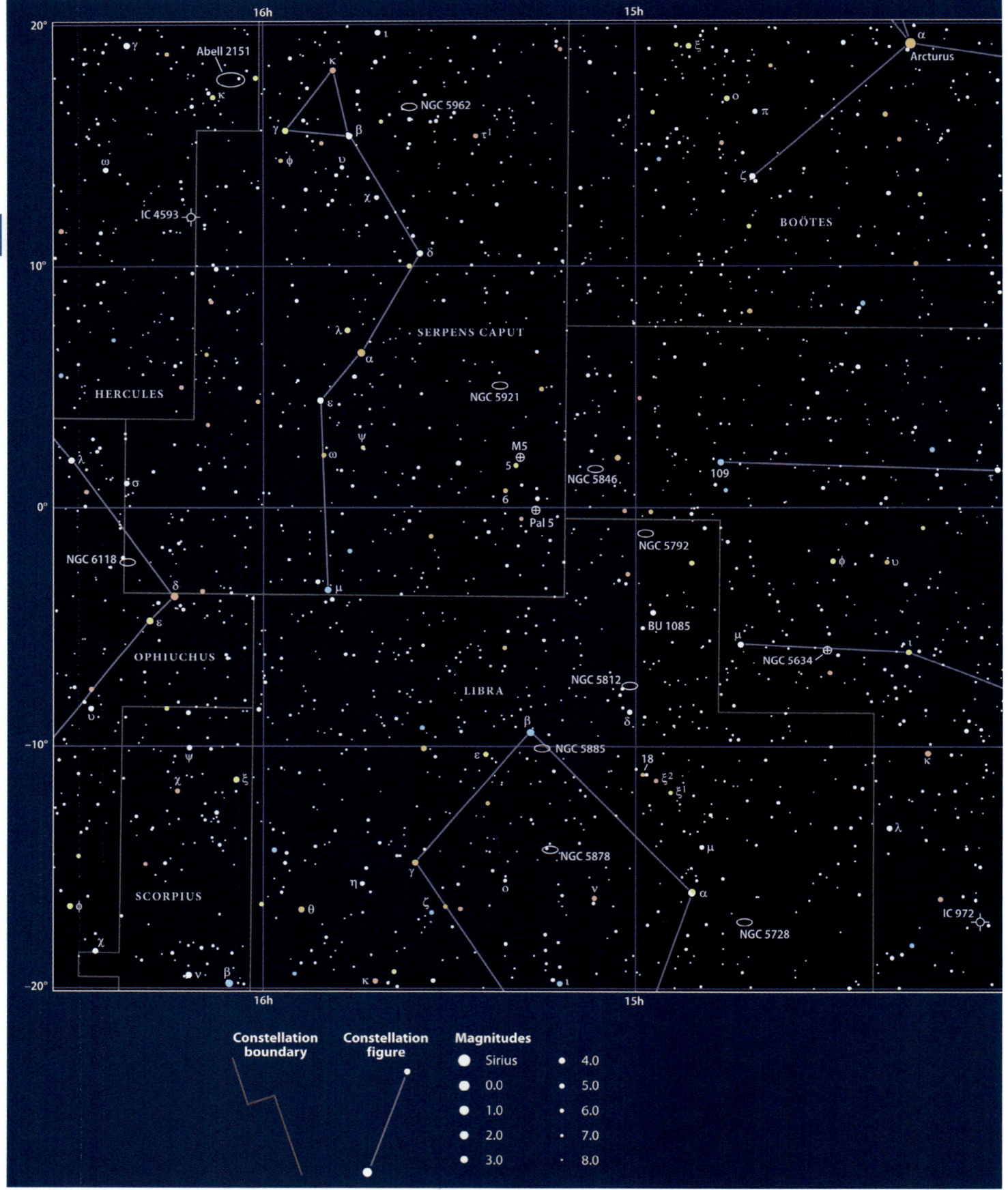

5.12 Star chart featuring the second-largest of all constellations, Virgo the Maiden. The many galaxies in the upper right

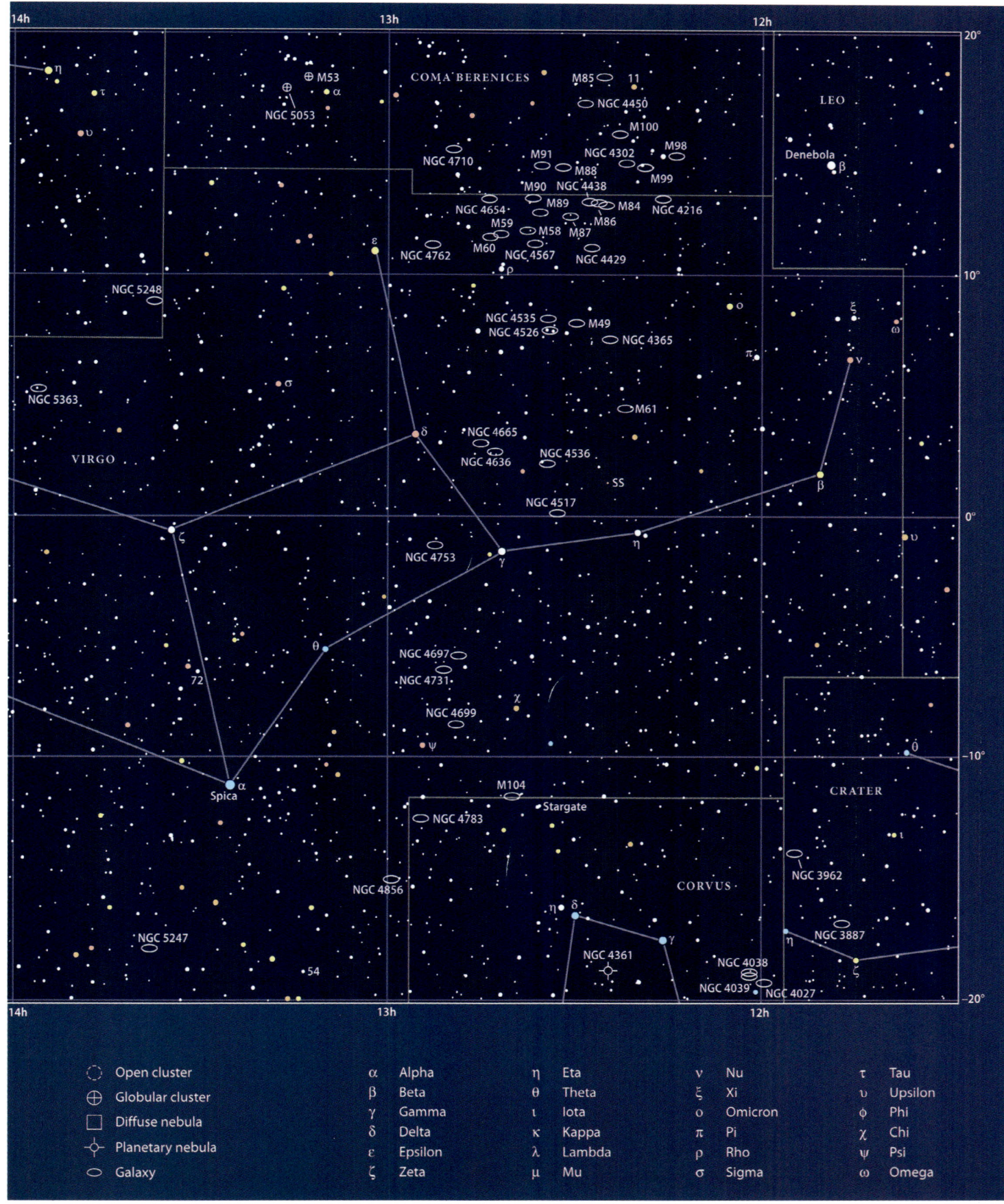

are the brightest members of the Coma-Virgo cluster of galaxies.

Courtesy of *Astronomy* magazine/Kalmbach Publishing

Star Maps CHAPTER **5** 81

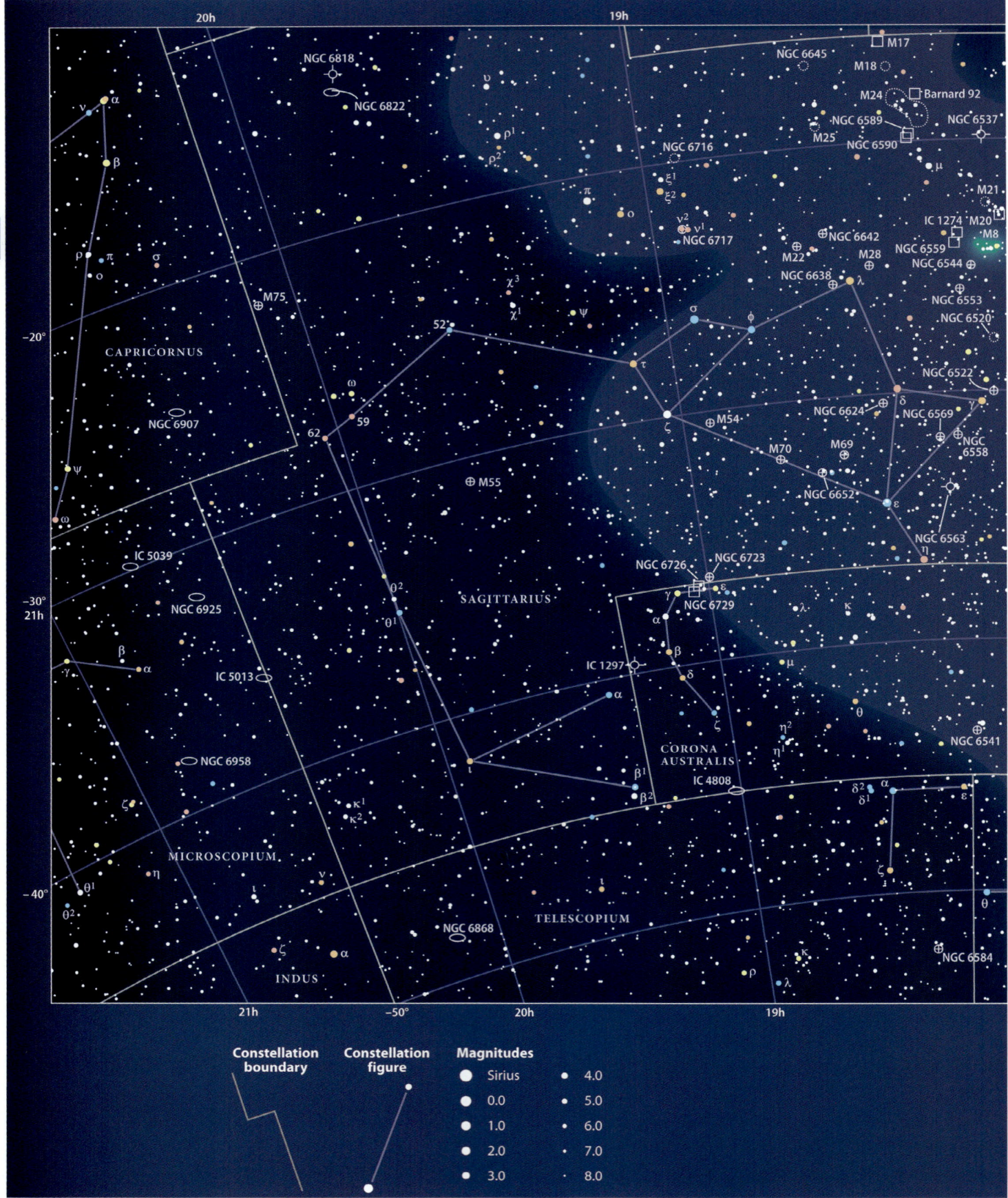

5.13 Star chart featuring Scorpius the Scorpion and Sagittarius the Archer, which lie toward the center of our galaxy from

our perspective on Earth. The large hazy region represents the Milky Way as we see it at night.

Name _____ Section _____ Date _____

Using *Atlas of the Stars* Charts

1 The charts label constellations with capital letters, and each chart shows all or parts of a group of constellations that are located in the same section of the sky. List five constellations from each two-page chart in Table 5.2. Note whether the whole constellation appears on the chart or just part of it.

TABLE 5.2 Constellation Data

	Figure 5.11		Figure 5.12		Figure 5.13	
	Constellation	Whole (W) or Part (P)	Constellation	Whole (W) or Part (P)	Constellation	Whole (W) or Part (P)
1						
2						
3						
4						
5						

2 Describe how constellations are shown on any of the charts. Draw an example in the space below.

3 Which is the largest constellation shown on any of the charts?

Exercise **5.2** Data Sheet

Star Maps **CHAPTER 5**

Exercise 5.2 Data Sheet

4 Note the difference between the lines that show constellation figures and those that show constellation boundaries. What stands out to you about the constellation boundary lines?

5 On the first chart (Fig. 5.11), which constellations border Canis Minor?

6 Look at any of the two-page charts, and describe how it represents star magnitudes. Give an example of a magnitude 0.0 star, magnitude 1.0 star, and magnitude 3.0 star from each chart.

7 How do the numbers of stars shown at a given brightness change as the magnitude gets fainter?

8 Look at the key for the deep-sky objects (open clusters, globular clusters, etc.). Which star chart shows the greatest number of open clusters? Which chart shows the most galaxies? Why do you think one chart shows more of one object while another chart shows more of a different object?

9 As you study the charts, you will see numerous objects marked with the letters "NGC" followed by a number. Describe the way these numbers run across the charts.

10 On Figure 5.11, find Beta (β) Orionis. What are its common name, color, and approximate magnitude? List two other stars that are the same color and nearly the same brightness.

Name _____ Section _____ Date _____

11 On Figure 5.11, what do you think the greenish "clouds" represent?

_____

12 Throughout the atlas, the charts show the Milky Way as a white area meandering through the stars. Does the Milky Way appear on each of the three charts in Figures 5.11–5.13? Why do you think the Milky Way does or does not appear on every chart in the atlas?

_____

_____

_____

_____

13 There are numbers at the edges of all three charts. Those with the letter "h" attached represent right ascension (R.A.), which is the sky's east-west coordinate. Those with degree symbols (°) represent declination (Dec.), which is the sky's north-south coordinate. What are the extents of R.A. and Dec. on each of the charts?

_____

_____

_____

14 Astronomers divide each hour of R.A. into 60 minutes, which is represented with the letter "m." So, for example, the R.A. halfway between 4h and 5h is 4h30m. One-third of the way from 7h to 8h is 7h20m, and so on. In Table 5.3, list the objects nearest to the following coordinates on Figure 5.11.

TABLE 5.3 Objects at Various Right Ascensions and Declinations

R.A.	Dec.	Object
4h35m	17°	
5h05m	−8°	
7h40m	5°	
6h45m	−17°	
6h	7°	
5h35m	−5°	
7h	−8°	
5h30m	0°	

Exercise 5.2 Data Sheet

Exercise 5.2 Data Sheet

15 Using Figure 5.12, list the approximate R.A. and Dec. of the objects in Table 5.4.

Table **5.4** Right Ascensions and Declinations of Various Objects

Object	R.A.	Dec.
Abell 2151		
Arcturus		
Denebola		
M87		
M104		
NGC 5053		
NGC 5885		
Spica		

16 Referring to Figure 5.13, list all the Messier objects (marked with an "M" and a number, for example M42) in the constellation Sagittarius.

Unit 2 Looking Up

Looking Up

This Season's Sky

LEARNING OBJECTIVES

Upon completion of this chapter, you should be able to:

1. Identify the major stars and constellations in the current season's sky.
2. Identify major myths associated with the constellations.
3. Explain how planets appear different than stars.
4. Define all glossary terms.

The Antennae Courtesy of NASA/ESA/SAO/CXC/JPL-Caltech/STScI

GLOSSARY TERM

Asterism recognizable pattern of stars that is not a constellation, but can contain stars from a single constellation or from several constellations

Seeing measurement (or an estimate using your eyes) of the quality of the air above an observing site; good seeing indicates that the air is steady and the stars are not twinkling much; poor seeing indicates that the air is unsteady, causing the stars to twinkle a lot

Transparency measurement (or an estimate using your eyes) that tells how clear the sky above an observing site is; perfect transparency indicates a sky free of clouds; zero transparency indicates a completely overcast sky

This lab contains four exercises, one for each season. So, no matter when the lab occurs, one of these exercises (with the accompanying star chart) represents the current night sky. In these labs, you will use these charts and information you have learned regarding how to read star charts and how to determine magnitudes and the level of light pollution at your observing site. You will learn the major constellations, some **asterisms**, classical mythology related to those star figures, and the season's brightest stars.

These charts have the following features:

- The constellation names are written in capital letters and the star names in lower-case letters.
- Star brightness is shown by the size of the dots representing the stars.
- Some of the brightest stars are labeled with Greek letters.
- Some deep-sky objects from Messier's list as well as the *New General Catalogue*.
- The outline of the Milky Way as it appears in the sky will be in the background.

These labs will ask you to determine **seeing** at your site. Professional astronomers build their observatories in places that have good seeing. The astronomical term *seeing* refers to the steadiness of the atmosphere. You will make an approximate guess of the seeing at your site by noting how much the stars twinkle. You will use a scale from 1 to 5, with 1 being the worst possible seeing—stars are twinkling wildly, even overhead. On the other end of the scale, 5 is the best seeing; the stars overhead are not twinkling at all, and those near the horizon are barely twinkling.

Another important factor to record is the **transparency** at your site. Transparency is simply a measurement of how clear the sky is. For this use a scale of 1 to 10, with 1 meaning it is totally cloudy. It is hoped you didn't pick such a night, but the weather does change, so maybe it surprised you. A transparency of 5 might mean that you can see a layer of thin clouds, but the brightest stars are visible. A 10 means the sky is completely clear and many faint stars are visible.

The Autumn Sky

Exercise 6.1

Fall is a great time for stargazing; evenings arrive earlier than in summer, and although the fall sky may not contain the most brilliant stars, great targets abound. Early in the fall season, the Milky Way arches from the northeast to the southwest and contains countless wonders. Set up a comfortable lawn or camp chair at a dark viewing site, and take some time to scan our galaxy with just your eyes or through binoculars.

Once you learn the stars and constellations in the fall sky, it is an easy transition to the other three seasons. In a short time, you will be locating numerous objects to share with others, and you won't even need a star chart.

 Procedure

Materials
- ❏ Flashlight (preferably one with a red filter) for reading the procedural instructions and star charts outside at night
- ❏ Autumn Sky star chart (provided)
- ❏ Binoculars
- ❏ Telescope (optional)

1. For this procedure, refer to the Autumn Sky star chart on page 93. The chart represents the fall sky of mid-September at about 9:00 p.m. your local time, late September at 8:00 p.m., and mid-October at 7:00 p.m. Make sure to use the chart within an hour of these times.

2. Upon arrival at your destination, your first task is to assess the quality of your observing site. Answer the first seven questions on the data sheet, page 94.

3. Steps 4 through 13 will guide you through the autumn sky. While observing the sky, refer to your star chart for help answering the remaining questions on the data sheet, pages 95–96.

4. Start by looking high overhead. This corresponds to the center of the circular star map. Find the three bright stars that form the Summer Triangle: Vega, in the constellation Lyra (the Harp); Altair, in Aquila (the Eagle); and Deneb, in Cygnus (the Swan), which remain visible in the autumn sky. On your star chart, look next to each of these three stars, and you will find the lowercase Greek letter Alpha (α). The alternate designation for Deneb using Bayer's system is Alpha Cygni. That star marks the tail of the Swan. Likewise, Vega is Alpha Lyrae, and Altair is Alpha Aquilae.

5. The swan's head is the star Albireo, but the chart is a little crowded in that area, so rather than trying to fit in the name, its label is simply the Greek letter Beta (β).

6. Look to its east for four stars that form the main part of the constellation Pegasus (the Winged Horse). Astronomers call this quartet the Great Square, and it is the easiest way to find the legendary horse. Pegasus is a huge constellation that spreads out toward the west from the Great Square.

7. As you examine the chart you will notice that the Great Square has two Alpha stars. The reason is that the northeast star of the Great Square does not belong to Pegasus. It is Alpheratz (Alpha Andromedae), the brightest star in the constellation Andromeda (the Princess). That constellation appears as two curving lines of stars that start at Alpheratz and bend toward the northeast.

8. Follow the northern curve from Alpha to the second unlabeled star. Andromeda's β star lies to its left. To its right on the star chart is the designation M31. This entry, the 31st on Messier's list, is the Andromeda Galaxy. The Andromeda Galaxy looks as if a faint piece of the Milky Way has broken off, but it is not part of our galaxy. Rather it is a separate galaxy that numbers nearly half a trillion stars.

9 Just north of the star pattern of Andromeda, look for the bright constellation Cassiopeia (the Queen). Depending on the time of night, the five luminaries of this star group appear either as the letter W or M. Cassiopeia lies directly in the midst of the Milky Way, so it is full of tightly packed star clusters you can see through binoculars.

10 If you view Cassiopeia as a W, that letter opens toward the north and the constellation Cepheus (the King). This star group will test your ability to find patterns because none of its stars are bright. Search for five stars that look like an outline of a house.

11 Cepheus lies midway between Cassiopeia's W and Polaris, the Alpha star of Ursa Minor (the Bear Cub). Polaris, ranked as the 47th-brightest star, is quite bright, but it is not the star's brightness that makes it famous. Rather, its position near the North Celestial Pole means that, of all the stars you can see, Polaris is the one that doesn't appear to move. Because it seems fixed, you can use it to find the direction north.

12 Next, return to Cassiopeia's W figure on the star chart. Look just to the east for the numbers 869 and 884. Astronomers refer to NGC 869 and NGC 884 as the Double Cluster. The pair is located slightly above the head of the constellation Perseus (the Hero). Although from a dark viewing site you can spot both clusters as fuzzy objects without optical aid, binoculars bring out their best.

13 At the other end of Perseus (and just rising in the east at chart time), you will find the sky's brightest cluster of stars: the Pleiades, also known as M45. With your naked eyes, you can see several Pleiads (individual stars in the cluster) in the form of a tiny dipper.

Name _____ Section _____ Date _____

The Autumn Sky

Courtesy of Richard Talcott and Roen Kelly, *Astronomy* magazine

This Season's Sky **CHAPTER 6**

Exercise **6.1** Data Sheet

93

Quality of Observing Site

1. How dark does the site seem to you? Explain how you came to your conclusion.

2. How far away from the nearest large city are you? How far from the nearest small city?

3. Are there any "light domes" visible? These bright areas of the sky (brightest near the horizon) originate with artificial lights.

4. Are there any clouds? Approximately how much of the sky do they cover?

5. Is the Moon visible? It is best to choose a moonless night to observe, but maybe the nights have been cloudy. What is the Moon's phase?

6. What is your estimate of your site's seeing? Explain your answer.

7. What is your estimate of your site's transparency? Explain your answer.

Name _____ Section _____ Date _____

The Autumn Sky (continued)

Observations

8 Attempt to find the Big Dipper in the sky. In the space below, draw the star pattern you think is the Big Dipper along with some of the stars around it. When you have finished, check with your instructor to verify that you have, in fact, found the Big Dipper.

9 As you face the Big Dipper, hold the star chart in front of you and turn it so the part of the horizon closest to the Big Dipper is at the bottom of the chart. How close to the horizon is the Dipper?

10 In which direction is the Big Dipper? _____

11 According to a North American Indian legend, the four stars of the Big Dipper's bowl represented a bear, and the three stars of the handle were hunters chasing the bear across the sky. Look closely at the hunter (star) in the middle of the handle. What can you see nearby?

12 The Big Dipper is not a constellation; it is an asterism. An asterism is an easily recognized group of stars that is not an official constellation. The stars of an asterism may come from a single constellation or from stars of several constellations. Of which constellation is the Big Dipper a part?

13 Two stars of the Big Dipper point to the North Star, Polaris. What is their position in the figure of the Dipper?

14 Polaris lies at the end of the handle of the Little Dipper, which is not labeled on the star chart; however, it is shown as a figure connected by lines. Can you see all of its stars? If not, how many can you see? Of which constellation is it a part?

Exercise 6.1 Data Sheet

Exercise 6.1 Data Sheet

15 On the opposite side of Polaris from the Big Dipper is a group of stars shaped like the letter W (or M, depending on what time of night you are observing). Which constellation is this?

16 High in the west, about three-quarters of the way up from the horizon, is the brightest star in the asterism called the Summer Triangle. Which star is this?

17 What color is this star? In which constellation is this star located?

18 Which star marks the southernmost point of the Summer Triangle?

19 In which constellation is this star located?

20 Which is the northernmost star of the Summer Triangle asterism?

21 In which constellation is this star located?

22 The "head" of the Swan is a faint star located almost in the middle of the Summer Triangle. Through a telescope, this star is a colorful double star. What is this star's name?

23 Using a telescope with an eyepiece that magnifies about 100 times, you will see Albireo split into two stars, one of which glows gold and the other sapphire blue. But not all human eyes have the same color response. What colors did you see in Albireo's two stars?

24 Look slightly east of the Summer Triangle. About two-thirds up from the horizon you will see four stars in the shape of a large box. Which constellation is this?

25 Which asterism does it contain?

26 The two westernmost stars of the box point to which bright star low in the southeast?

27 Look low between north and west and you will see another bright star. What is its name?

28 Taurus (the Bull) is rising low in the east. Riding on its back is the brightest open star cluster in the sky, the Pleiades, also known as M45. Open star clusters are loose collections of dozens to hundreds of stars that all formed out of the same nebula. Using only your naked eye, take a close look at this object. Note how many stars you can count.

29 Using binoculars, again look at M45. What type of binoculars did you use? (example: 7 × 50)?

30 How many stars did you count in the Pleiades using binoculars?

96 *Unit 2* Looking Up

The Winter Sky

Exercise 6.2

The winter sky offers the brightest stars of any season, which makes finding the constellations that contain them easy. At this time of year, the Milky Way arches from north to south. Along that starry path, celestial treasures visible through binoculars abound.

Learning the major constellations and bright stars in the winter sky is easy, even from a brightly lit backyard. And if you get cold quickly, just 5 or 10 minutes viewing each night will be enough. In a short time, you will be sharing these objects with others, and you won't even need a star chart.

 Procedure

Materials
- ❏ Flashlight (preferably one with a red filter) for reading the procedural instructions and star charts outside at night
- ❏ Winter Sky star chart (provided)
- ❏ Binoculars
- ❏ Telescope (optional)

1. For this exercise, refer to the Winter Sky star chart on page 99. The chart represents the winter sky of mid-December at about 9:00 p.m. your local time, late December at 8:00 p.m., and mid-January at 7:00 p.m. Make sure to use the chart within an hour of these times.

2. Upon arrival at your destination, your first task is to assess the quality of your observing site. Answer the first seven questions on the data sheet, page 100.

3. Steps 4 through 17 will guide you through the winter sky. While observing the sky, refer to your star chart for help answering the remaining questions on the data sheet, pages 101–102.

4. Toward the south, find the constellation Orion (the Hunter), which rules the sky. Orion looks like a giant butterfly, with the three stars of his belt marking the butterfly's body and wings to either side.

5. Locate Orion's brightest star, Rigel (Beta Orionis, also labeled with the Greek letter β on the star chart). It marks his left foot (the star to the bottom-right, as viewed in the sky). You will notice that Orion is an exception to the rule that the brightest star is labeled Alpha, because Rigel is the constellation's brightest star, even though Bayer labeled it Beta.

6. To Rigel's upper-left, past Orion's belt, is Betelgeuse (Alpha Orionis, also labeled with the Greek letter α). Although modern descriptions list Betelgeuse as one of Orion's shoulders, the original Arabic name more closely means "the armpit of the mighty one."

7. As you view Rigel and Betelgeuse in the sky, you will notice they have quite different colors: Rigel is bright blue, and Betelgeuse shines with a coppery tone. Stars shine with different colors because their surfaces radiate at different temperatures. Blue stars typically have surface temperatures above 50,000°F (27,760°C). The surface of a red star like Betelgeuse raises the thermometer's reading to only 3,000°F (1,650°C).

8. The region of Orion's Sword, located below his belt, is seen by a sharp-eyed observer at a dark viewing site as slightly fuzzy. This is the Orion Nebula. Binoculars improve the view, but a small telescope with an eyepiece that gives a magnification of about 100x works best. This object, 42nd on Messier's list, is a vast star-forming region.

9. Because Orion is so easy to find, you can use it as a starting point to locate other stars and constellations. For example, if you draw a line down from Orion's Belt,

you arrive at Sirius (Alpha [α] Canis Majoris), the night sky's brightest star. Because of its position in Canis Major (the great dog), the ancient Greeks referred to Sirius as the dog star.

10. If you have binoculars, point them at Sirius then move them until the star sits at the top of the field of view. Near the center you will see the open star cluster M41, which is composed of dozens to hundreds of stars that all formed out of the same nebula.

11. Canis Major isn't Orion's only dog—there is also Canis Minor (the small dog). To find it, make an equilateral triangle using Betelgeuse, Sirius, and Canis Minor's brightest star, Procyon. These three stars form a bright asterism (a recognizable star pattern that is not a constellation) called the Winter Triangle. Once you find Procyon, draw a line up to Beta (β) Canis Minoris and you have outlined the small dog—just two stars!

12. Now draw a line from Orion's Belt in the opposite direction away from Sirius. You will soon see a v-shaped group of stars called the Hyades, which marks the head of Taurus (the bull). Taurus' fierce eye is the reddish star Aldebaran (Alpha Tauri).

13. Continue the line upward, and you will see the Pleiades (M45). This naked-eye star cluster (sometimes called the Seven Sisters) appears best through binoculars that magnify 10 to 15 times. Telescopes tend to magnify the cluster too much, reducing the number of stars you will see at one time.

14. If you extend a line straight up through Orion's body and head, you will find Auriga (the Charioteer). The brightest star in this constellation is yellow Capella (Alpha Aurigae). Capella means "mother goat," and early observers called the triangle of faint stars just to the southwest of çapella the "kids." On the star chart, they are marked with the Greek letters Epsilon [ϵ], Zeta [ζ], and Eta [η].

15. Auriga is located in the heart of the Milky Way, so deep-sky objects abound here. Three open clusters—M36, M37, and M38—look nice through binoculars and spectacular through a telescope.

16. Return to Orion one last time. Draw a line from Rigel to Betelgeuse. Continue the line about one and one-half times that distance to find Gemini (the Twins). The two stars marking the Twins' heads are Castor (Alpha Geminorum) and Pollux (Beta Geminorum). In the sky, an easy way to tell which is which is to remember a first-letter match: Castor lies closer to Capella, and Pollux is on the same side as Procyon.

17. Finally, locate M35, a bright, open star cluster in Gemini. It is located slightly more than 2° (four full Moon widths) northwest of Eta Feminorum. Place Eta in the center of your binocular field of view, and M35 will be visible.

Name _____ Section _____ Date _____

The Winter Sky

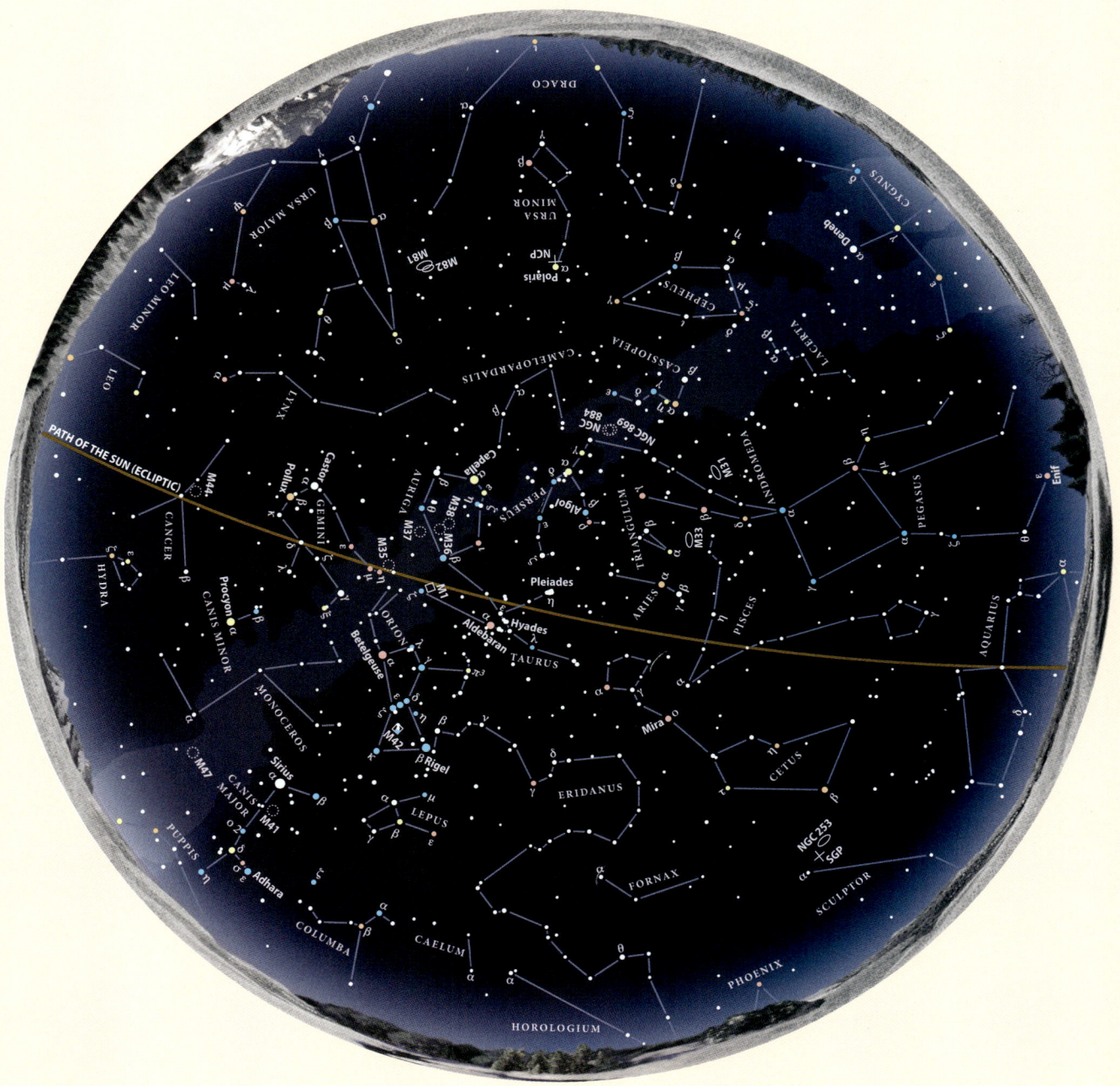

Courtesy of Richard Talcott and Roen Kelly, *Astronomy* magazine

Exercise **6.2** Data Sheet

This Season's Sky CHAPTER **6**

Exercise 6.2 Data Sheet

Quality of Observing Site

1 How dark does the site seem to you? Explain how you came to your conclusion.

2 How far away from the nearest large city are you? How far from the nearest small city?

3 Are there any "light domes" visible? These bright areas of the sky (brightest near the horizon) originate with artificial lights.

4 Are there any clouds? Approximately how much of the sky do they cover?

5 Is the Moon visible? It is best to choose a moonless night to observe, but maybe the nights have been cloudy. What is the Moon's phase?

6 What is your estimate of your site's seeing? Explain your answer.

7 What is your estimate of your site's transparency? Explain your answer.

Name _____ Section _____ Date _____

The Winter Sky (continued)

Observations

8 Find the Big Dipper in the sky and on your star chart. In which direction is it located?

9 How is the star in the bend of the Dipper's handle described?

10 Of which constellation is the Big Dipper a part?

11 What is another name for Polaris?

12 Polaris is at the end of the handle of which asterism?

13 Briefly define an asterism.

14 In which constellation is the Little Dipper located?

15 Locate the two brightest stars in the bowl of the Little Dipper. Observers sometimes call them the "Guardians of the Pole" because of how they circle the North Star. What is the name of the brighter of these two stars?

16 On the opposite side of Polaris from the Big Dipper, which star group is shaped like the letter W (or M)?

17 Which constellation is just setting to the west of this group?

18 Two curved lines of stars extending from this constellation form another constellation. Which one is it?

Exercise 6.2 Data Sheet

Exercise 6.2 Data Sheet

19 Look toward the south. Also, turn your star chart so that south is at the bottom. What bright star is closest to the overhead point?

20 In which constellation is this star located?

21 What is the small group of three stars near this bright star called?

22 The constellation Orion (the Hunter) dominates the southern sky at this time of year. What are the names of its two brightest stars?

23 If you extend a line downward from Orion's Belt, you see the brightest of all nighttime stars. What is its name?

24 In which constellation is it located?

25 Just below Orion is a faint constellation. What is its name?

26 If you follow Orion's Belt upward, what is the first constellation you encounter?

27 What kind of animal does this constellation represent?

28 What is this constellation's brightest star?

29 Continue the line from Orion's Belt through this star pattern to a small, tightly packed group of stars. What is its name?

30 To the north and east of Sirius is which other bright star?

31 In which constellation is it located?

32 Continue to the north from this star group. Which two bright stars do you encounter?

33 In which constellation are they located?

34 Look back at the Big Dipper. Imagine poking a hole in the bottom of its bowl and letting all the water run out. What constellation will the water hit?

35 What is the name of the asterism that marks this animal's head?

The Spring Sky

Exercise 6.3

At last, spring. Finally, it is warm enough for even fair-weather observers to break out their binoculars and telescopes. If you are new to astronomy, however, you will want to become familiar with the constellations and bright stars that populate this season's night sky before looking through the eyepiece. Then, when you read that a celestial target is in Boötes or near Regulus, you will already have a frame of reference.

Spring is an ideal time to begin learning the sky. Many of the constellations are large, and helpful indicators like the Big Dipper lead you to the brightest stars.

 Procedure

Materials

- ❏ Flashlight (preferably one with a red filter) for reading the procedural instructions and star charts outside at night
- ❏ Spring Sky star chart (provided)
- ❏ Binoculars
- ❏ Telescope (optional)

1. For this exercise, refer to the Spring Sky star chart on page 105. The chart represents the spring sky of mid-March at about 9:00 p.m. your local time, late March at 8:00 p.m., and mid-April at 7:00 p.m. Make sure to use the map within an hour of these times.

2. Upon arrival at your destination, your first task is to assess the quality of your observing site. Answer the first seven questions on the following data sheet, page 106.

3. Steps 4 through 15 will guide you through the spring sky. While observing the sky, refer to your star chart to help answer the remaining questions on the data sheet, pages 107–108.

4. High in the north, the sky's most recognizable star pattern—the Big Dipper—is easy to spot after sunset. The Big Dipper is part of the third-largest constellation, Ursa Major (the Great Bear). All of the other stars are faint, however, so concentrating on the Big Dipper is the best way to start.

5. Find the bend of the Big Dipper's handle. Sharp-eyed observers can tell that two stars occupy this space, although only one is shown on the star chart. Arabian astronomers 10 centuries ago called this pair the Horse and Rider. The brighter of the two is Mizar, which also is labeled with the Greek letter Zeta [ζ] on the star chart.

6. Alcor, which sits a little to the east-northeast of Mizar, glows about one-fifth as brightly as Mizar. Astronomers classify this pair as an "optical" double star, as in "optical illusion." The two luminaries are not part of the same system. In fact, Alcor sits three light-years (approximately 18 trillion miles) more distant than Mizar.

7. Because the Big Dipper is so easy to find, it can help you locate many constellations and stars in the spring sky. For example, use the Pointer Stars, which sit at the end of the Big Dipper's scoop. Their names are Dubhe and Merak, but on the chart, their labels are α (Alpha) and β (Beta). Use them to find perhaps the most well-known single star in the sky: Polaris, the North Star (the Alpha star in Ursa Minor). To find this celestial marker, draw a line from Beta through Alpha in Ursa Major, and extend that line about five times the distance between those two stars.

8. Polaris marks the end of the Little Dipper's handle. It is also the brightest star in Ursa Minor (the Bear Cub), and it marks the tip of the Bear's tail. Unlike the Big Dipper, however, the other six stars of the Little Dipper are faint. You will need a dark viewing location to see them all.

9 Return to your starting point, the Big Dipper. Follow beyond the curve of the handle to two brilliant stars: Arcturus (the Alpha star in the constellation Boötes) and Spica (the Alpha star in Virgo). (You can remember this with the phrase "Arc to Arcturus and speed on to Spica.") Arcturus' name comes from a combination of terms that mean "the bear's guard," which signifies its position near Ursa Major (the Great Bear). The main part of Boötes looks like a thin kite or an ice-cream cone. Spica sits in the southern reaches of the constellation Virgo (the Maiden). This sprawling star group ranks second in size among the 88 constellations that cover the sky.

10 These two luminaries, Arcturus and Spica, illustrate some of the color differences between stars. Arcturus appears orange, while Spica is bright blue. Stars have different colors because nuclear reactions in their cores heat their surfaces to different temperatures. Arcturus has a relatively cool surface temperature of 4,300 Kelvins (7,300°F). Spica's surface burns at 11,000 K (19,300°F). For comparison, the Sun is in the middle of the temperature range. On its surface, a thermometer would read 6,000 K (10,300°F).

11 Return to the Big Dipper one last time. To find the next constellation, imagine poking a hole in the Dipper's bowl, letting all the water run out, and waiting for a loud mythical roar. That is what you might hear as the water falls on the back of Leo (the Lion).

12 Leo's main figure has two parts: a backward question mark and a triangle. The question mark, or sickle-shaped, figure represents the front of the lion. Dotting the question mark with a distinctive blue-white color is bright Regulus (Leo's Alpha star).

13 East of the question mark, a right triangle marks the lion's back and tail. The star farthest west of Regulus is Denebola (the Beta star in Leo). It lies 36 light-years away. Although Denebola carries the "Beta" label (usually signifying the second-brightest star) in Leo, Algieba (the Lion's Gamma [γ] star) slightly outshines it. Algieba is an orange and yellow double star that you will have no difficulty splitting through a 3-inch telescope.

14 Located about 20° west of Leo's question mark is one of the faintest constellations in the sky: Cancer (the Crab). This faint star pattern is identified only so you can point binoculars toward its center. There you will find the Beehive Cluster, labeled on the star chart as M44. This object is the 44th on Messier's list. Through 10×50 binoculars most observers can count three dozen stars in it.

15 Below Virgo you will find the small constellation Corvus (the Crow). Its four moderately bright main stars and "crooked box" shape make it easy to find. There is another way to be sure you are looking at Corvus: see if its top two stars point upward to Spica.

Name _____ Section _____ Date _____

The Spring Sky

Exercise 6.3 Data Sheet

Courtesy of Richard Talcott and Roen Kelly, *Astronomy* magazine

This Season's Sky **CHAPTER 6**

Exercise 6.3 Data Sheet

Quality of Observing Site

1 How dark does the site seem to you? Explain how you came to your conclusion.

2 How far away from the nearest large city are you? How far from the nearest small city?

3 Are there any "light domes" visible? These bright areas of the sky (brightest near the horizon) originate with artificial lights.

4 Are there any clouds? Approximately how much of the sky do they cover?

5 Is the Moon visible? It is best to choose a moonless night to observe, but maybe the nights have been cloudy. What is the Moon's phase?

6 What is your estimate of your site's seeing? Explain your answer.

7 What is your estimate of your site's transparency? Explain your answer.

Name _____   Section _____ Date _____

The Spring Sky (continued)

Observations

8 Find the Big Dipper in the northeastern sky. Follow its curved handle downward toward the east. Rotate your star chart so that you are facing east. (Place east at the bottom.) To which bright star has the Big Dipper's handle led you?

_____

9 In which constellation is this star located?

_____

10 Extend the curved line from the Big Dipper's handle past that bright star to another that is nearly as bright. What is the name of this star?

_____

11 In which constellation is this star found?

_____

12 In which small constellation can you find southwest of this bright star?

_____

13 This constellation, and another small one to its west, seems to be riding on the back of a long constellation. In fact, this is the largest constellation in the sky. What is its name?

_____

14 The object labeled M44 is in the constellation Cancer the Crab. What two bright stars does M44 lie between?

_____

_____

15 Which two constellations lie between the stars Arcturus and Vega?

_____

16 Which well-known winter constellation has mostly set in the west at chart time?

_____

17 The constellations Gemini, Cancer, Leo, Virgo, and Libra lie along the ecliptic, an imaginary line in the sky (but printed on the map) that shows the Sun's apparent path through the constellations. What is the group called that contains these five (and seven more) constellations?

_____

18 Find the label NGP on the map. Those letters stand for "North Galactic Pole," and that is where the north pole of our Milky Way Galaxy points. In which constellation is the NGP located?

_____

Exercise 6.3 Data Sheet

This Season's Sky **CHAPTER 6**

Exercise 6.3 Data Sheet

19 A large constellation winds its way between the two Dippers. What is it?

20 The constellation Auriga sets in the northwest soon after chart time. What is its brightest star?

21 Which three Messier objects does Auriga contain?

22 The stars Spica and Denebola, along with one other star, lie at the points of an equilateral (equal-sided) triangle that forms an asterism called the Spring Triangle. What is the other star?

23 In which constellation is it located?

24 The brightest stars on the map are color-coded according to their temperatures. For example, Arcturus is an orange star and Spica is blue. Which do you think is the hotter star? Explain your answer.

25 Of which constellation is the Big Dipper a part?

26 What are the names of the two stars that point to Polaris?

27 What is another name for Polaris?

28 Polaris is at the end of the handle of which asterism?

29 Locate the two brighter stars in the bowl of the Little Dipper. Observers sometimes call them the "Guardians of the Pole" because of how they circle the North Star. How are these two stars labeled?

Unit 2 Looking Up

The Summer Sky

Exercise 6.4

Perhaps the best time to begin observing the sky is summer. Many bright stars populate the sky during this season, guiding you to the constellations that contain them. The brightest part of the Milky Way arches from north to south. In that star-packed region, deep-sky treasures visible through binoculars abound. Plus, it is warm outside, which may keep you observing longer.

Once you learn the main constellations and bright stars in the summer sky, it is an easy transition to the other three seasons. In a short time, you will be locating many great objects to share with others, and you won't even need a star chart.

 **Procedure**

Materials
- ❏ Flashlight (preferably one with a red filter) for reading the procedural instructions and star charts outside at night
- ❏ Summer Sky star chart (provided)
- ❏ Binoculars
- ❏ Telescope (optional)

1 For this exercise, refer to the Summer Sky star chart on page 111. The chart represents the summer sky of mid-June at about 9:00 p.m. your local time, late June at 8:00 p.m., and mid-July at 7:00 p.m. Make sure to use the chart within an hour of these times.

2 Upon arrival at your destination, your first task is to assess the quality of your observing site. Answer the first seven questions on the data sheet, page 112.

3 Steps 4 through 17 will guide you through the summer sky. While observing the sky, refer to your star chart for help answering the remaining questions on the data sheet, pages 113–114.

4 High in the northwest in the evening, locate the seven stars that form the Big Dipper. This well-known group daily circles the North Star, Polaris (Alpha [α] Ursae Minoris). The two stars at the end of the Dipper's bowl—the Pointer Stars—point to Polaris. The Big Dipper, however, is not a constellation; it is part of Ursa Major the Great Bear.

5 Notice the description of Polaris in parentheses on the star chart. You will see a Greek letter, which is its symbol, and the possessive form of the constellation name. Only a few hundred stars have proper names, but about a thousand have Greek letters associated with them.

6 Follow the curve of the Big Dipper's handle south to Arcturus (Alpha Boötis), which is the brightest star in the constellation Boötes (the Herdsman). Arcturus glows orange, indicating it is a cool star. It ranks as the fourth-brightest nighttime star, and the brightest in the northern half of the entire sky.

7 Overhead, three bright stars form summer's second most identifiable star picture (next to the Big Dipper)—the Summer Triangle. Each is the Alpha star in its constellation. The brightest is Vega, in Lyra (the Harp); next brightest is Altair, in Aquila (the Eagle); the third is Deneb, the star marking the tail of Cygnus (the Swan).

8 Cygnus contains the asterism of the Northern Cross. Albireo (Beta [β] Cygni) lies at the base of the cross. This star is one of the top showpieces for small telescopes, because Albireo is a colorful double star. The separation of the two components is wide enough that even 10×50 binoculars (steadily held or mounted on a tripod) will separate the pair. Color perception is unique among humans, but most observers see these stars as gold and sapphire blue.

9 Lyra is located just outside of Cygnus' Milky Way region. In the same binocular field of view as Vega is Epsilon (ε) Lyrae, a wide double star with two equally bright components. Amateur astronomers know Epsilon Lyrae as the Double Double because each member is itself a double star. Each pair can be split by using high magnification through even a 2.4-inch (60-millimeter) telescope.

10 Slightly west of the Summer Triangle is Hercules (the Hero). This constellation contains one of amateur astronomy's main sights—the Hercules Cluster (M13), the finest globular star cluster in the northern sky. M13 lies along one side of the Keystone, a crooked box of four medium-bright stars that mark Hercules' body. The Keystone is located about two-thirds of the way from Arcturus to Vega. You can glimpse M13 with your unaided eyes under a dark sky, but it looks better through binoculars, where it appears half the width of the full Moon.

11 Hercules stands head to head with another giant in the sky, Ophiuchus (the Serpent Bearer). Ophiuchus is a man entwined by a large snake, Serpens—the only constellation divided into two separate parts. In his left hand, Ophiuchus holds the snake's head, Serpens Caput. To Ophiuchus' east is the snake's tail, Serpens Cauda.

12 Ophiuchus, in turn, stands on the body of Scorpius (the Scorpion). Because Scorpius and neighboring Sagittarius lie in the southern part of the sky, thick air layers dim their glory from northern latitudes. However, you still should be able to find Antares (Alpha Scorpii), the bright star marking the Scorpion's heart. Antares is a red supergiant 400 times larger than the Sun.

13 Next to the Scorpion's stinger lie two open star clusters, M6 and M7. M6, popularly known as the Butterfly Cluster because of its shape, appears slightly elliptical, and its brightest star is an orange giant that varies in brightness.

14 From a dark viewing site, M7 is easy to spot with your naked eyes as a bright knot in the southern Milky Way. Through binoculars, M7 appears more than twice as wide as the full Moon. M7's central stars are arranged in an X, while the outliers form a triangle.

15 The next constellation, Sagittarius, lies east of Scorpius toward the center of our galaxy. Sagittarius has many deep-sky objects within its borders. If you have a telescope, start with the Lagoon Nebula (M8), a milky glow three Moon-diameters long with a dark rift down its center.

16 Near M8 sits M22, a fine globular cluster. Under a dark sky, binoculars show it as a woolly ball about two-thirds the Moon's diameter.

17 Finally, move north along the Milky Way into Scutum to encounter the Wild Duck Cluster (M11), one of the summer sky's best open clusters. English observer William Henry Smyth (1788–1865) called M11 the Wild Duck Cluster because it is V-shaped, like a flight of wildfowl. Small telescopes show the V has a brighter star at its apex.

Name _____ Section _____ Date _____

The Summer Sky

Courtesy of Richard Talcott and Roen Kelly, *Astronomy* magazine

Exercise **6.4** Data Sheet

This Season's Sky **CHAPTER 6** 111

Exercise 6.4 Data Sheet

Quality of Observing Site

1 How dark does the site seem to you? Explain how you came to your conclusion.

2 How far away from the nearest large city are you? How far from the nearest small city?

3 Are there any "light domes" visible? These bright areas of the sky (brightest near the horizon) originate with artificial lights.

4 Are there any clouds? Approximately how much of the sky do they cover?

5 Is the Moon visible? It is best to choose a moonless night to observe, but maybe the nights have been cloudy. What is the Moon's phase?

6 What is your estimate of your site's seeing? Explain your answer.

7 What is your estimate of your site's transparency? Explain your answer.

Unit 2 Looking Up

Name _____   Section _____   Date _____

The Summer Sky (continued)

Observations

8 Look slightly east of the overhead point. What is the asterism formed by the three brightest stars in that part of the sky called?

_____

9 Which star is the brightest of the three? _____

10 In which constellation is it located? _____

11 Which star of the three is the southernmost? _____

12 In which constellation is it located? _____

13 What is the name of the dimmest star in this threesome? _____

14 What direction is it from the brightest of the three stars? _____

15 In which constellation is it located? _____

16 There is a moderately bright star about in the middle of this group of three brighter stars. It is a famous double star because of the striking color of the two components. What is its name?

_____

17 Look toward the northwest in the sky. Locate the Big Dipper. Describe the star in the bend of the Dipper's handle.

_____

_____

18 The Big Dipper is an asterism. In which constellation is it located? _____

19 Which two stars in the Big Dipper point to the North Star?

_____

20 What is the common name of this pair? _____

21 The North Star is part of another constellation. Which one? _____

22 Give another name for the North Star. _____

23 Which constellation lies between the Little Dipper and the bright star Vega?

_____

24 On the opposite side of the North Star from where the Big Dipper lies is a W-shaped group. What is this constellation is named?

_____

25 Now draw a line from the North Star to Deneb. Which constellation lies to the northeast of that line?

_____

Exercise 6.4 Data Sheet

Exercise 6.4 Data Sheet

26. By following the curve of the Big Dipper's handle, you can find two bright stars. Which is the first (and brightest) one you see?

27. What color is this star?

28. In which constellation is it located?

29. Which is the second star the Dipper's handle leads you to?

30. What color is it?

31. In which constellation is it located?

32. Which constellation lies overhead at chart time?

33. A bright reddish star lies about two-fifths of the way up from the southern horizon. What is its name?

34. In which constellation is it located?

35. Which constellation is located one-third of the way from Arcturus to Vega?

36. Which huge constellation lies between Lyra and Scorpius?

37. Which constellation is located midway between Spica and Antares?

38. Which constellation of the classical zodiac is slightly east of Scorpius?

39. A line from Deneb to Altair extended its own length (35°) ends just east of which constellation?

40. A line from Vega to Altair (35°) extended 25° ends just to the east of the Alpha and Beta stars in which constellation?

41. The asterism Job's Coffin is a small, diamond-shaped group of stars just outside the Summer Triangle, about 13° east-northeast of Altair. Of which constellation is this asterism a part?

42. During summer in the Northern Hemisphere, the Milky Way arches overhead around chart time. It starts in southern Scorpius toward the south and passes through Cassiopeia in the northeast. Describe the Milky Way as you see it at your observing site.

Looking Up
Outdoor Sky Observations

LEARNING OBJECTIVES

Upon completion of this chapter, you should be able to:

1. Identify and explain the sky conditions during the observing session.
2. Locate some of the sky's major constellations and bright stars.
3. Describe an assigned number of objects viewed through binoculars during an outdoor observing session.
4. Assess the performance of the binoculars used.
5. Define all glossary terms.

7

Crab Nebula Courtesy of NASA/ESA/J. Hester/A. Loll

GLOSSARY TERM

Constellation one of 88 arbitrary configurations of stars; the (officially recognized) area of the celestial sphere containing one of these configurations

7

Exercise 7.1

In this lab, you will learn some of the **constellations** and bright stars in the night sky during the season your lab is in session. Your instructor will assign the monthly exercise that best fits the class schedule. In addition to the objects visible to your naked eyes, the lab contains several objects for you to observe through binoculars. As you view each object, take enough time to notice its details so you can write descriptions later.

Almost all of the deep-sky objects in these exercises (open clusters, globular clusters, nebulae, and galaxies) will appear different when seen with your naked eye, through your binoculars, or through a telescope than they do in the photographs. The reason for this is that most of these photographs are time-lapsed images taken with very powerful telescopes.

The astronomical term *seeing* refers to the steadiness of the atmosphere. You can make an approximate guess of the seeing at your site by noting how much the stars twinkle. Using a scale from 1 to 5, 1 is the worst possible seeing—stars are twinkling wildly, even overhead. On the other end of the scale, 5 means the stars overhead are not twinkling at all, and those near the horizon are barely twinkling.

Transparency at your site is simply a measurement of how clear the sky is. For this use a scale of 1 to 10, with 1 meaning it is totally cloudy. A transparency of 5 might mean that you can see a layer of thin clouds, but the brightest stars are visible. A 10 means the sky is completely clear and many faint stars are visible.

January–February: A Hero for the Ages

January and February's binocular highlights (Fig. 7.1) feature Cassiopeia the Queen and her daughter, Andromeda the Princess. These two women play pivotal roles in one of the sky's most famous myths: the legend of Andromeda and Cetus. According to the story, Poseidon, Greek god of the sea, punished Cassiopeia for her constant bragging by

7.1 Star chart to locate January and February's binocular highlights. Courtesy of Richard Talcott and Roen Kelly, *Astronomy* magazine

116 Unit 2 Looking Up

kidnapping Andromeda and chaining her to a rock by the sea. There, Poseidon summoned his vicious sea monster, Cetus, to devour her. Just as it looked as though all was lost, the story's hero, Perseus, swooped out of the sky on the back of Pegasus the Flying Horse and saved Andromeda.

Procedure

Materials
- ❏ Flashlight (preferably one with a red filter) for reading the procedural instructions and star chart outside at night
- ❏ January–February star chart (provided)
- ❏ Binoculars

1. Upon arrival at your destination, your first task is to assess the quality of your observing site. Answer the first seven questions on Data Sheet 7.1 (page 119).

2. In the sky, find Perseus standing protectively over Andromeda—although trying to see the figure of a mighty warrior in the stars is a stretch. Perseus looks more like two jagged arcs of stars curving away from the W pattern of Cassiopeia.

3. Now locate Perseus' brightest star, Mirfak (Alpha [α] Persei), which is located about two-thirds of the way along a line that stretches from Pegasus in the northwest to the bright star Capella in Auriga. Mirfak shines at **magnitude** 1.8. Astronomers classify it as a **giant star**. It lies about 500 light-years away.

4. Examine Mirfak through binoculars. It is surrounded by dozens of fainter stars scattered in small clumps and patterns. While most of the stars appear white or blue white, a few show slight tinges of yellow or orange.

5. Search for the orange star Sigma (σ) Persei, one of three suns forming a small triangle south of Mirfak. Also, look for two whitish **double stars** to Mirfak's north. Together, the stars gathered into this football-shaped area form the Alpha Persei Association. A stellar **association** contains mostly hot, blue-white and white stars, like many of the sky's open star clusters. Typically, however, the stars in an association gather more loosely than those in **open clusters**. In the case of the Alpha Persei group, approximately 50 stars are bound by their mutual—but weak—gravitational field. All lie about the same distance away as Mirfak, having formed from a common cloud of interstellar gas and dust 50 million years ago.

6. On the star chart, look for two open star clusters that appear as two overlapping circles in the constellation Perseus. Their combined nickname is the "Double Cluster" (Fig. 7.2). Their labels, NGC 869 and NGC 884, correspond to their entry numbers in the *New General Catalogue*, which lists nearly 8,000 **NGC objects**. Even from suburban skies, your unaided eyes can spot the Double Cluster as a small, faint smudge of light between Perseus and Cassiopeia.

7. Now find the Double Cluster through your binoculars by extending a line from Gamma (γ) Cassiopeiae, the center star of the W, through Delta (δ) Cas and continuing east. Maintain a straight course, and you will see both clusters as two tiny knots of stars. The one closer to Cassiopeia, NGC 869, appears more densely packed. Most of the stars look either white or blue white, but you might notice a few yellow and orange stars as well. To make subtle star colors stand out more vividly, defocus your binoculars slightly.

8. Finally, take aim at open cluster M34. It is located about halfway between the stars Almach (Gamma Andromedae) and Algol (Beta [β] Persei). A great target for binoculars, M34 covers an area as large as the full Moon. Look for the brightest of its 100 stars **twinkling** in the hazy glow of fainter, **unresolved** suns.

GLOSSARY TERMS

Magnitude measure of brightness of a celestial object; the smaller (or more negative) the magnitude, the brighter the object

Giant star any star approximately 10 to 100 times larger than the Sun

Double star any pair of stars that appear close to one another on the celestial sphere; may be physically linked by gravity to one another or may be chance alignments; also called binary star

Association large, loose grouping of young stars of similar type; a much looser group than an open cluster

Open cluster young system of stars usually containing between a few hundred and a few thousand stars

NGC object any object found in J. L. E. Dreyer's *A New General Catalogue of Nebulae and Clusters of Stars*, first printed in 1888, or in any of the two supplements, printed in 1895 and 1908

Twinkling also called scintillation, the motion of a star's (or other celestial object's) light due to its passage through our turbulent atmosphere, which acts as a series of lenses bending the light first one way, then another

Unresolved observing description that indicates detail in an object was not seen or that no separation was seen between two close stars

9 Answer the remaining questions on Data Sheet 7.1 (page 120) at the end of your viewing session (or after you return indoors).

7.2 Double Cluster in Perseus, formed by (**A**) NGC 869, and (**B**) NGC 884.

Courtesy of Anthony Ayiomamitis

Name _____   Section _____   Date _____

January - February

Quality of Observing Site

1 How dark does the site seem to you? Explain how you came to your conclusion.

2 How far away from the nearest large city are you? How far from the nearest small city?

3 Are there any "light domes" visible? These bright areas of the sky (brightest near the horizon) originate with artificial lights.

4 Are there any clouds? Approximately how much of the sky do they cover?

5 Is the Moon visible? It is best to choose a moonless night to observe, but maybe the nights have been cloudy. What is the Moon's phase?

6 What is your estimate of your site's seeing? Explain your answer.

7 What is your estimate of your site's transparency? Explain your answer.

Exercise 7.1 Data Sheet

Exercise 7.1 Data Sheet

Observations

8 What model of binoculars did you use?

9 Was the unit easy to handle? Were you happy with the views?

10 How long did you observe?

11 Which of the objects discussed in the procedural steps on pages 117–118 did you observe? Describe each briefly.

12 Which of the objects discussed did you not see? Why?

13 If you used your binoculars to further explore the sky, which objects not discussed on pages 117–118 did you observe? Describe a maximum of three of them.

March–April: Visit the Charioteer

Exercise 7.2

Head outside some March or April evening, and scan the sky for the binocular highlights (Fig 7.3) discussed in this exercise. If you live in mid-northern latitudes, you will see a lone beacon cresting near the **zenith**—the point in the sky straight overhead—as the sky darkens. Capella lies in Auriga (the Charioteer) and marks one of seven points in the winter oval of brilliant stars. Tracing back to ancient Rome, the name *Capella* translates to "mother goat," a reference to the position it holds in the Charioteer's picture. Near Capella lie three stars that represent the goat's kids.

7.3 Star chart to locate March and April's binocular highlights.
Courtesy of Richard Talcott and Roen Kelly, *Astronomy* magazine

GLOSSARY TERM

Zenith point in the sky directly above an observer; the highest point in the sky for an observer

Looking Up

Giovanni Battista Hodierna (1597–1660), astronomer at the court of the Duke of Montechiaro in Sicily, discovered all three Messier open clusters in Auriga through a 1-inch telescope around 1654.

Today, Capella is a known **binary star** system lying about 42 light-years away. Each of Capella's suns is a yellow star, like our Sun. All three have similar surface temperatures, but our Sun is about one-tenth as large as either of the Capella giants. Spotting the two component stars through even the largest telescopes can be difficult. Only 60 million miles (97 million kilometers)—less than the distance between the Sun and Venus—separate the two.

GLOSSARY TERM

Binary star any pair of stars that appear close to one another on the celestial sphere; may be physically linked by gravity to one another or may be chance alignments; also called double star

Procedure

1. Upon arrival at your destination, your first task is to assess the quality of your observing site. Answer the first seven questions on Data Sheet 7.2 (page 123).
2. Locate Capella in Auriga (the Charioteer).

Materials

- ❏ Flashlight (preferably one with a red filter) for reading the procedural instructions and star chart outside at night
- ❏ March–April star chart (provided)
- ❏ Binoculars

Outdoor Sky Observations **CHAPTER 7**

GLOSSARY TERMS

Field of view angle (in degrees) you can see through any binoculars; also refers to the actual area in view when you look through the eyepieces

Asterism unofficial, recognizable grouping of stars either part of a single constellation (the Big Dipper) or several constellations (the Summer Triangle)

Messier object one of the objects on French comet-hunter Charles Messier's list of 109 objects that, he thought, astronomers of his time could confuse with comets

Condensed concentrated, especially when it refers to the light from a celestial object

Note

March is traditionally the best month to spot all 109 Messier objects in a single dusk-to-dawn observing session. The best time is around the new Moon. The Sun's position in the sky around the vernal equinox reveals all these objects sometime during the night.

3. Just south of Capella look for the Kids—three 3rd-magnitude stars in the shape of an isosceles triangle: Epsilon (ε) shines with a yellowish-white hue, Eta (η) is blue, and Zeta (ζ) Aurigae appears orange.

4. Turn slightly more than a binocular **field of view** south of the triangle to see a pattern of five faint stars. Four form a parallelogram while the fifth lies just below. This is not a star cluster, but an **asterism**—one of those shapes occasionally encountered. Given March's windy days, the pattern is reminiscent of a box kite with a tail being whipped about. With your unaided eyes from a dark viewing site, you might see the kite asterism's oblong glow slightly west of the constellation's center, where it is located.

5. Look just north of the box kite for a dim glow among the stars. That is M38, one of Auriga's three open clusters, all of which are **Messier objects**. Do not be surprised if at first you can't see M38 through your binoculars, however. This cluster can be difficult to pick out from among the surrounding stars.

6. While looking for M38, you might notice M36, a second fuzzy glow situated between two faint field stars about half a binocular field of view to the east. M36 is smaller and more **condensed** than M38, so you should be able to spot it more easily. If you have good eyes and 70mm or larger binoculars, you might see a few faint stars. Telescopes reveal that the cluster's brightest stars fall into a pattern resembling a crooked Y.

7. The third and brightest Messier open cluster in Auriga is the Salt-and-Pepper Cluster, M37 (Fig 7.4). It is located about a binocular field of view east of M36. Because its brightest stars shine at only 9th magnitude, M37 appears as a faint, misty patch through binoculars.

8. Answer the remaining questions on Data Sheet 7.2 (page 124) at the end of your viewing session (or after you return indoors).

7.4 Salt and Pepper Cluster (M37) in Auriga.

Courtesy of Anthony Ayiomamitis

Name _____   Section _____   Date _____

March - April

Quality of Observing Site

1 How dark does the site seem to you? Explain how you came to your conclusion.

_____
_____
_____
_____

2 How far away from the nearest large city are you? How far from the nearest small city?

_____
_____

3 Are there any "light domes" visible? These bright areas of the sky (brightest near the horizon) originate with artificial lights.

_____
_____

4 Are there any clouds? Approximately how much of the sky do they cover?

_____
_____
_____

5 Is the Moon visible? It is best to choose a moonless night to observe, but maybe the nights have been cloudy. What is the Moon's phase?

_____
_____

6 What is your estimate of your site's seeing? Explain your answer.

_____
_____
_____
_____
_____

7 What is your estimate of your site's transparency? Explain your answer.

_____
_____
_____
_____
_____

Exercise 7.2 Data Sheet

Exercise 7.2 Data Sheet

Observations

8 What model of binoculars did you use?

9 Was the unit easy to handle? Were you happy with the views?

10 How long did you observe?

11 Which of the objects discussed in the procedural steps on pages 121–122 did you observe? Describe each briefly.

12 Which of the objects discussed did you not see? Why?

13 If you used your binoculars to further explore the sky, which objects not discussed on pages 121–122 did you observe? Describe a maximum of three of them.

May–June: The Big Dipper

Exercise 7.3

The Big Dipper is visible every clear night in the Northern Hemisphere, and it will be when you venture out to see the binocular highlights (Fig. 7.5) discussed in this exercise. Beginners learn to use the Big Dipper as a handy signpost for finding other stars in the sky—"Follow the Pointers to Polaris" or "Arc to Arcturus."

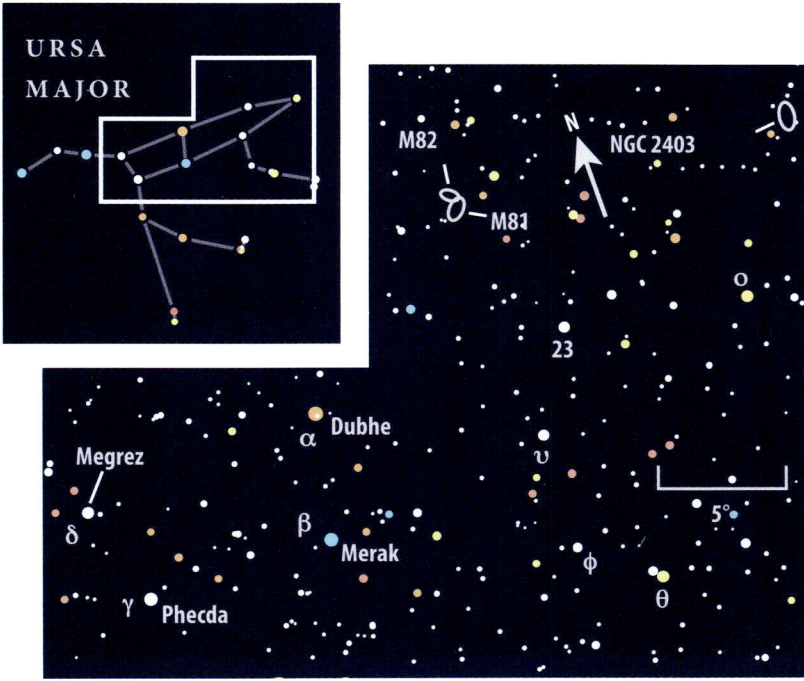

7.5 Star chart to locate several of May and June's binocular highlights.

Courtesy of Richard Talcott and Roen Kelly, Astronomy magazine

Procedure

Materials
- ❏ Flashlight (preferably one with a red filter) for reading the procedural instructions and star chart outside at night
- ❏ May–June star chart (provided)
- ❏ Binoculars

1. Upon arrival at your destination, your first task is to assess the quality of your observing site. Answer the first seven questions on Data Sheet 7.3 (page 129).

2. Locating the stars in the sky, examine each of the four stars that make up the bowl of the Big Dipper: Dubhe (Alpha Ursae Majoris), Merak (Beta), Phecda (Gamma), and Megrez (Delta). All shine nearly white, except for Dubhe at the bowl's northwest corner, which radiates an orangish glow because it is a type-K giant star—larger, but cooler, than our own Sun.

3. Draw a line from Phecda, at the bowl's southeastern corner, to Dubhe. Extend that line an equal distance to the northwest until you come to a small, right triangle of stars. The star marking the triangle's right angle is 24 Ursae Majoris. Note, however, that most star charts do not name the other stars.

4. If you look carefully, you might see a faint blur to the triangle's southeast. That is not a star, but rather, is Bode's **Galaxy** (M81). Although finding M81 can prove daunting at first, give it a try and you are bound to spot it. Through 7×35 binoculars, the galaxy's oval shape appears small and dim, but unmistakable. Larger binoculars increase the contrast between the galaxy's bright center and the dimmer surrounding

GLOSSARY TERM

Galaxy collection of up to thousands of billions of stars, dust, and gas held together by gravity

halo; however, binoculars won't resolve the features astronomers refer to as "arms" that make this object a spiral galaxy.

5. Look just north of M81, and try to make out a dimmer splinter of grayish light. If you are able to see it, you have found the Cigar Galaxy, M82 (Fig. 7.6), a dramatic object astronomers now classify as a starburst galaxy. Within this object, stars are forming and dying at an astounding rate. If the sky is quite dark, 7×35 binoculars will reveal this galaxy's famous cigar shape. Unlike M81, however, which shows a brighter core, M82 appears uniformly dim from end to end. M81 and M82 are each about 12 million light-years away, but about 100,000 light-years separate them. Together, they form one of the most striking galaxy pairs in the sky. Other galaxies belong to the M81 group; unfortunately, only one can be seen with binoculars, its spiral galaxy NGC 2403 in the constellation Camelopardalis, the others are too faint.

6. The next target is NGC 2403. This visible member galaxy lies a good distance away, across the border within the faint constellation Camelopardalis.

7. To find NGC 2403, start at M81, and move about one binocular field of view southwest, to a slender triangle formed by 5th-magnitude stars Rho (ρ), Sigma^1 (σ^1), and Sigma^2 (σ^2) Ursae Majoris. This distinctive triangle points toward the west-northwest, directly at a lone 5th-magnitude star about a field of view away.

8. From there, move another binocular field of view southwest to a larger right triangle of three 6th-magnitude stars. The target lies halfway between the triangle's right angle and its southern corner. You can spot NGC 2403 through 10×50 binoculars. You will see its tiny oval glow against the background sky, even from a backyard under a light-polluted sky.

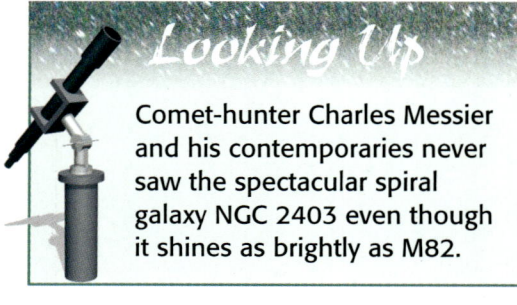

Looking Up

Comet-hunter Charles Messier and his contemporaries never saw the spectacular spiral galaxy NGC 2403 even though it shines as brightly as M82.

7.6 Cigar Galaxy (M82), an irregular galaxy in the constellation Ursa Major.

Courtesy of Adam Block/Mount Lemmon SkyCenter/University of Arizona

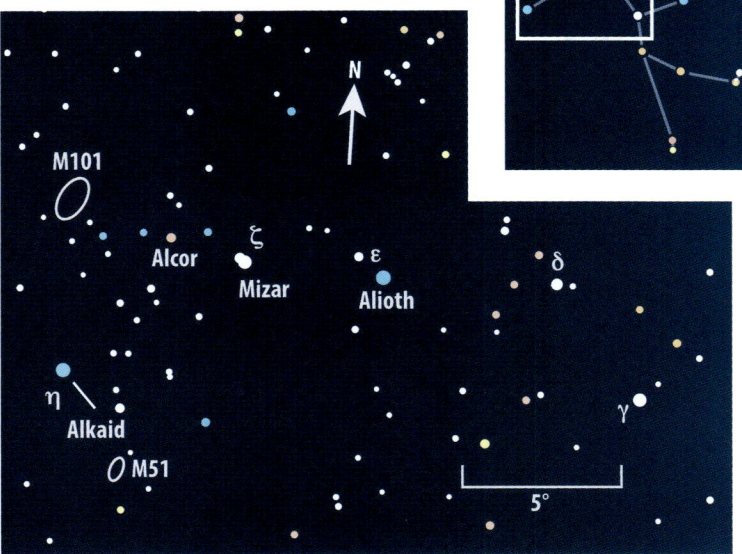

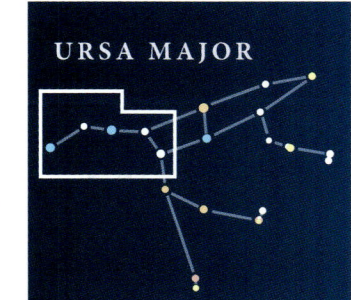

7.7 Star chart to locate May and June's remaining binocular highlights. *Courtesy of Richard Talcott and Roen Kelly, Astronomy magazine*

9. The Big Dipper is not a constellation by itself, but rather is part of Ursa Major (the Great Bear). Figure 7.7 shows the Great Bear and the rest of May and June's binocular highlights. The Dipper's bowl forms the Bear's back and belly. Although artists often portray the three stars in the Dipper's handle—Alioth, Mizar, and Alkaid—as its long tail, bears don't have long tails.

10. A better explanation comes from a Native American legend that depicts the tail stars as three hunters chasing the bear. If you look carefully, you can see that the hunter in the middle is carrying a pot in which to cook the bear after they catch it.

11. The "pot" is the star Alcor (80 Ursae Majoris). With Mizar (Zeta UMa), they form one of the sky's best-known double-star teams. They are a good naked-eye test and are an easy target for even the smallest binoculars. Mizar shines at 2nd magnitude, while Alcor glows at 4th. They may look like a physical pair, but Alcor and Mizar are actually frauds. Mizar lies 78 light-years away, while Alcor is 81 light-years away. Mizar turns out to be a true close-set binary star, but you will need a telescope to see its companion.

12. Don't confuse Mizar's companion for a third, fainter star that forms a tiny triangle with Alcor and Mizar.

13. From Mizar, trace a zigzag of four faint stars eastward, away from the Big Dipper's curving handle. At the fourth star, turn slightly northeast toward a diamond of four fainter stars. There you will see a faint smudge next to the diamond's eastern point. That is the spiral galaxy M101—a challenging target for small

Looking Up

Although Mizar's companion star is not physically related to either Alcor or Mizar, it holds an interesting footnote in astronomical history. In 1722 German mathematician Johann Liebknecht (1679–1749) thought he saw this star move against the background from one night to the next. He concluded it was not a star, but a new planet. In his excitement, he christened it Sidus Ludoviciana ("Ludwig's Star") after Ludwig V, then king of Germany. It quickly became apparent that Liebknecht was mistaken, but the star is still called Sidus Ludoviciana.

GLOSSARY TERM

Averted vision technique observers use for viewing faint objects in which they focus, not on the object itself, but a little off to its side

telescopes, let alone binoculars. Even with **averted vision**, M101 probably will seem little more than a dim glow, but consider that you are looking at a system of billions of stars. The light from these stars left 27 million years ago, long before our earliest ancestors evolved on the grassy savannahs of eastern Africa.

14. If M101 proves a little too difficult, try your luck with a brighter galaxy. Aim your binoculars toward Alkaid, the star at the end of the Dipper's handle. Look half a field of view to Alkaid's west-southwest for a 4th-magnitude star, and from there move an equal distance farther south to a pentagon of dimmer suns. If you look carefully just inside the pentagon's eastern corner, you might spot a dim glow; that is the Whirlpool Galaxy, M51 (Fig. 7.8).

15. Through 70mm and larger binoculars, you might notice M51 appears a little lopsided. You aren't looking at just one galaxy, but two. Charles Messier missed M51's companion, so you won't find it in his catalog. Today it is known as NGC 5195, which glows at about 10th magnitude.

16. Answer the remaining questions on Data Sheet 7.3 (page 130) at the end of your viewing session (or after you return indoors).

7.8 Whirlpool Galaxy (M51) in Canes Venatici.

Courtesy of Adam Block/Mount Lemmon SkyCenter/University of Arizona

Name _____ Section _____ Date _____

May-June

Quality of Observing Site

1 How dark does the site seem to you? Explain how you came to your conclusion.

_____
_____
_____

2 How far away from the nearest large city are you? How far from the nearest small city?

_____
_____

3 Are there any "light domes" visible? These bright areas of the sky (brightest near the horizon) originate with artificial lights.

_____
_____

4 Are there any clouds? Approximately how much of the sky do they cover?

_____
_____

5 Is the Moon visible? It is best to choose a moonless night to observe, but maybe the nights have been cloudy. What is the Moon's phase?

_____
_____

6 What is your estimate of your site's seeing? Explain your answer.

_____
_____
_____
_____

7 What is your estimate of your site's transparency? Explain your answer.

_____
_____
_____
_____

Exercise 7.3 Data Sheet

Observations

8 What model of binoculars did you use?

9 Was the unit easy to handle? Were you happy with the views?

10 How long did you observe?

11 Which of the objects discussed in the procedural steps on pages 125–128 did you observe? Describe each briefly.

12 Which of the objects discussed did you not see? Why?

13 If you used your binoculars to further explore the sky, which objects not discussed on pages 125–128 did you observe? Describe a maximum of three of them.

July–August: A Tail's Tale

Exercise 7.4

The constellation Scorpius is unique, because Scorpius actually looks like what it is supposed to represent. No trying to force a centaur aiming an arrow out of a star pattern that looks more like a teapot. That is neighboring Sagittarius. The stars of Scorpius actually trace the outline of a scorpion, complete with two stars for its poisonous stinger at the end of a hook-shaped body.

The bright orange star Antares (Alpha Scorpii), marking the Scorpion's heart, also attracts attention. Even though it lies about 600 light-years away, Antares is the 15th-brightest star in the night sky.

Immersed in the glow of the Milky Way, Antares makes a good starting point to search for many hidden treasures in July and August (Fig. 7.9).

7.9 Star chart to locate July and August's binocular highlights. *Courtesy of Richard Talcott and Roen Kelly, Astronomy magazine*

Procedure

1. Upon arrival at your destination, your first task is to assess the quality of your observing site. Answer the first seven questions on Data Sheet 7.4 (page 135).

2. Find Antares in the constellation Scorpius.

3. Next find **globular cluster** M4. You don't need to move to see it when you are staring at Antares, it is just a half degree to the west. Look for a dim, round glow that resembles a distant cotton ball. At 7,200 light-years away, M4 (Fig. 7.10) is relatively close to Earth for a globular cluster. This makes it appear bigger than most and easy to identify through 7×50 binoculars. Because M4 is loosely structured, its lack of a condensed center can confound observers who must battle light pollution.

Materials

- ❏ Flashlight (preferably one with a red filter) for reading the procedural instructions and star chart outside at night
- ❏ July–August star chart (provided)
- ❏ Binoculars

GLOSSARY TERM

Globular cluster spherical collection of old stars orbiting outside the main bodies of galaxies

7.10 Globular cluster M4 glowing softly near the constellation Scorpius' brightest star, Antares.
Courtesy of George Seitz/Adam Block/NOAO/AURA/NSF

4 Scorpius also houses a second Messier globular cluster that proves more challenging through binoculars: M80. This cluster's small, dim disk rests halfway between Antares and Graffias (Beta Scorpii). Although its starlike core is typical of globular clusters, M80 is tiny compared to M4, so identifying it from surrounding stars can be difficult.

5 Now place Antares at the western edge of your binoculars' field of view, and look on the opposite side for a close-set pair of 5th-magnitude stars lying across the invisible border in Ophiuchus. The globular cluster M19 is located just a degree to the south. Looking a little larger and slightly brighter than M80, M19 should reveal itself as a fuzzy "star" through 7×50 binoculars.

6 M62 sounds like it should be a twin of M19, but when you look for it about half a field to the south it is more difficult to see. M62 may appear featureless through 8x42 binoculars; however, 15×70 binoculars reveal a brighter, stellar core.

7 Next, proceed southward along the Scorpion's crooked body. Stop partway at Mu (μ) Scorpii, a wide double star, and at Zeta Scorpii, another wide double star. Although both are just chance line-of-sight pairings, the area between them is dazzling through binoculars.

8 Make sure to pause at open cluster NGC 6231, which is located just north of Zeta. There are 120 searing blue and blue-white stars crowded inside this cluster. They look faint through binoculars because they are being seen from 5,900 light-years away. If you could reduce that to the distance of the Pleiades Cluster (440 light-years), the brightest stars in NGC 6231 would outshine Sirius, the brightest star in our sky.

9. Continue along the Scorpion's curved body to the tip of its stinger, marked by the stars Shaula (Lambda [λ] Sco) and Lesath (Upsilon [υ] Sco). If you place the pair on the southern edge of your binoculars' field of view, you should see two distinct clumps of stars to the north and northeast (above and to the left): M6 and M7, respectively.

10. M6 is distinguishable, because the brightest stars in M6 seem to form a rectangle, although more inventive eyes can imagine a butterfly's outline among the stars. Look for two wings outstretched from the butterfly's centered body. The butterfly appears to be headed southeast.

11. M7 (Fig. 7.11) is larger and brighter than M6, so it should be more obvious. Even through the smallest binoculars, M7 will burst into a striking assortment of stars covering an area larger than the full Moon. Several of its 80 stars show subtle hues of yellow and blue, with the brightest a yellow beacon lying close to the group's center.

12. Answer the remaining questions on Data Sheet 7.4 (page 136) at the end of your viewing session (or after you return indoors).

7.11 Ptolemy's Cluster (M7) lying just off the two stars that form the stinger of Scorpius the Scorpion.

Courtesy of Allan Cook/Adam Block/NOAO/AURA/NSF

Name _____ Section _____ Date _____

July - August

Quality of Observing Site

1 How dark does the site seem to you? Explain how you came to your conclusion.

2 How far away from the nearest large city are you? How far from the nearest small city?

3 Are there any "light domes" visible? These bright areas of the sky (brightest near the horizon) originate with artificial lights.

4 Are there any clouds? Approximately how much of the sky do they cover?

5 Is the Moon visible? It is best to choose a moonless night to observe, but maybe the nights have been cloudy. What is the Moon's phase?

6 What is your estimate of your site's seeing? Explain your answer.

7 What is your estimate of your site's transparency? Explain your answer.

Exercise 7.4 Data Sheet

Exercise 7.4 Data Sheet

Observations

8 What model of binoculars did you use?

9 Was the unit easy to handle? Were you happy with the views?

10 How long did you observe?

11 Which of the objects discussed in the procedural steps on pages 131–133 did you observe? Describe each briefly.

12 Which of the objects discussed did you not see? Why?

13 If you used your binoculars to further explore the sky, which objects not discussed on pages 131–133 did you observe? Describe a maximum of three of them.

September–October: Strum the Harp

Exercise 7.5

The brilliant stellar sapphire Vega (Alpha Lyrae) sparkles high in the west and is the starting point for this exercise's binocular highlights (Fig. 7.12). Famous as the fifth-brightest star in the night sky, Vega's glow punches through even severe light pollution, making it difficult to miss.

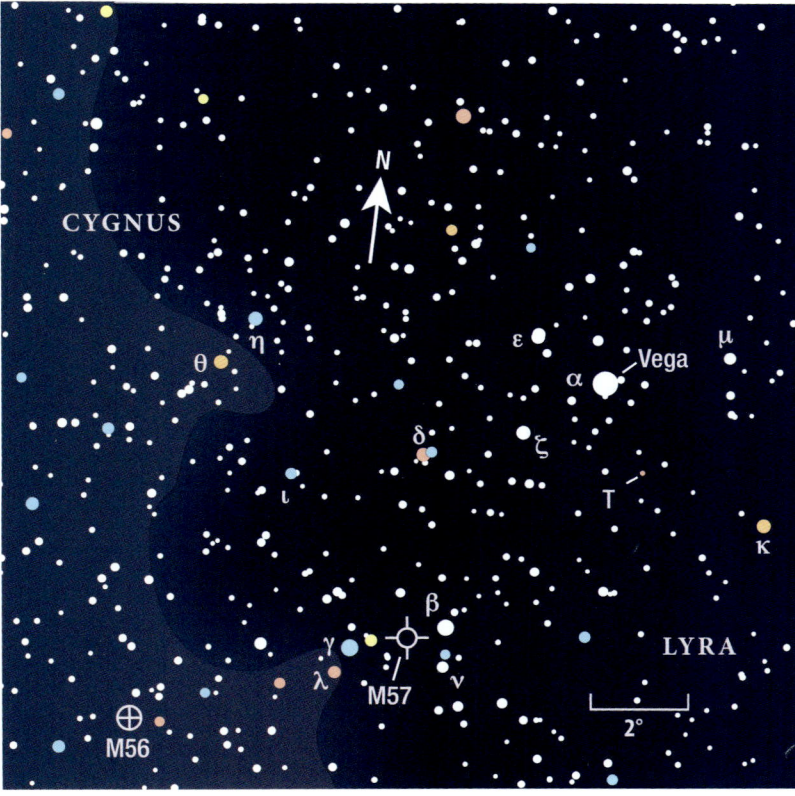

7.12 Star chart to locate some of September and October's binocular highlights.
Courtesy of Richard Talcott and Roen Kelly, *Astronomy* magazine

At your next opportunity, look at Vega through your binoculars. As you observe its blue-white luster, consider that the light you are seeing left there 25 years ago. Even though that works out to be 147 trillion miles (237 trillion kilometers), Vega is just a "town" or two away on the cosmic distance scale. We are practically neighbors. Lyra, Vega's constellation, symbolizes the lyre, or harp, owned by the mythological musician Orpheus.

Procedure

1. Upon arrival at your destination, your first task is to assess the quality of your observing site. Answer the first seven questions on Data Sheet 7.5 (page 141).

2. Locate Vega in Lyra the Harp. Vega marks a portion of the harp's handle, while four fainter stars in a parallelogram frame its body. Light pollution may hide those four stars, so if you can't make them out by naked eye, use your binoculars. Each is worth a closer look. For example, Sheliak (Beta Lyrae), at the parallelogram's southwestern corner, is more than a faint star. Sheliak is actually an eclipsing binary that is perfect for binocular study. In just under 13 days, an

Materials

- ❏ Flashlight (preferably one with a red filter) for reading the procedural instructions and star chart outside at night
- ❏ September–October star chart (provided)
- ❏ Binoculars

unseen companion star causes Sheliak to flicker from magnitude 3.3 to 4.3. Compare Sheliak's brightness to that of nearby stars that do not vary to confirm its 13-day cycle.

> **GLOSSARY TERM**
>
> **Planetary nebula** outer, gaseous layers of a red giant star, which have been gently blown off into space and glow because the gas is excited by radiation from the central, collapsing star

3. The most-photographed **planetary nebula** is the Ring Nebula (M57), which lies along the southern edge of the harp's parallelogram. To spot it, look midway between Sheliak and Sulaphat (Gamma Lyrae) for three faint stars that create a tiny right triangle. The "star" at the right angle is actually the Ring. Although it takes at least 50-power binoculars to make out its classic smoke-ring shape, M57 may be seen as a faint, starlike point through binoculars as small as 7×35.

4. Now scan southeast of Sulaphat toward the star Albireo (Beta Cygni), and pause about halfway in between. There you will find a conspicuous asterism shaped like the number 7. If you look just to the 7's southeast, you should spot a smudge that doesn't quite look like a star. That is globular cluster M56 (Fig. 7.13). Although 100,000 stars make up the cluster, M56 is too far away to resolve through binoculars. That feat requires at least a 6-inch telescope.

5. Even the smallest pocket binoculars will reveal Delta (δ) Lyrae, located at the Harp's northeastern corner, as two close-set stars. Some sharp-eyed stargazers don't need any optical aid to see them. The pair's brighter star, magnitude 4.3 Delta^2, looks orangish, while magnitude 5.6 Delta^1 is bluish white. The two Delta stars belong to a scattered open cluster nicknamed the Delta Lyrae Cluster and are cataloged as Stephenson 1.

6. Fifteen stars belong to Stephenson 1, with most too faint for binoculars. Although both Deltas belong to this cluster, studies suggest that Delta^1 is about 1,200 light-years away, while Delta^2 is about 200 light-years closer to us—a classic example of an optical binary star.

7. The last stop is Epsilon (ε) Lyrae, just to Vega's northeast. If you have acute vision, you may be able to split Epsilon into two stars by naked eye. The stars are separated by 3.5', which is near the naked eye's resolution limit. Of course, through binoculars, Epsilon easily resolves into two points of light. The northernmost of the pair is labeled Epsilon^1, while the other is Epsilon^2. Many observers also know this system as the Double Double, because Epsilon^1 and Epsilon^2 are each close-set pairs of stars.

7.13 Globular cluster M56 in the constellation Lyra, best seen with binoculars having front lenses of 50mm or more.

Courtesy of Anthony Ayiomamitis

Unfortunately, it takes at least 80× magnification to see all four Epsilon affiliates. Although you can see only two stars, the added beauty of Vega to the southwest and Zeta Lyrae to the southeast creates a stunning binocular scene.

8 Figure 7.14 shows the remaining binocular highlights for September and October. Altair, in the constellation Aquila (the Eagle), marks the Summer Triangle's southern tip and sparkles radiantly through binoculars. You would never know it just by looking, but Altair rotates rapidly, spinning around once in less than 10 hours. Compare that to our Sun, which takes about 25 days to rotate once.

9 Located on either side of Altair are two fainter stars: Alshain (Beta Aquilae) to the southeast, and Tarazed (Gamma Aquilae) to the northwest. These giant stars are classified as spectral type K and are larger—but cooler—than either Altair or our Sun. Can you detect their subtle orange coloring through your binoculars? They contrast nicely with white Altair.

10 If you are viewing under dark skies, you might notice the hazy lane of our galaxy, the Milky Way, flowing past Altair. Scan southeastward a little more than a binoculars' field of view diameter to the 3rd-magnitude star Delta Aquilae, and then an equal distance farther southeastward, to 3rd-magnitude Lambda Aquilae. Lambda and two fainter, adjacent stars form an arc that hooks counterclockwise, directly toward the Scutum Star Cloud.

Scutum is a faint constellation that is difficult to pick out on its own. But through binoculars, the Scutum Star Cloud stands out as one of the Milky Way's

7.14 Star chart to locate September and October's remaining binocular highlights.

Courtesy of Richard Talcott and Roen Kelly, Astronomy magazine

Outdoor Sky Observations **CHAPTER 7** 139

richest regions. The star cloud shines bright enough that it is visible to the naked eye, even if light pollution blots out the more subtle surrounding Milky Way.

11 While you look around, notice several pockets where the star density increases. The Wild Duck Cluster (M11) stands out as a bright smudge of starlight just to the west of the Eagle's tail-feather stars (Fig. 7.15). Because most of M11's stars are too faint for binoculars, they blend into a small, round mist of starlight, except for a lone 8th-magnitude sun that shines above the fray.

12 A second smudge of condensed starlight is located about half a binocular field of view southwest of M11. Look for it just to the southwest of a slender right triangle of stars formed by Alpha, Delta, and Epsilon Scuti. Although this open cluster, cataloged as M26, is not nearly as bright and obvious as M11, it should still be easily visible through handheld binoculars on clear, moonless nights. M26 includes about 30 stars. None shines brighter than 10th magnitude, so they are not bright enough to be seen through most binoculars. Instead, M26's light produces a hazy glow covering about a quarter of a degree, or about half of the Moon's diameter in our sky.

13 Answer the remaining questions on Data Sheet 7.5 (page 142) at the end of your viewing session (or after you return indoors).

7.15 Wild Duck Cluster (M11) lies in the constellation Scutum the Shield.

Courtesy of Anthony Ayiomamitis

Name _____ Section _____ Date _____

September–October

Quality of Observing Site

1. How dark does the site seem to you? Explain how you came to your conclusion.

2. How far away from the nearest large city are you? How far from the nearest small city?

3. Are there any "light domes" visible? These bright areas of the sky (brightest near the horizon) originate with artificial lights.

4. Are there any clouds? Approximately how much of the sky do they cover?

5. Is the Moon visible? It is best to choose a moonless night to observe, but maybe the nights have been cloudy. What is the Moon's phase?

6. What is your estimate of your site's seeing? Explain your answer.

7. What is your estimate of your site's transparency? Explain your answer.

Exercise 7.5 Data Sheet

Observations

8 What model of binoculars did you use?

9 Was the unit easy to handle? Were you happy with the views?

10 How long did you observe?

11 Which of the objects discussed in the procedural steps on pages 137–140 did you observe? Describe each briefly.

12 Which of the objects discussed did you not see? Why?

13 If you used your binoculars to further explore the sky, which objects not discussed on pages 137–140 did you observe? Describe a maximum of three of them.

November–December: In the Queen's Court   Exercise 7.6

After the Big Dipper and Orion, the most recognizable pattern of stars visible from the Northern Hemisphere may be the "W" of Cassiopeia, the mythical Queen of Ethiopia. As a circumpolar constellation (one that never sets) for many northerners, Cassiopeia's five stars are visible on clear nights. Cassiopeia is the starting point for this exercise's binocular highlights (Fig. 7.16). During November, Cassiopeia rides high above Polaris and looks resplendent in the Milky Way's gentle glow. Set your binoculars on this target.

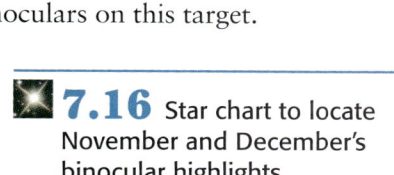 **7.16** Star chart to locate November and December's binocular highlights.

Courtesy of Richard Talcott and Roen Kelly, *Astronomy* magazine

Procedure

Materials

- ❏ Flashlight (preferably one with a red filter) for reading the procedural instructions and star chart outside at night
- ❏ November–December star chart (provided)
- ❏ Binoculars

1. Upon arrival at your destination, your first task is to assess the quality of your observing site. Answer the first seven questions on Data Sheet 7.6 (page 147).

2. One of observers' favorite open clusters in Cassiopeia is M52 (Fig. 7.17). To find it, draw a line from Schedar (Alpha Cassiopeiae) to Caph (Beta Cas), the westernmost stars in the W pattern, and continue it an equal distance to the northwest. There, you should spot a slender, four-star diamond pattern. M52 lies to the diamond's south. Although about 200 stars are located in M52, few are bright enough to be visible through binoculars; the rest blend into a little cloud of misty starlight.

3. After you have viewed M52, return to the star Caph in the Cassiopeia W and look around. Can you spot three pairs of stars just to its south-southwest? Although none of these is a true binary star system, each pair forms the corner of a slim triangular asterism that reminds some people of a skinny hang glider. The brightest star in the asterism, Rho Cassiopeiae, is

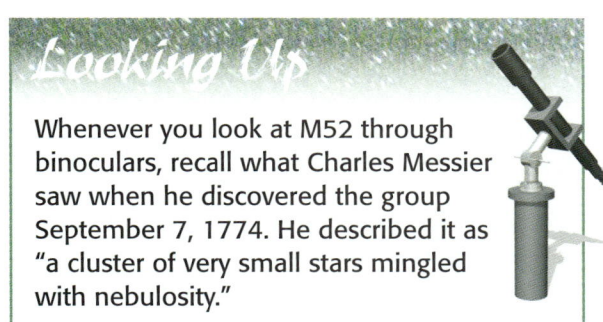

Whenever you look at M52 through binoculars, recall what Charles Messier saw when he discovered the group September 7, 1774. He described it as "a cluster of very small stars mingled with nebulosity."

7.17 Open cluster M52, located in a rich part of the Milky Way in the constellation Cassiopeia the Queen.

Courtesy of Anthony Ayiomamitis

one of the two at the nose of the hang glider, while 5th-magnitude Sigma Cas—the brighter of the two stars—is at the end of the southern wing.

4 You might also spot a faint stain on the leading edge of the glider's southern wing. That is NGC 7789, a distant open cluster discovered in the fall of 1783 by Caroline Herschel (1750–1848), Sir William's sister. Don't expect to see a lot of stars through your binoculars, however. Instead, NGC 7789 looks like a faint, round glow with perhaps one or two elusive points shining through.

5 One of Cassiopeia's sights lies near the star Zeta Cas. Through binoculars, 4th-magnitude Zeta's distinct aquamarine hue contrasts nicely against Schedar's orangish tint.

6 Zeta is located at the end of a semicircular asterism of seven fainter stars that astronomer Garrett P. Serviss described as an "array of stars . . . in a broken half-circle, which may suggest the notion of a crown." Together, they resemble a backward "3" or a miniature of the spring constellation Corona Borealis. Because older depictions of the constellation show Zeta representing the Queen's head, "Cassiopeia's crown" seems an appropriate name for this little asterism.

7 Go deeper into the constellation by following the zigzag path along the Cassiopeia five-star "W" to Ruchbah (Delta Cas), at its lower-left

Looking Up

"Here the Milky Way is so rich that an observer hardly needs any guidance; he is sure to stumble upon interesting sights for himself," is how American astronomy popularizer Garrett P. Serviss (1851–1929) describes the binocular view of Cassiopeia in his 1888 classic, *Astronomy with an Opera Glass*.

Serviss' book introduced the idea that people could use binoculars to view the heavens. Of course, binoculars made in the 1800s don't compare to the quality found in even the least expensive binoculars today. But the low quality of his equipment didn't deter Serviss from viewing the binocular universe from his home in Brooklyn, New York.

corner. Aim half a binocular field of view to the southwest of Ruchbah toward 5th-magnitude Phi (o) Cas. You will see a second, fainter star just to Phi's southwest as well as a tiny smear of dim starlight to the north. Together, they form the Owl Cluster, also known as NGC 457 (Fig. 7.18). The two brightest stars mark the owl's eyes, while the fainter stars outline its body and outstretched wings.

8 Next, view open cluster M103, located to the other side of Ruchbah. This group is seen as a small, triangular patch of starlight nestled in a Milky Way field when viewed with 10×50 binoculars. You should be able to see four or five separate stars here. The remaining 170 stars pool their faint light to create what looks like a tiny arrowhead. Others think M103 resembles a handheld fan.

9 Scan the area around M103, and you will notice the hazy glow of NGC 663 a little east of the halfway point between Ruchbah and Segin (Epsilon Cas). You may see it before you even notice M103. NGC 663 looks like an unresolved blur of light through 10×50 binoculars, while 16×70 binoculars add a few feeble points of light.

10 Answer the remaining questions on Data Sheet 7.6 (page 148) at the end of your viewing session (or after you return indoors).

7.18 Owl Cluster, also known as NGC 457 in the constellation Cassiopeia the Queen, is one of the brightest open star clusters in the northern half of the sky.

Courtesy of Anthony Ayiomamitis

Name _____ Section _____ Date _____

November–December

Quality of Observing Site

1 How dark does the site seem to you? Explain how you came to your conclusion.

2 How far away from the nearest large city are you? How far from the nearest small city?

3 Are there any "light domes" visible? These bright areas of the sky (brightest near the horizon) originate with artificial lights.

4 Are there any clouds? Approximately how much of the sky do they cover?

5 Is the Moon visible? It is best to choose a moonless night to observe, but maybe the nights have been cloudy. What is the Moon's phase?

6 What is your estimate of your site's seeing? Explain your answer.

7 What is your estimate of your site's transparency? Explain your answer.

Exercise 7.6 Data Sheet

Exercise 7.6 Data Sheet

Observations

8 What model of binoculars did you use?

9 Was the unit easy to handle? Were you happy with the views?

10 How long did you observe?

11 Which of the objects discussed in the procedural steps on pages 143–145 did you observe? Describe each briefly.

12 Which of the objects discussed did you not see? Why?

13 If you used your binoculars to further explore the sky, which objects not discussed on pages 143–145 did you observe? Describe a maximum of three of them.

Looking Up

Sketching Techniques

LEARNING OBJECTIVES

Upon completion of this chapter, you should be able to:

1. Explain why sketching will help make you a better observer.
2. Identify the materials needed to sketch celestial objects.
3. Determine and explain the stages necessary to make a sketch of a particular object.
4. Compare and contrast asterisms and constellations.
5. Locate some of the sky's major asterisms.
6. Define all glossary terms.

Star-forming region LH 95 in the Large Magellanic Cloud
Courtesy of NASA/ESA/Hubble Heritage (STScI/AURA) – ESA/Hubble Collaboration

If you are not ready to capture the planets, stars, and galaxies with a camera, there is an alternative: sketching. Drawing with pencil on paper enables you to create a pictorial record of your observations and become a better observer in the process. When considering sketching, remember, you don't have to be an artist. Be happy with how you sketch and know—with absolute conviction—that your results will get better. Over time you will be amazed at the improvement in your ability to see minute features. In this lab, you will learn how to look for the details in celestial objects with your naked eyes and while using a telescope or binoculars. You will also sketch **asterisms** from real-time sky targets.

Preparation and Materials

Sketching Forms and Pencils

Much of what you sketch will be circular (whole solar or lunar disks and planets), and all the noncircular objects will lie in the circular field of view of your binoculars' or telescope's eyepiece. In this lab, you will be given circles in which to sketch. For future sketches, you can draw circles either with a compass or computer software. The advantage of the latter is the ability to quickly print as many circles as needed. A slight disadvantage: the circle will have a different "look" than the rest of the sketch.

Circles may be different sizes, but don't make them too large or your sketches will take too long. Use a circle 4 inches (10 centimeters) in diameter. Smaller sizes can be used, but beginning sketchers may not feel confident using smaller circles. Drawing tiny details requires a fine touch. Two 4-inch circles also easily fit on a page, leaving space for comments.

When you are ready to head outside, take your sketching forms, regular #2 pencils, pencil sharpener, and a good quality soft eraser.

Table or Clipboard

If possible, it is best to sit while observing and sketching. When you are observing, comfort is important, especially when you are looking for minute details and transferring them to paper. Ideally, you should set up a short table in a comfortable position near the telescope eyepiece. If this is impractical, you will need a clipboard or some sort of sturdy notebook to back up your paper.

Red Flashlight

You will need a red flashlight, preferably one that you can dim, for observing faint objects. A red light degrades your **dark adaption** less than any other color. Holding a red flashlight (Fig. 8.1) with your hand or teeth for any length of time is tiring. If you can fasten the light to your sketching board or table, or suspend the light above it, you will be more relaxed at the eyepiece.

8.1 Examples of red flashlights.

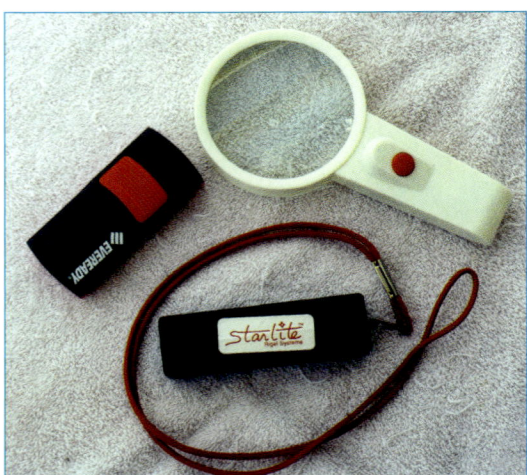

> **GLOSSARY TERM**
> **Asterism** unofficial, recognizable grouping of stars either part of a single constellation (the Big Dipper) or several constellations (the Summer Triangle)

> **Note**
> More advanced sketchers will benefit from an Eberhard Faber Ebony art pencil, which will produce better results than a regular #2 pencil.

> **GLOSSARY TERM**
> **Dark adaption** process by which your eyes get used to the darkness; exposure to bright light at night can temporarily ruin your dark adaption

Check Your Understanding

1.1 What is the first step in sketching celestial objects?

1.2 Why will sketching make you a better observer?

1.3 What supplies will you need during a sketching session?

Techniques

When you begin sketching celestial objects you will, by necessity, spend more time looking at objects rather than moving quickly from one object to another. To make a sketch, it is necessary to memorize small areas of detail before transferring what you see to paper. The more objects you sketch, the more details you will notice as part of your normal observing routine. In essence, you are training your eyes (or eye, because most observers use one eye exclusively) to see objects in a new way. We have all observed objects and wished we had a larger telescope. Sketching helps you see the maximum detail possible using your current equipment.

Taking Your Time

To practice training your eye, pick a celestial object that you can see well through your telescope or binoculars and can observe over a month's time. Observe and sketch the object once a week for four weeks. Use the same binoculars or telescope and eyepiece, and observe from the same site. Take 15 or 20 minutes during each session to sketch the object. Most observers who sketch at the telescope use dark pencils on white paper. This approach is intuitive, but it takes some effort to convert sketches of deep-sky objects to how they looked through the eyepiece, with the blacks and whites reversed. Just take your time. After one month, you will have four sketches to compare, and you should see improvements in the drawings.

Keeping a Log

Before you begin sketching, you should always make notes of the date and time the sketch was made, the telescope and eyepieces used, and a brief description of sky conditions. Your sketches will be more valuable to you if you keep a log of this relevant information. Where you put these details is up to you. Your method may consist of nothing more than

> **Note**
>
> Because it is so bright, the Moon requires that you *ruin* your dark adaption to sketch it. Sit near a white light bright enough to trigger your daytime vision, which is better for seeing both detail and color than your nighttime vision. A single incandescent bulb usually emits enough light to do this. Use this technique only for solo observing.

> **GLOSSARY TERM**
>
> **Feature** distinct property or part of an object

filling in the blank areas on your sketching form. Or, you can record the details on a separate page in an observing notebook, or on a computer disk using word-processing software.

Sketching Solar System Objects

If you are sketching the Sun (Fig. 8.2), the Moon, or a planet such as Jupiter (Fig. 8.3), try to identify the center of the object's circle.

Next, look at the object as you would a clock. Use the clock's face as a reference, and note where **features** begin and end. Choose a pair of features and note:

- Are they the same height and width?
- Do they appear tilted at the same angle?
- Are they the same lightness and darkness?
- Are their borders similar or different?
- Do they have the same degree of smoothness?
- Are there any color differences?
- Where are they located with respect to each other?

Draw light outlines of your selected pair of features first. Start with obvious features such as sunspots, large lunar craters, or planetary bands. When you are satisfied with the way one pair of features looks, move on to the next. Move on to smaller, fainter features, and finally, carefully shade in areas until you are satisfied with the details.

> **CAUTION**
>
> To observe the Sun, sunspots, and other solar features, ALWAYS use an approved solar filter.

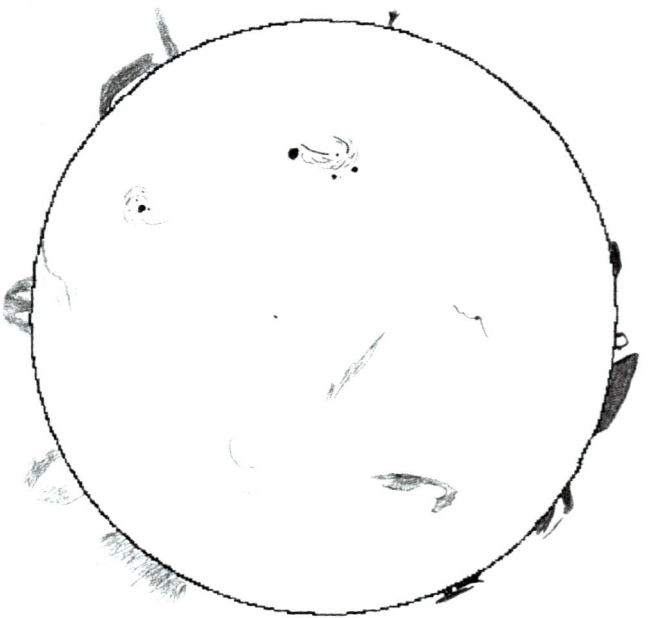

8.2 Solar features seen on September 14, 2000, between 23h11m UT and 23h27m UT through a 4-inch refractor at $f/30$ and a magnification of $60\times$.

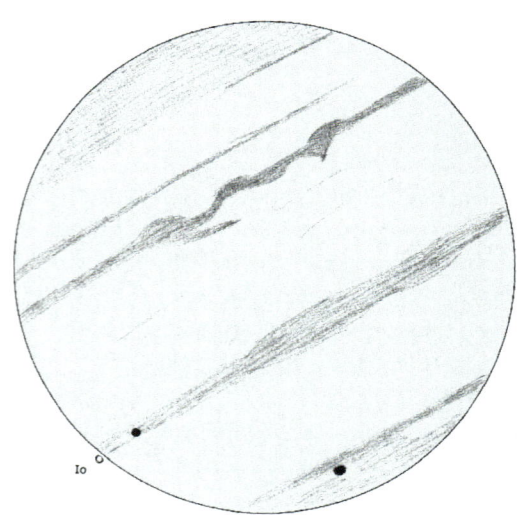

8.3 Shadow transits of the jovian moons Io and Ganymede seen on January 8, 2001, at 2h53m UT through a 4-inch refractor at $f/15$ and a magnification of $250\times$.

Sketching Deep-Sky Objects

Not all objects in the night sky will be circular in shape. You might sketch a galaxy (Fig. 8.4), a nebula, or a comet. For these objects, you need a slightly different set of techniques.

A good first target for beginning sketchers is the Ring Nebula (M57) in the constellation Lyra (Fig. 8.5). It is a bright planetary nebula that is small enough for a good first sketch.

Begin by comparing its size to that of the entire field of view. Is it half as large as the field of view? One-third as large? One-fifth? Try to capture this ratio in your drawing.

Next, note the shape of the nebula (it is not quite circular), the density of both the ring and the central area, and the positions of the field stars.

Now, sketch the positions of the brightest stars first, using them as **reference points** (Fig. 8.6A). Once you are satisfied that the bright stars on your sketch match what you are seeing through the eyepiece, plot some of the fainter stars.

Next, add the target object's rough outline (Fig. 8.6B). For star clusters, start with the brightest stars and work to the fainter ones. Finally, add shading to complete the sketch (Fig. 8.6C). It is easy to go overboard at this point, so work slowly, and frequently compare your sketch to the object.

Some observers prefer to use pencils with soft leads when sketching comets, nebulae, and galaxies. To simulate the glow of nebulosity, first draw the outline of the object. Then, using a clean finger (or an ever-so-slightly moist one), gently **smudge** the outline into a blob that accurately represents the object. An example of smudging is shown in Figure 8.7, a sketch of the sky's brightest globular cluster, Omega Centauri (NGC 5139). Sketching this object can be difficult because so many stars are visible, especially through large telescopes. For this type of sketch, taking time and paying special attention to the center of the object helps in creating an accurate image.

> **Note**
> Some sketchers wait until they have drawn the object before they add any faint stars. Pick whichever sequence seems to work best for you.

GLOSSARY TERMS

Reference point prominent, easy-to-locate point, sometimes used as a starting point

Smudge noun: a blurry spot or streak; verb: to smear or make indistinct

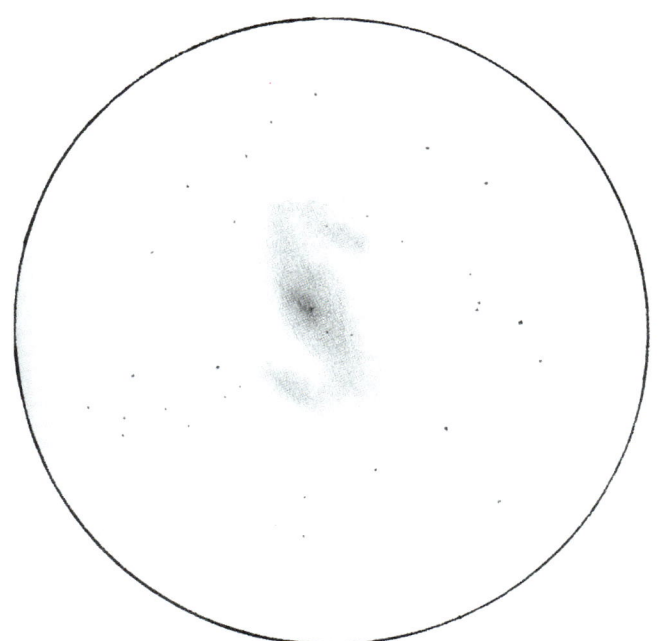

8.4 Bode's Galaxy (M81) in the constellation Ursa Major, seen through a 17.5-inch f/4.5 Newtonian reflector equipped with a 32mm eyepiece (magnification = 63×).

Courtesy of David J. Eicher

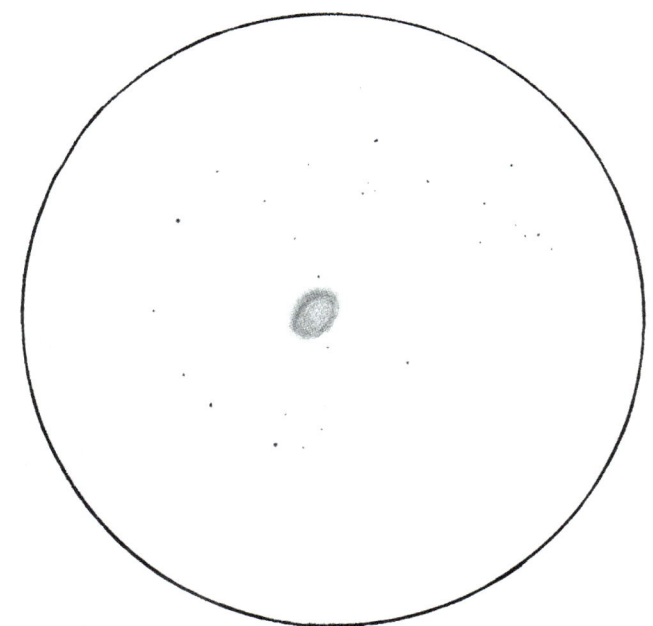

8.5 Ring Nebula (M57) in the constellation Lyra seen through a 24-inch Newtonian reflector at f/4 with a 13mm eyepiece (magnification = 188×).

Courtesy of David J. Eicher

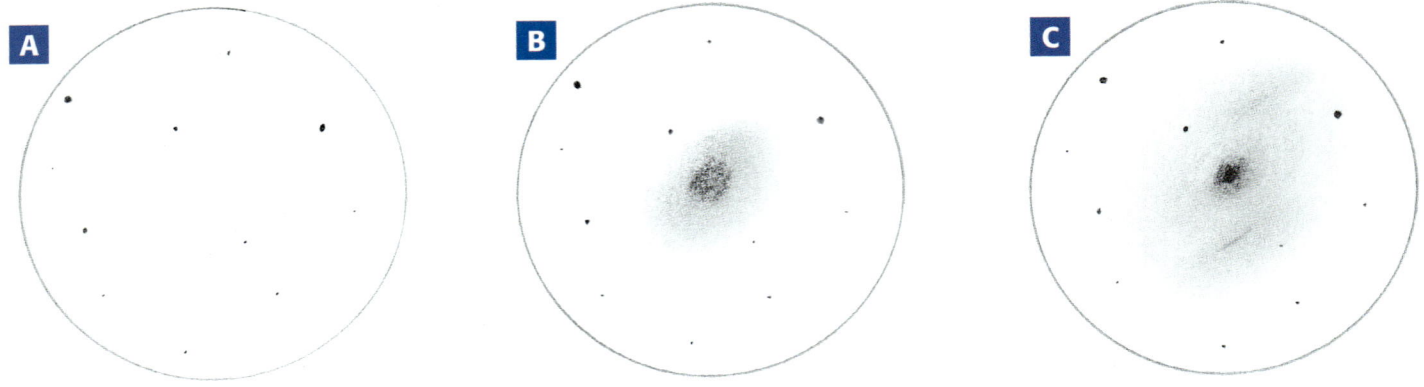

8.6 Tried-and-true steps for sketching a deep sky object: (**A**) step 1, sketch brightest stars; (**B**) step 2, add a rough outline of the target object; (**C**) step 3, add shading to finish.

Courtesy of David J. Eicher

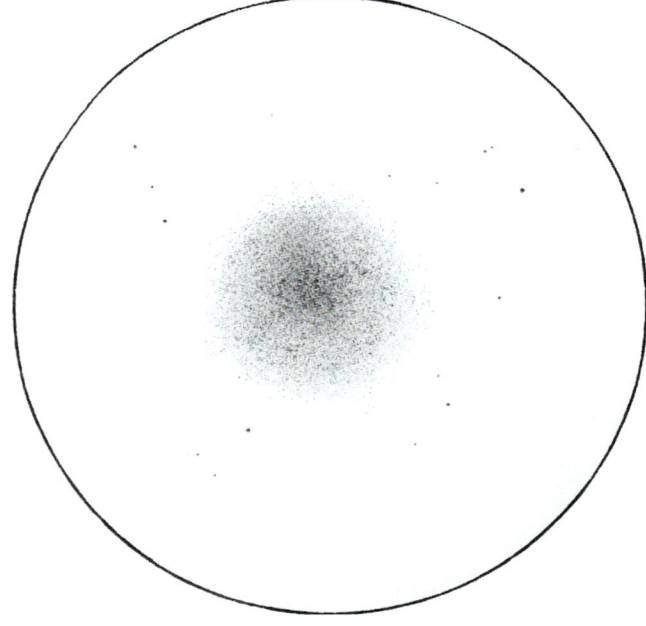

8.7 Omega Centauri (NGC 5139) seen through a 17.5-inch *f*/4.5 Newtonian reflector equipped with a 32mm eyepiece, giving a magnification of 63.

Courtesy of David J. Eicher

Incorporating Color

Once you are comfortable with black-and-white sketching, you may want to create some color drawings. Start with a small selection of colored pencils, and expand your color palette as your sketching comfort level increases. For planetary sketching, you will use a variety of earth-tone colors. Different hues of blue and green pencils will come in handy for reproducing subtle colors in planetary nebulae.

To show different increments of the same color with a single pencil, press harder or softer with the pencil. If you want to produce an even darker shade, first lightly color the area with black, and then lightly color over the black with the color of your choice. To make a lighter tint, first color firmly with white, and then add a color lightly over the white. Experiment with how much black or white and color you need to get the final tone you want. This procedure changes depending on the color you are adding. If an under-coloring with black doesn't give you the look you want, try blending your top color over brown, gray, purple, or other dark colors.

The use of regular pencils on white paper is only one option for sketching. Some observers choose to sketch with white pencil on black paper. Pen-and-ink drawings take a delicate touch but can be striking. Blendable chalk is another option. Some observers use only black and white chalk, blending them to obtain various shades of gray. Other sketchers prefer colored chalk.

After Observing

After your observing session, you may wish to continue working on your sketch. First, if necessary, clean up your sketch. Remove stray pencil marks or unwanted smudges. A light touch-up to fill in shaded areas or to add color is fine as long as you don't add any details you didn't see through the telescope.

Some observers make nice, round stars with a black pen or do their smudging at this time. Some also spray a layer of fixative on each sketch to protect it from handling and the Sun's ultraviolet light.

After each session, file your sketches and their descriptions in a safe location. You also may choose to scan them and save them in a file on your computer's hard drive. With a scanner hooked to your computer, you can scan your original sketch and almost instantly produce a **negative image** of it. You will soon have an archive to reference and share with others, without using expensive imaging equipment. Your sketches will create a time-lapse "movie" of sorts that will also show your improvement as an artist.

> **GLOSSARY TERM**
>
> **Negative image** opposite of a normal image; a negative image shows light areas as dark and dark ones as light; also reverses color, with red areas appearing cyan, greens appearing magenta, and blues appearing yellow

Check Your Understanding

2.1 Why do you want a nearby light turned on when you sketch the Moon?

2.2 Which features should you draw first when sketching a solar system object, like the Sun?

2.3 What are the first objects whose positions you should record when sketching a deep-sky object?

2.4 List three details you should be aware of when comparing two similar features.

2.5 With regard to sketching, what is the main difference between deep-sky objects and solar system objects?

2.6 Briefly explain three reasons a sketcher should keep a log.

Exercise 8.1 Sketching Asterisms

Procedure

1. Depending on the season in which you are observing, use one of the following tables to seek out visible asterisms. Table 8.1 lists autumn asterisms, Table 8.2 lists winter asterisms, Table 8.3 lists spring asterisms, and Table 8.4 lists summer asterisms.

2. Upon arrival at your destination, your first task is to assess the quality of your observing site. Answer the first eight questions on the data sheet, page 159.

3. Sketch five asterisms from the list, using the following data sheets.

4. Were there any asterisms you couldn't see? Why?

_____
_____
_____

Materials

- ❏ Seasonal asterisms table (provided)
- ❏ Sketching pencils
- ❏ Flashlight with red filter (for reading procedural instructions and asterisms table outside)
- ❏ Asterisms data sheets (provided)
- ❏ Telescope
- ❏ Binoculars

TABLE 8.1 Autumn Asterisms

Observation Method	Name	Description
Naked eye	The Circlet (of Pisces)	γ, β, Θ, ι, 19, λ, and κ Piscium
Telescope	Delphinus Minor	Group of five or six 7th- and 8th-magnitude stars that look like the constellation Delphinus (the Dolphin); to find it, look about halfway along a line drawn from Scheat (Beta [β] Pegasi) to Markab (Alpha [α] Pegasi) (in Pegasus)
Naked eye	The Great Square (of Pegasus)	α, β, and γ Pegasi and α Andromedae
Naked eye	The Head (of Cetus)	α, γ, ξ², μ, and λ Ceti
Naked eye	Job's Coffin	α, β, γ, and δ Delphini
Binoculars	Kemble's Cascade	Chain of stars (the Waterfall) that ends at open cluster NGC 1502 (the Pool) (in Camelopardalis)
Naked eye	The Kids (near Capella)	ε, ζ, and η Aurigae
Naked eye	The Little Dipper	α, δ, ε, ζ, η, γ, and β Ursae Minoris
Naked eye	The Northern Fly	35, 39, and 41 Arietis
Telescope	The Owl	Open cluster NGC 457; the two brightest stars are the Owl's eyes (in Cassiopeia)
Naked eye	The Segment (of Perseus)	η, γ, α, δ, ε, and ζ Persei
Binoculars	Triangulum Minor	Small triangle similar in shape to the constellation Triangulum (the Triangle) made up of the stars 6, 10, and 12 Trianguli (in Triangulum)
Naked eye	The "W" (of Cassiopeia)	ε, δ, γ, α, and β Cassiopeiae
Naked eye	The Water Jar (of Aquarius)	γ, η, π, and ζ Aquarii

TABLE 8.2 Winter Asterisms

Observation Method	Name	Description
Telescope	The 37	Open cluster NGC 2169, whose stars, to some, form the numbers "3" and "7"; others see the letters "L" and "E" (in Orion)
Naked eye	The Belt (of Orion)	δ, ε, and ζ Orionis
Naked eye	The Butterfly (of Orion)	ζ, ε, δ, γ, α, β, and κ Orionis
Telescope	The Cheshire Cat	Group of stars southwest of open cluster M38 that forms a smiley face (in Auriga)
Binoculars	Davis' Dog	The following 5th-magnitude stars: 50, 51, 53, 65, 67, 69, and 72 Tauri (in Taurus)
Naked eye	The Heavenly "G"	α Aurigae, α and β Geminorum, α Canis Minoris, α Canis Majoris, β Orionis, α Tauri, and α Orionis.
Naked eye	The Hyades	α, γ, δ, and ε Tauri.
Binoculars	The Lambda-Lambda	The star Lambda (λ) Orionis combined with open cluster Collinder 69, whose stars might just form a Greek letter Lambda (in Orion)
Naked eye	The Little Dipper	α, δ, ε, ζ, η, γ, and β Ursae Minoris
Naked eye	The Pleiades	17, 19, 20, 23, 27, and η Tauri
Naked eye	The "V" (of Taurus)	ε, δ, γ, θ, and α Tauri
Naked eye	The Winter Triangle	α Canis Majoris, α Canis Minoris, and α Orionis

TABLE 8.3 Spring Asterisms

Observation Method	Name	Description
Naked eye	The Bier	α, β, γ, and δ Ursae Majoris
Naked eye	The Big Dipper	α, β, γ, δ, ε, ζ, and η Ursae Majoris
Telescope	The Broken Engagement Ring	C-shaped grouping of 10 stars near Merak (Beta [β] Ursae Majoris) (in Ursa Major)
Telescope	The Engagement Ring	Circle of stars that includes Polaris (Alpha [α] Ursae Minoris), which is the "diamond" of the ring (in Ursa Minor)
Naked eye	The Guardians of the Pole	β and γ Ursae Minoris
Naked eye	The Head (of Hydra)	δ, ε, ζ, η, ρ, and σ Hydrae
Naked eye	The Kite (of Boötes)	α, ε, δ, β, γ, and ρ Boötis
Naked eye	The Little Dipper	α, δ, ε, ζ, η, γ, and β Ursae Minoris
Naked eye	The Pointer Stars	α and β Ursae Majoris
Naked eye	The Sail	β, δ, γ, and ε Corvi
Binoculars/Telescope	The Shark	Group of 12 stars that, to a creative mind, looks like a shark; to find it, draw a line from Eta (η) Ursae Minoris to Epsilon (ε) Ursae Minoris; the shark lies one-third of the way from Eta to Epsilon (in Ursa Minor)
Naked eye	The Sickle (of Leo)	α, η, γ, ζ, μ, and ε Leonis
Naked eye	The Spring Triangle	α Boötis, α Virginis, and β Leonis
Binoculars/Telescope	The Stargate	Group of six stars 5° north of Delta (δ) Corvi and 1° southeast of the Sombrero Galaxy (M104) (in Corvus)
Naked eye	The Trapezoid (of Boötes)	β, γ, δ, and μ Boötis
Naked eye	The "Y" (of Virgo)	α, γ, δ, ε, η, and β Virginis

TABLE 8.4 Summer Asterisms

Observation Method	Name	Description
Binoculars	The Coathanger	Also known as Collinder 399 and Brocchi's Cluster, this is probably the sky's most famous binocular asterism; made up of 10 stars between 5th and 7th magnitude about midway between Alpha (α) Vulpeculae and Alpha Sagittae, the Coathanger stretches 1.5°, or three full Moon diameters (in Vulpecula)
Naked eye	The Family (of Aquila)	α, β, and γ Aquilae
Naked eye	The Fish Hook (of Scorpius)	σ, α, τ, ϵ, μ^1, ζ^2, η, θ, ι^1, κ, λ, and υ Scorpii
Naked eye	The Head (of Draco)	β, γ, ξ, and ν Draconis
Naked eye	The Keystone (of Hercules)	ϵ, ζ, η, and π Herculis
Naked eye	The Little Dipper	α, δ, ϵ, ζ, η, γ, and β Ursae Minoris
Telescope	Little Orion	Also called Leiter 9, this is a small group of stars that looks like the constellation Orion (the Hunter) nearly halfway from Alpha (α) to Sigma (σ) Cygni (in Cygnus)
Binoculars	The Little Queen	Five stars that look like the main ("W") part of the constellation Cassiopeia (the Queen); located 3.5° east of Chi (χ) Draconis; the brightest star shines at magnitude 6.8 (in Draco)
Naked eye	The Lozenge	β, γ, ξ, and ν Draconis
Naked eye	The Milk Dipper (of Sagittarius)	ζ, τ, σ, ϕ, and λ Sagittarii
Naked eye	The Northern Cross	α, β, γ, δ, and ϵ Cygni
Naked eye	The Parallelogram (of Lyra)	β, γ, δ, and ζ Lyrae
Naked eye	The Summer Triangle	α Lyrae, α Cygni, and α Aquilae
Naked eye	The Teapot (of Sagittarius)	δ, λ, ϕ, σ, τ, ζ, ϵ, and γ Sagittarii
Binoculars/Telescope	The Zig Zag	This group of 12 stars of 8th and 9th magnitude is located 2° west of Omega (ω) Herculis (in Hercules)

5 Today astronomers know that the stars that create constellations or asterism patterns do not all lie at the same distance. To observers on Earth, however, the sky appears flat; that is, we do not perceive depth (distance). So, anyone can use stars in a given area to create a pattern meaningful to them. Take some time to examine the sky. While scanning the sky, notice other patterns that remind you of known asterisms. For example, you might see a flowerpot, your uncle's cane, or Thor's hammer.

6 Pick two patterns you've invented, and explain them on pages 165–166. Include sketches of each.

Name _____   Section _____   Date _____

Sketching Asterisms

Quality of Observing Site

1 Note your observing times:

Start _____   End _____

2 How dark does the site seem to you? Explain how you came to your conclusion.

_____
_____
_____
_____

3 How far away from the nearest large city are you? How far from the nearest small city?

_____
_____

4 Are there any "light domes" visible? These bright areas of the sky (brightest near the horizon) originate with artificial lights.

_____

5 Are there any clouds? Approximately how much of the sky do they cover?

_____

6 Is the Moon visible? It is best to choose a moonless night to observe, but maybe the nights have been cloudy. What is the Moon's phase?

_____
_____

7 What is your estimate of your site's seeing? Explain your answer.

_____
_____
_____

8 What is your estimate of your site's transparency? Explain your answer.

_____
_____
_____

Sketching Techniques   **CHAPTER 8**

Asterism #1

1 Name of the asterism: _____

2 List the bright stars in the asterism, if any: _____

_____

3 Constellation(s) the asterism is part of: _____

_____

4 How did you make your observations? (*check one*)   ☐ Naked eye   ☐ Binoculars   ☐ Telescope

5 Sketch the asterism:

6 Additional notes:

_____

_____

_____

_____

Name _____ Section _____ Date _____

Sketching Asterisms (continued)

Asterism #2

1 Name of the asterism: _____

2 List the bright stars in the asterism, if any: _____

3 Constellation(s) the asterism is part of: _____

4 How did you make your observations? (*check one*)   ☐ Naked eye   ☐ Binoculars   ☐ Telescope

5 Sketch the asterism:

6 Additional notes: _____

Sketching Techniques CHAPTER 8

Exercise 8.1 Data Sheet

"Invented" Asterism #2

1. Name of your asterism: _____

2. List the bright stars in the asterism, if any: _____
   _____

3. Constellation(s) the asterism is part of: _____
   _____

4. How did you make your observations? (*check one*) ☐ Naked eye   ☐ Binoculars   ☐ Telescope

5. Sketch the asterism:

6. Explain why you chose to invent this asterism. What about the star pattern appealed to you?
   _____
   _____
   _____

7. Additional notes:
   _____
   _____
   _____
   _____

Unit 2 Looking Up

Looking Up
The Magnitude System and Light Pollution

LEARNING OBJECTIVES
Upon completion of this chapter, you should be able to:

1. Explain how the magnitude of an object corresponds to its light output.
2. Compute how much brighter one star is than another by knowing the magnitude of each.
3. Identify several ways light pollution affects areas other than astronomy.
4. Estimate the limiting magnitude, or sky quality.
5. Define all glossary terms.

Light Pollution Courtesy of Laurent Laveder/Science Source

GLOSSARY TERMS

Magnitude measure of brightness of a celestial object; the smaller (or more negative) the magnitude, the brighter the object

Apparent magnitude how bright a star appears from Earth

Absolute magnitude how bright a star would look from a standard distance of 10 parsecs (32.6 light-years)

Parsec a distance equal to approximately 3.26 light-years

TABLE **9.1** Magnitude System

Magnitude Difference	Brightness Ratio
0.1	1.10
0.2	1.20
0.3	1.32
0.4	1.45
0.5	1.58
0.6	1.74
0.7	1.91
0.8	2.09
0.9	2.29
1.0	2.5118865
2.0	6.3
2.5	10
3.0	15.8
4.0	39.8
5.0	100
6.0	251
7.0	631
7.5	1,000
8.0	1,585
9.0	3,981
10.0	10,000
12.5	100,000
15.0	1,000,000

The Magnitude System

The first observer to describe and catalog differences in the brightnesses of stars was the Greek astronomer Hipparchus, who lived in the second century BCE. He divided his list of approximately 850 stars into six brightness ranges, or **magnitudes**. He called the brightest stars of the 1st magnitude and the faintest stars of the 6th magnitude. His system remained almost unchanged for more than 1,800 years.

Then Galileo arrived. In addition to discovering the phases of Venus, the large moons of Jupiter, and many other things, he noted that his telescope did not just simply magnify, but also "revealed the invisible."

Writing in 1610, Galileo stated, "Indeed, with the glass you will detect below stars of the sixth magnitude such a crowd of others that escape natural sight that it is hardly believable." Then he coined a term that nobody had heard before. He called the brightest of the stars below naked-eye visibility "7th magnitude."

Following the invention of the telescope, astronomers had to expand the magnitude system. They were discovering many stars fainter than those listed by Hipparchus as 6th magnitude. In addition, observers noted that stars of the 1st magnitude varied greatly in brightness.

Around the end of the eighteenth century, most astronomers used a loose system that defined two stars that differed by one magnitude as having a brightness difference of approximately 2.5. Remember, with magnitudes the *smaller* the number, the *brighter* the object; so, fainter objects have larger, more positive, magnitudes.

In 1856 English astronomer Norman R. Pogson (1829–1891) suggested that all observations be calibrated with a ratio between magnitudes of 2.5118865. For the purposes in this lab, however, we will define a star of a certain magnitude as 2½ times brighter than a star one magnitude fainter. In the mid-nineteenth century, the concept of using magnitudes equal to and less than zero also came into being. Astronomers recognized that the brightest stars were much brighter than 1st magnitude, to say nothing of the bright planets, the Moon, and, of course, the Sun.

This definition of magnitudes leads to the fact that a difference of five magnitudes equals a 100-fold difference in brightness. So, the star Sirius (α CMa) at magnitude –1.46 is 100 times as bright as Wasat (δ Gem) at magnitude 3.54. Note that in star charts, the decimal in magnitudes often is not shown.

Table 9.1 lists magnitude differences and their brightness ratios. If you are looking for a value not listed, multiply the ratios of the magnitude differences which, when added, give the desired difference. For example, if you are looking for the brightness ratio between Antares (α Sco) at magnitude 1.2 and Ras Algethi (α Her) at magnitude 3.5. That is a magnitude difference of 2.3, which is not listed in the table.

To find the value, multiply the ratios of 6.3 (a magnitude difference of 2) and 1.32 (a magnitude difference of 0.3); so, Antares is 6.3 × 1.32 = 8.316, or, rounding this off, it is eight times as bright as Ras Algethi. Here is another example: on a certain date, Jupiter blazed at magnitude –2.5 near the star Sadalmelik (Alpha [α] Aquarii), which shines at magnitude 2.9. How much brighter was Jupiter than Sadalmelik? The total magnitude difference is 2.9 – (–2.5) = 5.4. To find the difference in brightness, multiply the brightness ratios of five magnitudes (100) and that of 0.4 magnitude (1.45). So, 100 × 1.32 = 145. Jupiter was 145 times brighter than Sadalmelik.

Apparent vs. Absolute Magnitude

When astronomers describe the brightness of a celestial object, they use **apparent magnitude** (designated **m**, and usually just called "magnitude"). This is a measure of how bright an object appears to us. Astronomers also use **absolute magnitude** (designated **M**), which allows comparison of celestial objects in terms of their real brightnesses (Fig. 9.1).

The absolute magnitude of a star is the brightness the star *would* have if it were at a distance of 10 **parsecs** (32.6 light-years). Comparing the brightnesses of all stars at this

9.1 Star A lies less than 10 parsecs from Earth, so its apparent magnitude is brighter than its absolute magnitude. Any star farther than 10 parsecs from Earth, such as Star B, has an absolute magnitude brighter than its apparent magnitude.

standard distance lets astronomers determine how bright each really is compared to all the others.

Put another way, absolute magnitude is a measure of the star's luminosity—the total amount of visible energy radiated by the star. Absolute magnitude tells astronomers much more about a star than apparent magnitude.

Note that for comets and asteroids, scientists developed a totally different absolute magnitude system. For those objects, absolute magnitude is the brightness of a comet or asteroid as it would appear to a theoretical observer standing on the Sun if the object were one astronomical unit (the average distance between Earth and the Sun) away.

The Brightness of the Night Sky

Even the darkest parts of the sky are not perfectly dark. So, how bright is the night sky? Observers measure sky brightness in terms of magnitudes per square **arcsecond**. That value lets them know how the background light spreads over an area of sky. This measurement is similar to one lighting engineers use to determine how bright the lights are in the laboratory room, and how much light is falling on your desk. In the 1980s astronomers performed a study of the sky's nighttime brightness at one of the darkest spots on Earth, Cerro Tololo Inter-American Observatory (CTIO) in Chile.

Astronomers measured the sky's brightness as the Moon went through its phases. The values in Table 9.2 are reasonable approximations for most dark sites. Remember that the larger the number, the darker the sky.

The difference is greatest when astronomers measure sky brightness through a blue filter because our atmosphere scatters blue light the most. Therefore, the sky background will be brighter when measured in blue, especially when there is a lot of light to scatter. This occurs at full Moon.

> **GLOSSARY TERM**
> **Arcsecond** angle measurement equal to 1/3600 of 1°

TABLE 9.2 Sky Brightness at Various Moon Phases

The Moon's Age In Days	Sky Brightness
0 (New Moon)	21.8
3	21.7
7 (First quarter)	21.4
10	20.7
14 (Full Moon)	20.0

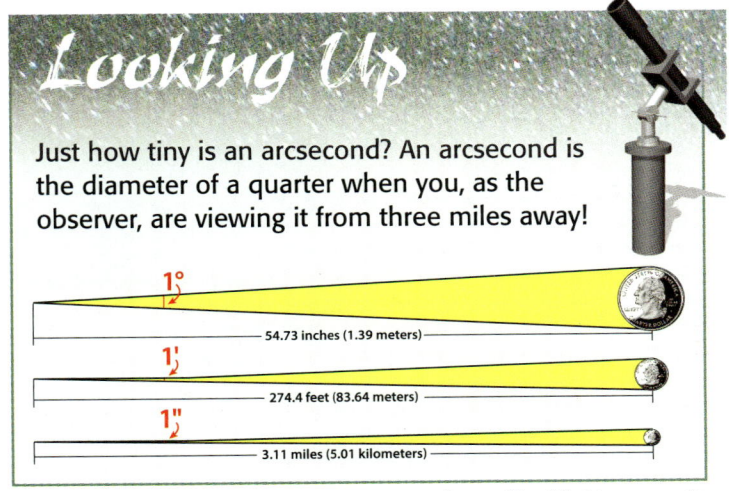

Just how tiny is an arcsecond? An arcsecond is the diameter of a quarter when you, as the observer, are viewing it from three miles away!

Courtesy of Roen Kelly, *Astronomy* magazine

The Magnitude System and Light Pollution **CHAPTER 9**

Estimating Limiting Magnitude

During an observing session, it is always a good idea to make an estimate of the sky quality in terms of how clear and dark it is. Observers call this exercise finding your **limiting magnitude**. Not only will this help you determine how good (or bad) the sky is at that particular time, it will also allow you to judge the quality of your recorded observations months or years from that session. Also, making repeated estimates will help make you a better observer because you will be more conscious of little details.

Many observers use the area around Polaris (Alpha Ursae Minoris) to calculate their limiting magnitude. Figure 9.2 shows the brightness of some of the stars to the nearest 0.1 magnitude. Note that the numbers do not contain decimal points, which a reader could misinterpret for stars. So, when you see 20, for example, the magnitude is 2.0.

Most observers take limiting magnitude estimates near the **zenith**, where sky conditions are usually the best. If, however, you are studying a particular object in depth or one that lies far from the zenith, you may want to make your estimate near the object rather than at the zenith. Just be sure to take your estimate at the same altitude (height above the horizon) as the object you are observing.

Atmospheric Extinction

When determining the magnitude of a celestial body, such as a star or comet, you must take into account a quantity known as atmospheric extinction, or how much our atmosphere dims an object. Table 9.3 combines a celestial body's altitude with an observer's elevation to provide the amount of extinction. For example, if you observe from an elevation of 1 kilometer (0.6 mile) above sea level and an object lies 10° above the horizon, the atmospheric extinction is 1.16 magnitudes. In other words, the object appears 1.16 magnitudes fainter than it would if our atmosphere had no effect.

> **GLOSSARY TERMS**
> **Limiting magnitude** faintest star visible at an observing site
> **Zenith** point in the sky directly above an observer; the highest point in the sky for an observer

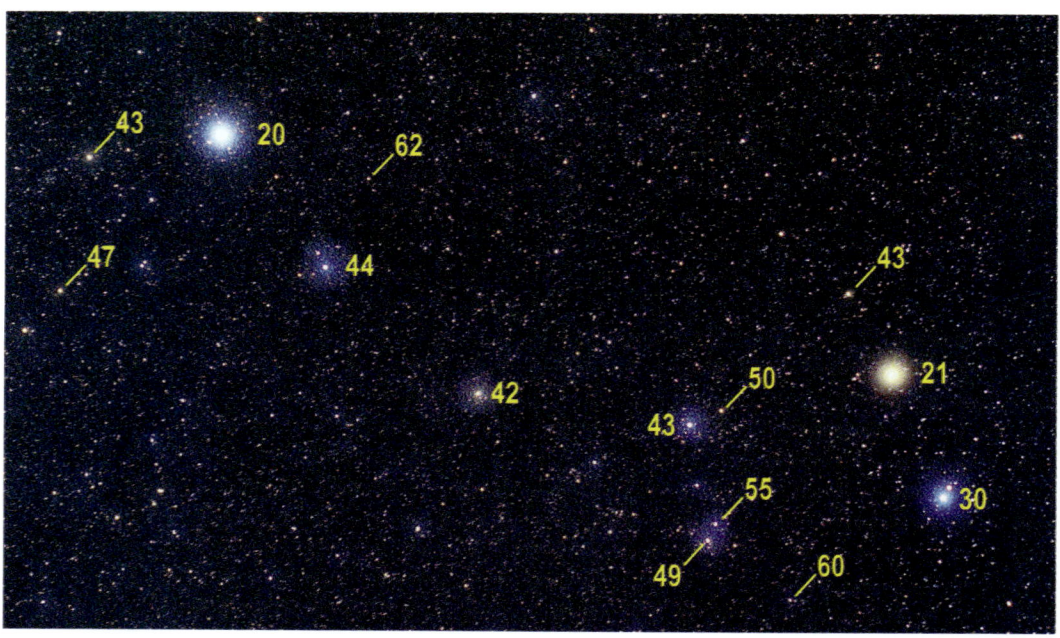

9.2 Area around Polaris (Alpha Ursae Minoris) used in calculating limiting magnitude. Magnitudes are written as whole numbers because decimal points look like stars. So, 43 = 4.3.

Courtesy of Bill and Sally Fletcher

TABLE 9.3 Atmospheric Extinction at Various Elevations

Altitude of Object	Observer at Sea Level	Observer at 500 meters (0.3 mile)	Observer at 1 km (0.6 mile)	Observer at 2 km (1.2 miles)	Observer at 3 km (1.9 miles)
89°	0.28	0.24	0.21	0.16	0.13
80°	0.29	0.24	0.21	0.16	0.13
70°	0.30	0.25	0.22	0.17	0.14
60°	0.32	0.28	0.24	0.19	0.15
50°	0.37	0.31	0.27	0.21	0.17
45°	0.40	0.34	0.29	0.23	0.19
40°	0.44	0.37	0.32	0.25	0.21
35°	0.49	0.42	0.36	0.28	0.23
30°	0.56	0.48	0.41	0.32	0.26
25°	0.64	0.54	0.47	0.37	0.30
20°	0.82	0.70	0.60	0.47	0.39
15°	1.08	0.92	0.79	0.62	0.51
10°	1.59	1.34	1.16	0.91	0.74
5°	2.91	2.46	2.13	1.66	1.36
1°	7.38	6.26	5.40	4.22	3.46

Check Your Understanding

1.1 How many brightness ranges did the first magnitude system have?

1.2 Why did astronomers have to expand the magnitude system after Galileo began using a telescope to explore the sky?

1.3 If a certain star is 7.5 magnitudes brighter than another star, how much brighter does it appear to us?

1.4 How far away would a star have to be for its apparent magnitude to equal its absolute magnitude?

1.5 Which filter transmits the same colors to which the human eye is sensitive?

1.6 How many magnitudes brighter does the night sky look to us at full Moon than at new Moon?

1.7 Give three reasons observers should determine their limiting magnitude.

Light Pollution

Those who love skywatching and sky-shooting eventually become aware of a problem: excess nighttime lighting, now generally called light pollution (Fig. 9.3). During the past several decades, nighttime outdoor lighting has increased at a high rate. Today, there are satellite images that no longer show the night side of our planet as dark. Figure 9.4 shows light pollution across the United States. The darkest areas are black and usually remote. Gray and mauve regions are ideal for finding pristine skies; these areas are typically rural and not accessible by road. Green sections have acceptable dark-sky conditions. If you

9.3 Light pollution in El Paso, Texas.

9.4 Map of light pollution across the United States. Courtesy of Craig Mayhew and Robert Simmon, NASA GSFC

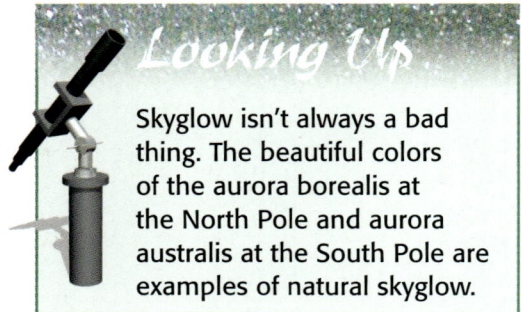

Skyglow isn't always a bad thing. The beautiful colors of the aurora borealis at the North Pole and aurora australis at the South Pole are examples of natural skyglow.

GLOSSARY TERMS

Light trespass condition that exists when unwanted light crosses a property line and interferes with a person's view, privacy, or sleep; examples include shopping centers, convenience stores, and service stations

Glare condition that occurs when bright lights or brightly lit areas lie next to unlit areas; glare never helps visibility and often creates dangerous safety problems

Skyglow type of light pollution that can be artificial (city lights) or natural (the aurora borealis)

Clutter badly designed group of lights that can affect drivers or pilots

don't have access to any of the black areas, as will be the case in many of the northeastern states, go for the green.

For much of the populations of North America and Europe the night sky is no longer black, or even dark. Rather, it is a bright yellow orange, aglow from poorly designed light fixtures and almost bereft of stars. For some people born in the 1980s or later, it is possible that the only object they have seen in the night sky is the Moon. In today's world, the most immediately endangered natural resource may be the dark night sky.

Nobody denies that some outdoor lighting is necessary for people's safety and security. Other lighting, such as that used for advertising, may not be necessary, but it is a consequence of living in our world today. If those lights are designed and maintained correctly, however, they can be a benefit without taking away from the night sky's beauty.

Defining the Problem

Outdoor light pollution manifests itself in three ways: light trespass, glare, and clutter. **Light trespass** occurs when unwanted light enters your property. It can ruin an imaging session or cause sleep deprivation if bright light enters through a bedroom window.

Glare (Fig. 9.5) results from high contrast between lit and unlit areas. Although it affects amateur astronomers, it is most often a problem for drivers. Bright streetlights, advertising signs, and poorly planned lights used by businesses are the primary causes. Figure 9.6 shows various kinds of poorly planned outdoor lights. The lights in Figure 9.6A send light up into the atmosphere, causing both glare and **skyglow**. The lights in Figures 9.6B and 9.6C are better, but still are not ideal.

Clutter refers to badly designed groups of lights that may lead to confusion. Although clutter generally affects automobile drivers, it also can pose a hazard to air traffic.

Several other lighting problems also must be considered. The first is the financial waste caused by bad lighting. In the United States alone, shining light when and where it is not needed and using inefficient light sources waste $2 billion, annually.

Bright lights also give people a false sense of security. A U.S. Department of Justice report to Congress concluded, "We can have little confidence that lighting prevents crime, particularly since we do not know if offenders use lighting to their advantage. . . . In short, the effectiveness of lighting is unknown." Bright lights may make us feel safer, but are we really? Do criminals need light to commit crime? Are they afraid of the dark, too?

9.5 Glare from streetlights.

  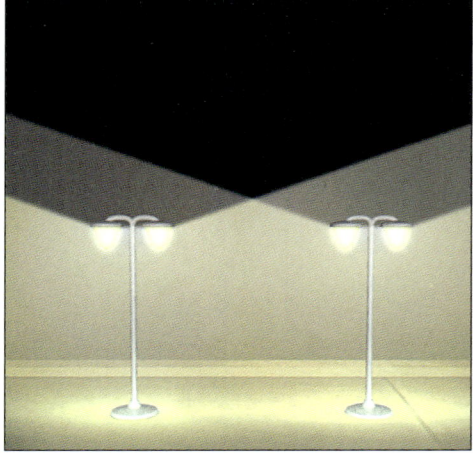

9.6 Examples of poorly planned outdoor lights: (**A**) globe post lamp, (**B**) industrial yard light, and (**C**) twin globe post lamps.

Courtesy of Roen Kelly, *Astronomy* magazine

Evidence also is growing that the loss of the night is throwing nature out of balance. Light pollution affects more than people. Florida's sea turtles have been facing a major threat from bright lights. Many species of birds, especially the small insect-eaters, migrate thousands of miles at night. Lights shining from skyscrapers, broadcast towers, lighthouses, monuments, and other tall structures attract them. The birds either flutter about until they drop from exhaustion or actually hit the object and die.

What the Eye Needs

Low lighting levels are acceptable once your eyes adapt to the darkness, a process called dark adaption. Problems arise when eyes must move from dark to light, then light to dark, back and forth until they really aren't functioning well in either situation.

Older adults may encounter a major problem when dealing with overly bright nighttime lighting. Changes within the eye due to age, such as diminished dark adaption speed, can prove dangerous. For example, entering a service station with brilliant lights and then driving out onto a highway with much dimmer or no lights places the elderly at a safety disadvantage. Older eyes are also more sensitive to glare, particularly from bluer light sources.

Below is a list of 11 things you can do to help combat light pollution.

1. **Light only what needs lighting.** This sounds simple, but it typically gets overlooked. Ask yourself, "Does this even require lighting?" If you have determined that something requires light, does it need to be lit at all times? If you go out to your shed only once a week, does it really need to be lit dusk to dawn every night?

2. **Install motion sensors.** Motion sensors turn your lights on automatically whenever there is activity outside your home or business. This typically reduces your use of electricity for lighting by more than 90 percent, easily paying for the cost of the sensor and its installation.

3. **Use only as much light as required.** Don't always install the highest available wattage. If you are not performing surgery on your patio, there is no need for operating room light levels.

The Astronomical League, the largest organization of amateur astronomers, recognizes that light pollution is an issue for many amateur astronomers. They offer an observing certificate with objects chosen specifically for urban stargazers.

4. **Turn down the lights you are using.** Brighter lights than what is recommended by engineers waste money because human eyes quickly adapt to the higher level of illumination. Install lower-wattage bulbs in the areas inside and outside of your home, such as hallways and yards, in which bright lights are not needed for specific activities.

5. **Pick the right kind of lamp for the job.** Selecting, for example, high-pressure sodium lights saves 40 percent in operating costs over metal halide lamps and 150 percent over mercury vapor lights.

> **GLOSSARY TERM**
>
> **Full-cutoff** light fixture with a shield that directs light to the intended target

6. **Use only full-cutoff light fixtures.** The best type of outdoor lighting falls under the classification "full-cutoff." Full-cutoff fixtures (Fig. 9.7) do not allow any light to escape upward, which means the edge of the lamp's shade, and assure lights shine down, preventing light from shining across property lines or up into the sky. They distribute their light in a directed pattern and provide acceptable light on the ground with less power. The first full-cutoff light fixture was General Electric's M100, which the company debuted in 1959.

Many states now mandate full-cutoff lights for building or highway construction. Today's full-cutoff fixtures generally employ high-pressure sodium lamps, but bright light-emitting diode (LED) lights also are becoming popular. High-pressure sodium lamps appeared in 1970, and are the dominant streetlights in the United States. The main characteristic is their orange-yellow glow. They use far less energy than mercury-vapor or metal-halide lamps. LED lamps emit white light and use even less energy than high-pressure sodium lamps.

7. **Shield existing lights.** If you are not quite ready to install new, full-cutoff lights, shades are available for many existing fixtures, which will convert them into night-sky-friendly lights for minimal cost.

9.7 Full-cutoff light fixtures.
Courtesy of Roen Kelly, *Astronomy* magazine

Looking Up

The International Dark-Sky Association

The International Dark-Sky Association (IDA) incorporated in 1988 as a tax-exempt, nonprofit organization for educational and scientific purposes. The IDA's goals are "to be effective in stopping the adverse environmental impact on dark skies by building awareness of the problem of light pollution and of the solutions, and to educate everyone about the value and effectiveness of quality nighttime lighting." The IDA believes in a united approach that is supportive of local and individual efforts. The group has accomplished much in some locations, but more needs to be done everywhere. The IDA believes it can succeed in preserving dark skies and in improving the nighttime environment for everyone. Quality outdoor lighting is the key.

The IDA is a membership-based organization, with about 11,000 members from each of the 50 states and from more than 70 other countries. These include more than 500 organization members.

Specific areas where the IDA is involved include: education on all aspects of the issues; a regular newsletter; information sheets, brochures, and leaflets; economic information; examples of good lighting design; a speaker's bureau; documentation of good and bad lighting using photos and video; presentation of Good Lighting awards; media contacts; press releases; and developing viable and effective sections and affiliates as resources around the world.

The IDA website (www.darksky.org) contains a vast amount of material, information sheets, slides, images, PowerPoint presentations, and links to other important sites.

8. **Install reflectors.** Often, you can use reflectors to outline a driveway instead of a string of lights. Reflectors are cheaper to purchase, cost nothing to run, and are unaffected by power outages.

9. **Get used to the dark.** Our eyes work quite well in the dark. If you take the trash out at night, do you need to turn your outdoor lights on? Chances are good that you will be able to find the trash can and make your way to the curb without the extra light.

10. **Educate those around you about light pollution.** This includes your friends, neighbors, and elected officials. With just a little bit of thought and effort, this is one type of environmental pollution that can be cleaned up without any side effects, and money will be saved in the process.

11. **Show your support.** Help promote groups such as the International Dark-Sky Association and manufacturers that produce approved fixtures. If you blog, write about light pollution. Don't just complain! Mention success stories whenever you can. Taking a proactive approach to light pollution is the best way to ensure that future nights will be safe, healthy, and dark.

Check Your Understanding

2.1 What are the three ways light pollution appears?

2.2 Why are older adults particularly sensitive to light pollution?

2.3 Of the 11 ways listed that you can use to help alleviate the problem of light pollution, which one do you think is the most important, and why?

2.4 What is the advantage of a full-cutoff light fixture?

Exercise 9.1   Star Brightness Comparison

Procedure

Use Table 9.1 on page 168 and Table 9.4 below to answer the following questions, rounding all the magnitudes and brightness differences to the nearest 0.1.

TABLE 9.4 The 20 Brightest Stars

Star	Designation	Magnitude
Sirius	Alpha Canis Majoris	−1.46
Canopus	Alpha Carinae	−0.72
Rigil Kentaurus	Alpha Centauri	−0.27
Arcturus	Alpha Boötis	−0.04
Vega	Alpha Lyrae	0.03
Capella	Alpha Aurigae	0.08
Rigel	Beta Orionis	0.12
Procyon	Alpha Canis Minoris	0.38
Achernar	Alpha Eridani	0.46
Betelgeuse	Alpha Orionis	0.50
Hadar	Beta Centauri	0.61
Acrux	Alpha Crucis	0.76
Altair	Alpha Aquilae	0.77
Aldebaran	Alpha Tauri	0.85
Antares	Alpha Scorpii	0.96
Spica	Alpha Virginis	0.98
Pollux	Beta Geminorum	1.14
Fomalhaut	Alpha Piscis Austrinus	1.16
Mimosa	Beta Crucis	1.25
Deneb	Alpha Cygni	1.25

1 How many times brighter is Rigel than Hadar? _____

2 How many times brighter is Aldebaran than Mimosa? _____

3 How many times brighter is Arcturus than Betelgeuse? _____

4 How many times brighter is Antares than Fomalhaut? _____

5 How many times brighter is Canopus than Altair? _____

6 How many times brighter is Vega than Spica? _____

7 How many times brighter is Sirius than Procyon? _____

8 How many times brighter is Rigil Kentaurus than Pollux? _____

9 How many times brighter is Achernar than Acrux? _____

10 How many times brighter is Sirius than Deneb? _____

Limiting Magnitude Evaluation

Exercise 9.2

Procedure

1 Partner with someone else in the lab.

2 With overhead lights on and/or a spotlight aimed at a projection screen, your instructor will project an image of the constellation Orion. In the space below, sketch the constellation as you see it. Record the brightest stars first to give you an outline. Next, fill in fainter and fainter stars until you record all that you see. Your goal is to be as accurate as possible.

3 When you and your partner have each completed your sketches, exchange drawings. Using Figure 9.8, evaluate the limiting magnitude of your partner's sketch.

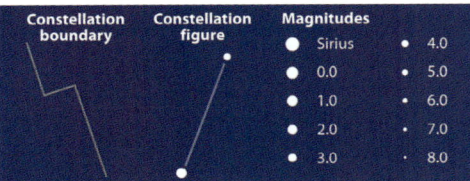

9.8 Constellation Orion the Hunter.
Courtesy of Roen Kelly, *Astronomy* magazine

The Magnitude System and Light Pollution   CHAPTER 9   179

4 Record your answer: _____.

5 Reclaim your own sketch.

    a Do you agree with your partner's estimate (above)?

    b If not, what limiting magnitude would you assign your sketch?

    c If your estimate was different from your lab partner's, list three possible reasons why.

Exercise 9.3 — Constellation Sketching with Light Pollution

For this activity, you must visit three locations with varying degrees of light pollution on three separate *clear* nights when the Moon is not in the sky. Possible locations include: for severe light pollution, a downtown office complex or a large shopping center parking lot; for moderate light pollution, an in-city public park or an unlit football field inside the city limits; for low light pollution, a location at least 30 miles from a large city or 10 miles from a small town.

You will sketch the same constellation at each location. Don't choose one that is too large (like Hydra) or too small (like Delphinus). Do choose one that lies near the zenith (the overhead point). And try to make the three sketches at times when the constellation is the same height above the horizon.

Procedure

Materials
- ❏ Sketching pencils
- ❏ Sketching pad
- ❏ Clipboard
- ❏ Flashlight (preferably one with a red filter)

1 For your first sketch on the data sheet, page 181, wait to arrive at your observing site until at least three hours after sunset. This timing is important because over the course of a month, Earth orbits one-twelfth the way around the Sun. The result of this motion is that a star, or group of stars, rises approximately 4 minutes earlier each night, or about 2 hours earlier each month. So a constellation that rises at 10:00 p.m. on the first of any month will rise at 8:00 p.m. on that month's last day.

2 To make your second and third sketches on page 182, you will need to observe "your" constellation when it is at the same position in the sky. To do this, take the time you made your first sketch and subtract 4 minutes for each day that has passed. For example: You begin with your first sketch on a clear night at 10:15 p.m. The next three nights are cloudy, so you don't go out. The following night is clear. Because four nights have passed, to start your next sketch when "your" constellation is in the same position as on your first night, you want to begin drawing at 10:15 − (4 minutes/day × 4 days) = 10:15 − 16 minutes = 9:59 p.m., or approximately 10:00 p.m.

Name _____   Section _____ Date _____

Constellation Sketches

Constellation chosen _____

Sketch 1

Date _____

Time _____

Location _____

Light pollution (*circle one*)

Low   Moderate   Severe

Exercise 9.3 Data Sheet

The Magnitude System and Light Pollution   **CHAPTER 9**

Exercise 9.3 Data Sheet

Sketch 2

Date _____

Time _____

Location _____

Light pollution (*circle one*)

Low   Moderate   Severe

Sketch 3

Date _____

Time _____

Location _____

Light pollution (*circle one*)

Low   Moderate   Severe

Night and Day

The Moon

LEARNING OBJECTIVES

Upon completion of this chapter, you should be able to:

1. Identify major lunar features.
2. Identify dates of current Moon phases and how they relate to lunar rising and setting times.
3. Determine which constellation's stars the Moon is in front of on given nights.
4. Explain why observing the Moon at certain times yields better results.
5. Calculate the scale of a photograph of the Moon.
6. Sketch the Moon and several lunar features.
7. Determine a scale for the lunar sketch on the Moon chart in the lab manual.
8. Compute the size of 10 lunar craters.
9. Estimate the sizes of several lunar features you draw and the same features on the Moon chart in your lab manual; compare the two results.
10. Define all glossary terms.

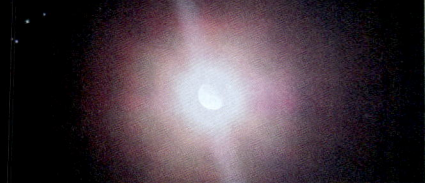

Moon and the Pleiades
Courtesy of John Chumack

GLOSSARY TERMS

Crater roughly circular depressions on the surface of many objects in the solar system; most craters were created through the impact of a meteorite, with the remainder being volcanic or caused when the surface collapsed

Lunation lunar month

Phase percentage of illumination of the Moon or other solar system object at a particular time in its orbit

Lunar month period of time between one new Moon and the next new Moon

New Moon lunar phase that occurs when the Moon lies between Earth and the Sun; at new Moon, we cannot see our nearest celestial neighbor

Full Moon lunar phase that occurs when the Moon lies 180° from the Sun; full Moon rises at sunset and is in the sky all night long

The Moon offers something for everyone. It is visible somewhere in the sky most nights, and its changing face presents features one night not seen the previous night. At times, the Moon is a thin crescent above the western horizon just after sunset. At other times, the Moon is full and shines at its brightest during the entire night. Sometimes, you may not see the Moon in the sky at all. Mountain ranges, vast volcanic plains, and more than 1,500 named **craters** make the Moon a target you will point your telescope at again and again.

During this lab, you will use charts to track the Moon's motion through one **lunation** (the period between similar phases, such as from one new Moon to the next—approximately one month). You also will sketch several lunar features, and the Moon's phase (appearance) on each clear night.

The Changing Face of the Moon

Phases

The changing position of the Moon with respect to the Sun causes the Moon, as seen from Earth, to cycle through a series of **phases** (Fig. 10.1). One complete set of phases is a **lunar month**. From **new Moon** (Fig. 10.2) to **full Moon** (Fig. 10.3), when the entire side of the Moon facing us is illuminated, is a continuous growth of the sunlit portion of the Moon, *but only as we see it*. The Moon is half in light and half in darkness at all times, as can be seen by the inner set of moons in Figure 10.1. It is simply its position relative to Earth and the Sun that determines how much of the bright, sunlit part we see, as indicated by the outer set of moons in Figure 10.1.

Traditionally, the lunar month begins at new Moon. It was originally called new Moon because people thought it was at these times that the Moon was being reborn. We can't actually see the new Moon from Earth, because the lit side lies in the same direction as the Sun and, in the sky, the Moon's position is also quite near the Sun's.

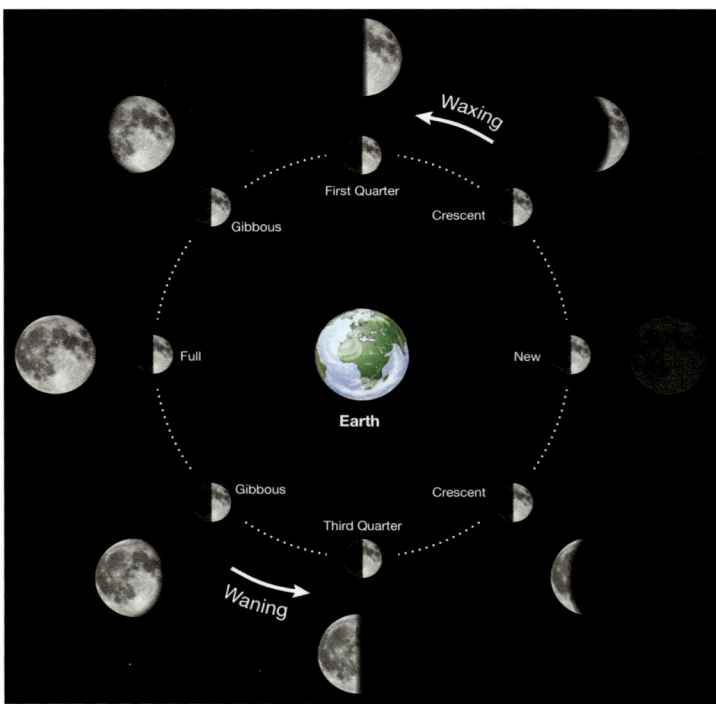

10.1 Phases of the Moon. The inner circle of Moons represents the view from space. The outer circle is what we see of the Moon from Earth.
Courtesy of Holley Y. Bakich

10.2 Moon's phase 23 hours and 17 minutes after new Moon at sunrise in the eastern sky of Athens, Greece.
Courtesy of Anthony Ayiomamitis

184 Unit 3 Night and Day

10.3 Full Moon.
Courtesy of John Chumack

10.4 First quarter Moon.
Courtesy of John Chumack

10.5 Earthshine is the light reflected from Earth onto the dark part of the lunar surface. This allows us to see that part of the Moon.
Courtesy of John Chumack

10.6 Last quarter Moon. Courtesy of John Chumack

From new Moon, our lone natural satellite progresses through **waxing crescent** until **first quarter** (Fig. 10.4). When the Moon is a thin crescent, sunlight reflected off Earth falls on its dark part. This subtle illumination is known as **earthshine** (Fig. 10.5). After first quarter, the Moon continues through **waxing gibbous** until it is a full Moon. Then it progresses through **waning gibbous** until **last quarter** (Fig. 10.6). It continues through **waning crescent** until it is back to new Moon to begin another lunar month.

Periods

The Moon orbits Earth once every 27 days, 7 hours, and 43.7 minutes. This is known as an **orbital period**. The Moon also has a **synodic period**, which takes 29 days, 12 hours, and 44.1 minutes to complete (Fig. 10.7). The synodic period is the time it takes for an object to reappear at the same place relative to two other objects. In this case, the two other objects are Earth and the Sun, and because Earth revolves around the Sun, it takes the Moon just a little bit longer to get to the same place.

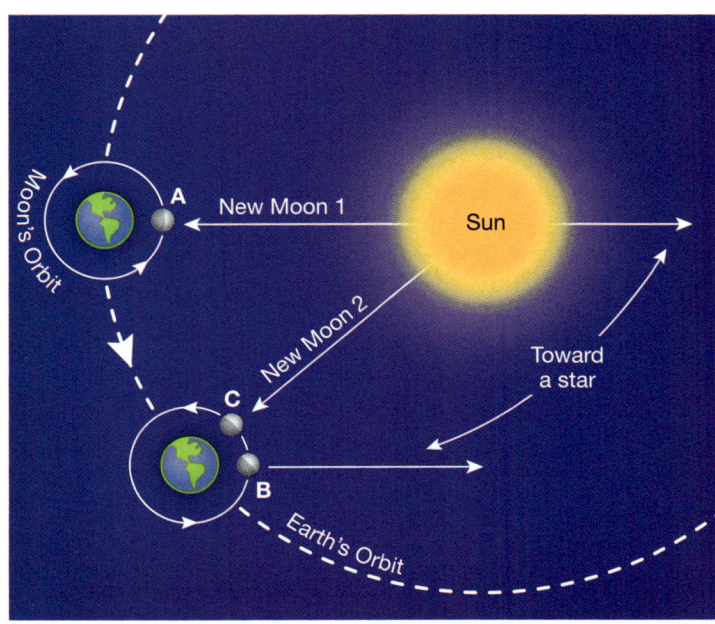

10.7 The Moon's orbital period (from A to B) is more than two days shorter than its synodic period (from A to C). The difference is due to Earth's motion around the Sun.

GLOSSARY TERMS

Waxing growing; for the Moon, this means the illuminated portion is getting larger from night to night

Crescent phase between either quarter Moon and new Moon

First quarter lunar phase that occurs when the Moon lies 90° east of the Sun; the first quarter Moon rises at midday and sets at midnight

Earthshine light reflected from Earth onto the dark part of the lunar surface, allowing us to see that part of the Moon

Gibbous major phase between either quarter Moon and the full Moon

Waning shrinking; for the Moon, this means the illuminated portion is getting smaller from night to night

Last quarter lunar phase that occurs when the Moon is located 90° west of the Sun; the last quarter Moon rises at midnight and sets at midday

Orbital period time it takes the Moon to orbit Earth once; 27 days, 7 hours, and 43.7 minutes; also Sidereal period

Synodic period time it takes the Moon to go from any phase to the same phase; 29 days, 12 hours, 44.1 minutes

Check Your Understanding

1.1 Why don't you ever see the new Moon in the sky?

_____

1.2 Compared to the full Moon, how bright is the last quarter Moon?

_____

1.3 Why is the Moon's synodic period longer than its orbital period?

_____
_____
_____

10

Features of the Moon

Looking Up

The word *mare* comes from the Latin meaning "sea." Early observers thought these dark regions contained water!

Of the 1,940 major named features on the Moon, 1,545 are **craters**. Other lunar features include:

- **mare** (*pl.*, **maria**)
- **rays**
- **rilles** or **rimae**
- **wrinkle ridges**.

You will need to know and locate the following 22 craters or crater pairs to complete Exercise 10.2, page 203.

When you look at the full Moon standing high in the southern sky, remember that astronomers have assigned the directions north and south to its top and bottom, respectively. The western part of the Moon is to your left, and the eastern part is to your right. This may sound wrong, because it means the western side lies nearest to Earth's eastern horizon, and vice versa, but it is correct. It may help to think of the Moon like you would a map of the United States. On such a map, the western states lie to the left, and the eastern states are on the right. For this lab, the Moon is divided into four **quadrants** (quarters).

Northwestern Quadrant

The northwestern quadrant is the upper left part of the Moon (Fig. 10.8). This quadrant is dominated by the dark Mare Imbrium (Sea of Rains). The lighter "sea" to the north is Mare Frigoris (Sea of Cold). Between them lies Plato Crater. To the south of Mare Imbrium straddling the northwest and southwest quadrants, you can find Copernicus Crater in Oceanus Procellarum (Ocean of Storms). Other interesting features include Archimedes Crater, Aristarchus Crater, and Reiner Crater.

GLOSSARY TERM

Mare/maria Basalt-filled impact basins common on the Moon's face

Ray bright streak originating at a crater formed by material ejected from the crater during its formation

Rille/rimae long narrow depression in the Moon's surface that resembles a channel

Wrinkle ridge ridge that formed when lava cooled

Quadrant one-quarter of the visible disk of a celestial object; can be named "upper-left," "northeast," or by some other designation

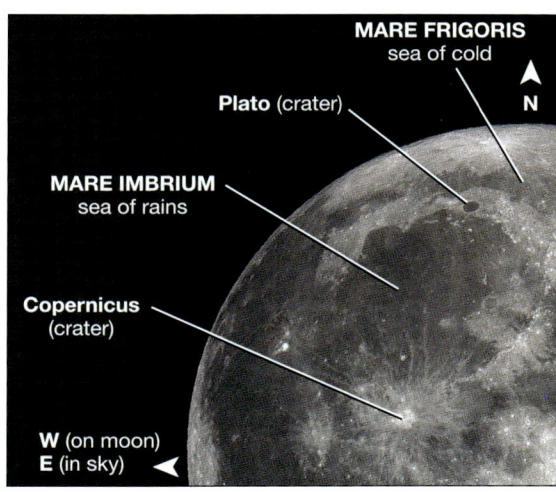

10.8 Northwestern quadrant of the Moon.

10.9 Archimedes Crater.
Courtesy of Damian Peach

Archimedes Crater

Archimedes Crater (Fig. 10.9) lies at 30° north latitude centered between the eastern and western outer edges. This 83 kilometer-wide crater lies just northwest of the Moon's largest mountain range, the Montes Apenninus.

Aristarchus Crater

Aristarchus Crater (Fig. 10.10) measures 40 km across. Nearby, look for Vallis Schröteri (Schröter's Valley), the Moon's largest sinuous valley. It begins at a small crater 25 km north of Herodotus and runs for 160 km.

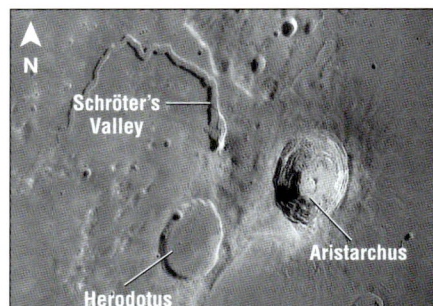

10.10 Aristarchus Crater.
Courtesy of Damian Peach

Copernicus Crater

Copernicus Crater (Fig. 10.11) is one of the Moon's most famous formations. It marks the center of a system of bright rays that extends for up to 800 km. Copernicus measures 93 km wide. Because of its great depth—3,750 meters—sunrise and sunset shadows here create dramatic relief. The central peak rises 1,200 m above the floor. Copernicus' outer wall gives it a peculiar hexagonal shape.

Plato Crater

Plato Crater (Fig. 10.12) lies at the Moon's top center for observers. Plato spans 101 km and has one of the darkest crater floors on the Moon. Polish astronomer Johannes Hevelius (1611–1687) called Plato the Greater Black Lake. One feature to observe within this crater is its slumped inner wall, especially on the western end. Even a small telescope will reveal the largest area, a triangular section that caved in millions of years ago.

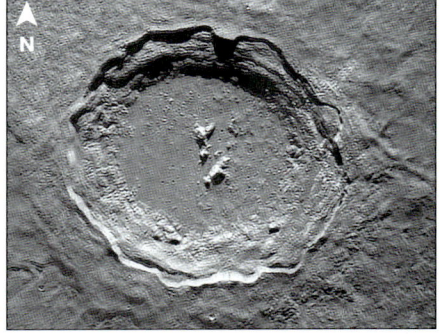

10.11 Copernicus Crater.
Courtesy of Damian Peach

Reiner Crater

Reiner Crater (Fig. 10.13), although only 30 km in diameter, features prominently within Oceanus Procellarum (the Ocean of Storms) on the Moon's western edge. Although its rim is nearly circular, Reiner appears oval. Just to Reiner's west, look for Reiner Gamma, a bright, flat albedo (reflective) feature that stretches for 70 km.

10.12 Plato Crater.
Courtesy of Damian Peach

10.13 Reiner Crater.
Courtesy of Damian Peach

10.15 Lacus Mortis.
Courtesy of Damian Peach

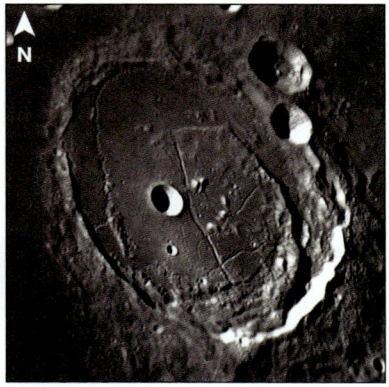

10.16 Posidonius Crater.
Courtesy of Damian Peach

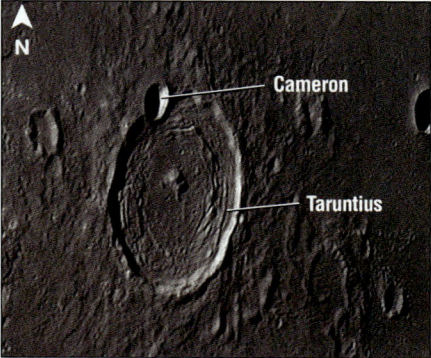

10.17 Taruntius Crater.
Courtesy of Damian Peach

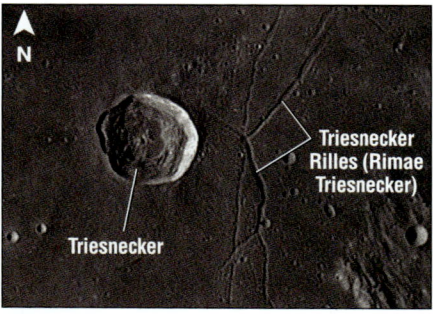

10.18 Triesnecker Crater.
Courtesy of Damian Peach

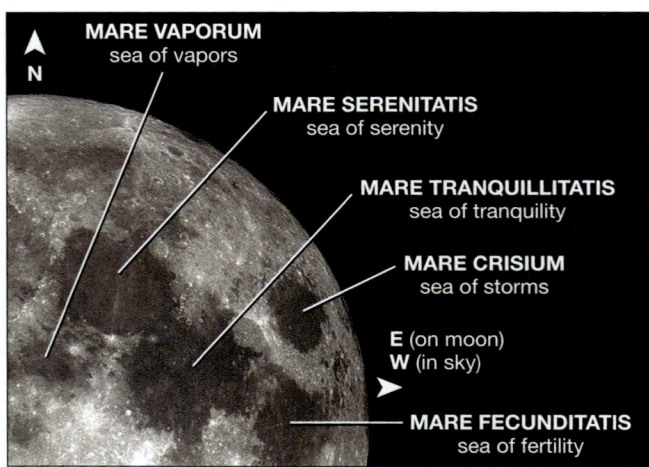

10.14 Northeastern quadrant of the Moon.

Northeastern Quadrant

The Moon's northeastern quadrant is the part to the upper right as we view it from Earth (Fig. 10.14). It has five maria: Mare Serenitatis (Sea of Serenity), Mare Tranquillitatis (Sea of Tranquility), Mare Crisium (Sea of Storms), Mare Vaporum (Sea of Vapors), and Mare Fecunditatis (Sea of Fertility). Features in this quadrant include Lacus Mortis, Posidonius Crater, Taruntius Crater, and Triesnecker Crater.

Lacus Mortis

Lacus Mortis, the Lake of Death (Fig. 10.15), spans 150 km and contains the 40 kilometer-wide Crater Bürg. Try to spot the rilles to the west of Bürg, which run for 60 miles (100 km). Lunar cartographers designated these collectively as Rimae Bürg.

Posidonius Crater

Posidonius Crater (Fig. 10.16) measures 95 km across. Located on its floor, within its wall, and just outside its boundaries are numerous smaller craters. The largest, Posidonius J, lies to the north. When you have finished with Posidonius, look to its west in vast Mare Serenitatis for a series of wrinkle ridges, which formed when the lava that covered this part of the lunar surface early in the Moon's history cooled.

Taruntius Crater

Taruntius Crater (Fig. 10.17) stretches 56 km and lies on the northwestern edge of massive, dark Mare Fecunditatis. Taruntius has a double wall and a prominent central peak. At the northwestern edge of the crater, 11 kilometer-wide Cameron Crater breaks the outer rim. When you observe Taruntius, look for breaks in the lower, inner wall.

Triesnecker Crater

Triesnecker Crater (Fig. 10.18) sits nearly centered on the Moon. It spans 26 km, but is definitely not circular. When most observers point their telescopes here, it is not to observe the crater but rather the Triesnecker Rilles (Rimae Triesnecker). This system is the Moon's best-known system of rilles, and it extends in a north-south orientation for more than 200 km. When you observe it, use a magnification of 200× or more for the best results.

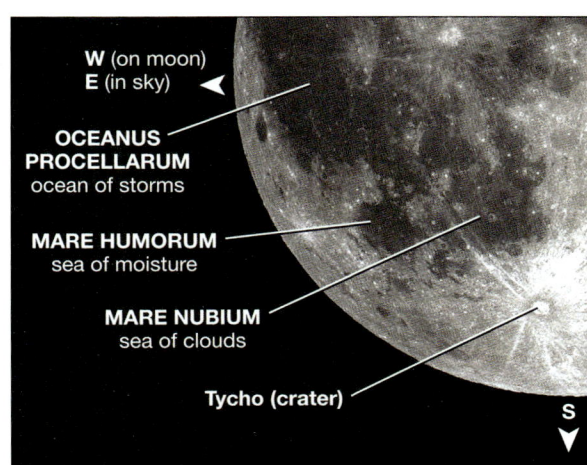

10.19 Southwestern quadrant of the Moon.

Southwestern Quadrant

The southwestern quadrant of the Moon is the part to the lower left as we view our lone natural satellite in the sky (Fig. 10.19). The large Oceanus Procellarum (Ocean of Storms) lies on the Moon's western edge. To its southeast are two maria: the smaller Mare Humorum (Sea of Moisture) and the larger Mare Nubium (Sea of Clouds). The largest visible crater in this area is Tycho Crater, close to the southern edge. Other craters to look for are Clavius Crater; Fra Mauro Crater; Gassendi Crater; Moretus Crater; Pitatus Crater; Ptolemaeus Crater; Rupes Recta, or the Straight Wall; and Schiller Crater.

Clavius Crater

Clavius Crater (Fig. 10.20) ranks as the third-largest crater on the Moon's nearside. It is visible to the naked eye and spans 225 km. But it is what is in Clavius that you should observe. Look for the crater chain of decreasing size that begins at Clavius' eastern wall. Oblong Rutherfurd Crater measures 54 km by 48 km. Following it are Clavius D (28 km), C (21 km), N (13 km), J (12 km), and JA (8 km).

Fra Mauro Crater

Fra Mauro Crater, which measures about 80 km in diameter, represents the remains of a walled plain (Fig. 10.21). To its south sit two craters whose walls overlap those of Fra Mauro. Bonpland Crater (60 km) is another old walled plain, and Parry Crater (48 km) has a floor that flooded with dark lava several billion years ago. Look carefully on Fra Mauro's floor for craterlets Fra Mauro E and Fra Mauro P.

Gassendi Crater

Gassendi Crater (Fig. 10.22), whose long axis measures 110 km across, has numerous clefts, hills, and central mountains that interrupt its floor. To the north, the crater Gassendi A has broken its wall. Together, both craters give the appearance of a diamond ring.

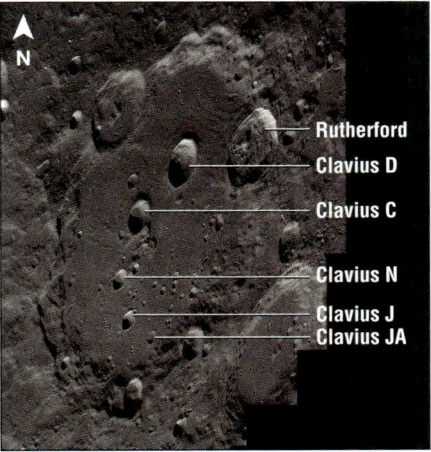

10.20 Clavius Crater.
Courtesy of David Tyler

10.21 Fra Mauro Crater.
Courtesy of Damian Peach

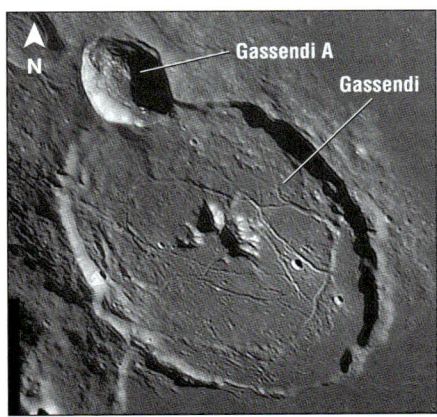

10.22 Gassendi Crater.
Courtesy of Damian Peach

Moretus Crater

Moretus Crater (Fig. 10.23) sits in a heavily impacted region near the Moon's south pole. When the Sun angle is low here, the central peak that rises 2.1 km above the surrounding floor should be easy to spot. Moretus measures 114 km wide. Note that Short Crater immediately to Moretus' south is deeper, so it still lies in shadow in this image.

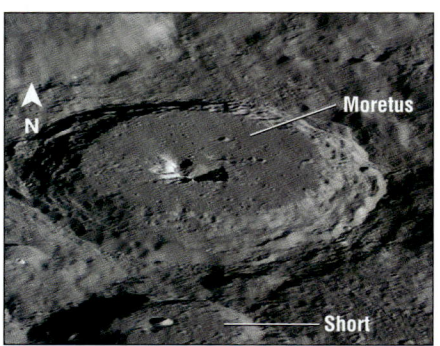

10.23 Moretus Crater.
Courtesy of Damian Peach

Pitatus Crater

Pitatus Crater (Fig. 10.24), which spans 97 km, contains many features strewn about its wide floor. A low central peak sits just to the northwest of the crater's center. Through an 8-inch telescope, look for the thin grooves called Rimae Pitatus on the western floor. More than 20 lettered craterlets surround Pitatus. Also be sure to observe the double-walled crater Hesiodus A directly to the west of Pitatus.

Ptolemaeus Crater

Ptolemaeus Crater (Fig. 10.25) is the largest and northernmost of a chain of three prominent craters that lie at the center of the Moon's near side. Alphonsus sits immediately to Ptolemaeus' south, followed by Arzachel. Ptolemaeus measures 153 km across. Scan its wide floor for numerous small craters. The easiest to see, Ammonius, measures 9 km across. Just north of Ammonius is the faint, saucer-like depression Ptolemaeus B.

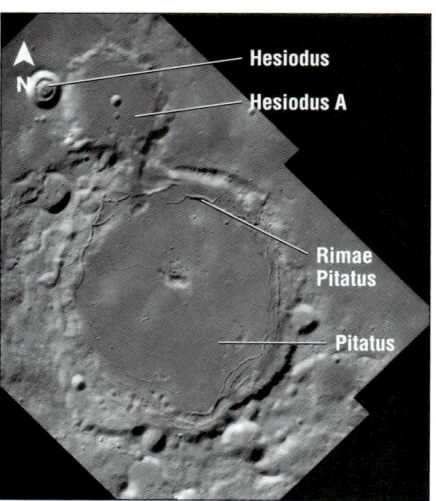

10.24 Pitatus Crater.
Courtesy of Damian Peach

Rupes Recta, the Straight Wall

Rupes Recta, the Straight Wall (Fig. 10.26), lies in Mare Nubium due east of Birt Crater. The Wall runs for 110 km, is 2.5 km wide, and rises 240 m to 300 m above the surrounding floor. Although through your telescope it may look like a sharp cliff, it actually slopes at a gentle 7°.

Schiller Crater

Schiller Crater (Fig. 10.27) is an elongated lunar impact feature that measures 179 km by 71 km. Note Schiller's terraced inner wall. Increase the magnification, and you will spot the double ridge along the center of the crater's floor toward the northwest.

Tycho Crater

Tycho Crater (Fig. 10.28) is one of the Moon's most famous features. It has an extensive system of rays that stretches more than 1,450 kilometers. Tycho's diameter is 85 km, and it has a depth of 4.8 km. The central peaks rise 1.6 km above the floor.

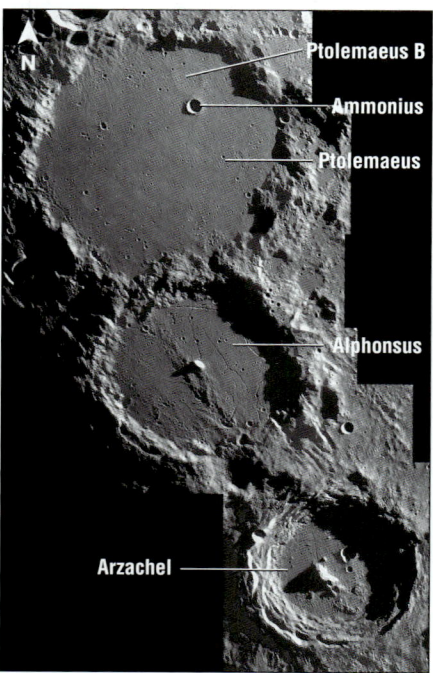

10.26 Rupes Recta, the Straight Wall.
Courtesy of Damian Peach

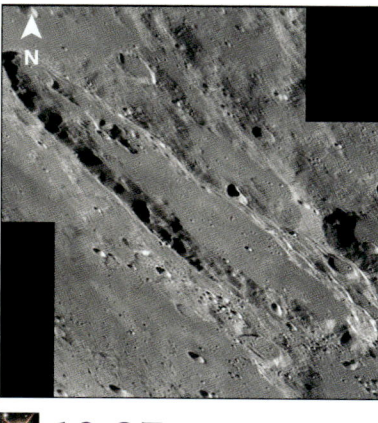

10.27 Schiller Crater.
Courtesy of Damian Peach

10.28 Tycho Crater.
Courtesy of Damian Peach

10.25 Ptolemaeus Crater.
Courtesy of Damian Peach

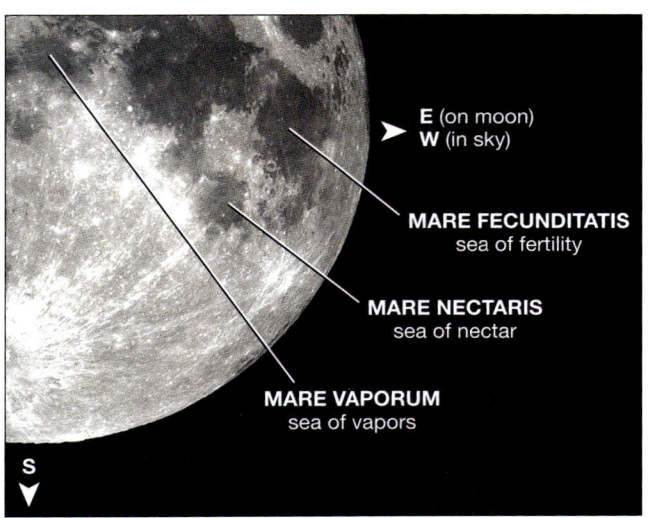

10.29 Southeastern quadrant of the Moon.

10.30 Albategnius Crater.
Courtesy of Damian Peach

10.31 Cyrillus and Theophilus Craters.
Courtesy of Anthony Ayiomamitis

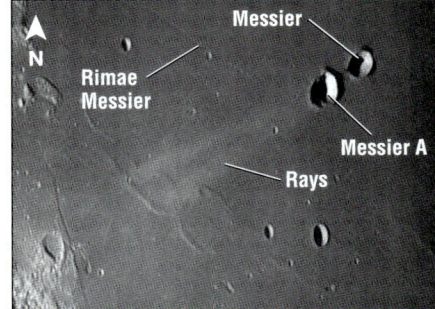

10.32 Messier and Messier A.
Courtesy of Damian Peach

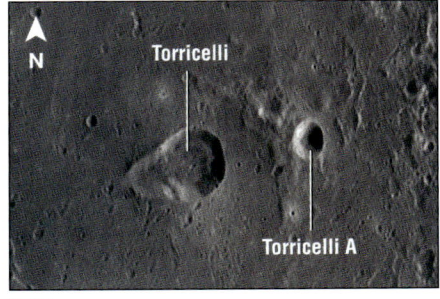

10.33 Torricelli Crater.
Courtesy of Damian Peach

Southeastern Quadrant

The Moon's southeastern quadrant is the lower right part of the Moon from our point of view (Fig. 10.29). This quadrant is dominated by the ray system of Tycho crater, but you can find a small "sea," Mare Nectaris (Sea of Nectar). To the north of Mare Nectaris are Mare Vaporum (Sea of Vapors) and Mare Fecunditatis (Sea of Fertility). Craters here include Albategnius Crater, Cyrillus Crater, Messier and Messier A, Theophilus Crater, and Torricelli Crater.

Albategnius Crater

Albategnius Crater (Fig. 10.30) is so large that lunar scientists often refer to it as a walled plain. It measures 136 km across and bears many scars of more recent meteor impacts. The most prominent are Klein (44 km), which sits to the southwest, and Albategnius B (20 km), just inside the northern rim. Note that Albategnius' outer wall has a rough hexagonal shape.

Cyrillus and Theophilus Craters

Theophilus Crater adjoins and sits to the upper right of Cyrillus Crater (Fig. 10.31). Theophilus measures 100 km across with a central triple peak towering 1,400 m above the crater floor. Cyrillus spans the same diameter as Theophilus, but its wall is not as intact. Look for the small, sharply defined crater Theophilus B at the northwestern edge of Theophilus' wall.

Messier and Messier A

Double crater Messier and Messier A (Fig. 10.32) sit on the Moon's eastern side only 2° south of its equator. Messier is an oblong crater measuring 9 km by 11 km. Messier A spans 13 km by 11 km. Two linear rays extend westward from Messier A for more than 100 km. If your sky is steady, look for the thin rille Rima Messier, which lies to the northwest of the craters.

Torricelli Crater

Torricelli Crater (Fig. 10.33) appears pear-shaped at first glance because its western wall is open and connects to a smaller crater to its west. Both craters lie in the upper part of a low-contrast circular formation named Torricelli R. The prominent crater to the east is 11 kilometer-wide Torricelli A. Torricelli Crater measures 23 km across.

Determining Feature Size

To determine the size of a feature on the Moon you will use a proportion to determine the size of the feature in question, both on your drawing and on the Moon chart (Fig. 10.39). First, you have to know the Moon's diameter: 3,474 kilometers. Next, measure the diameter of the Moon chart in millimeters. Then divide the number of millimeters into the Moon's diameter in kilometers. Your answer (call it M) is the scale for the Moon chart, in kilometers per millimeter. That is, each millimeter you measure on the Moon chart actually measures M kilometers. Now you can determine the size of any feature on the chart or on your drawing. For example, if you measure a feature on one of your illustrations to be 4 mm by 3 mm, on the Moon the feature actually measures 4 × M kilometers by 3 × M kilometers.

10.39 Example for determining feature size.

Courtesy of John Chumack

Calculation

The width of the image of the Moon (above) is 134.5 mm. The width of Mare Serenitatis on the image is 21 mm. To find its actual width:

$$M = \frac{\text{actual diameter in km}}{\text{image diameter in mm}} = \frac{3{,}474 \text{ km}}{134.5 \text{ mm}} = 25.829 \text{ km/mm}$$

Size of Mare Serenitatis: 21 mm × 25.829 km/mm = 542.409 km

Lunar Calculations

Exercise 10.1

Because this lab takes approximately one month to complete, start immediately on the date your instructor chooses. Briefly observe the Moon (3 to 5 minutes) every clear night for the entire period. Make all your observations in the early evening between 7:00 p.m. and 9:00 p.m., and try to make your observations at approximately the same time each night. Record which night the Moon is no longer visible at your chosen observing time. Try not to miss a night, and be sure to record those nights when you couldn't see the Moon because of clouds.

Procedure

Materials

- ❑ Astronomical Applications data sheet (available online, http://aa.usno.navy.mil/data/docs/RS_OneDay.php)
- ❑ Lunar chart (provided)
- ❑ Planisphere
- ❑ Sketching pencils
- ❑ Eraser
- ❑ Clipboard (to support your papers)
- ❑ Flashlight (if needed, to write and sketch in a dark location)

Determining Current Observing Conditions

1 Fill in the dates of the major phases for the current lunation in Table 10.1 below. Begin with the last new Moon. You can find this information on the *U.S. Naval Observatory Astronomical Applications Department Sun and Moon Data for One Day* online almanac data sheet.

TABLE 10.1 Dates of Moon Phases during Current Lunation

Phase	Date
New	
First quarter	
Full	
Last quarter	

2 Each night, before heading outside to observe, complete Table 10.2 on the data sheet, pages 199–200, including the current phase of the moon, the percentage of the Moon's face that is illuminated, the time of moonrise and moonset, and the constellation in which the Moon currently resides. Refer to the *U.S. Naval Observatory Astronomical Applications Department Sun and Moon Data for One Day* online almanac data sheet and your planisphere to find the answers.

The Moon CHAPTER 10   **197**

Recording the Moon's Features

1. Each night, you will draw a sketch of the Moon's features in the circles on the data sheets, pages 200–202. Sketches do not have to be masterpieces, but, because you want to be accurate, each will take some time to complete. Carefully gauge the relative size of all features.

2. Label the following mountain ranges (if visible): Montes Apennines, Altari Scarp.

3. Label the following maria (if visible): Mare Crisium, Mare Frigoris, Mare Humorum, Mare Imbrium, Mare Nectaris, Mare Nubium, Mare Serenitatis, Mare Tranquillitatis, Mare Vaporum, Ocean Procellarum, Sinus Iridium, Sinus Roris.

4. Label the following craters (if visible): Aristarchus, Aristoteles, Copernicus, Cyrillus, Grimaldi, Hercules, Plato, Theophilus, Tycho, Walter.

Name _____ Section _____ Date _____

Recording Lunar Data

TABLE 10.2 Daily Moon Data for 30 Days

Day	Date	Current Phase	Illumination	Moonrise Time	Moonset Time	Constellation Moon is in
1						
2						
3						
4						
5						
6						
7						
8						
9						
10						
11						
12						
13						
14						
15						
16						
17						
18						
19						
20						

Continues on next page

Exercise 10.1 Data Sheet

Exercise 10.1 Data Sheet

Day	Date	Current Phase	Illumination	Moonrise Time	Moonset Time	Constellation Moon is in
21						
22						
23						
24						
25						
26						
27						
28						
29						
30						

Sketching Lunar Features

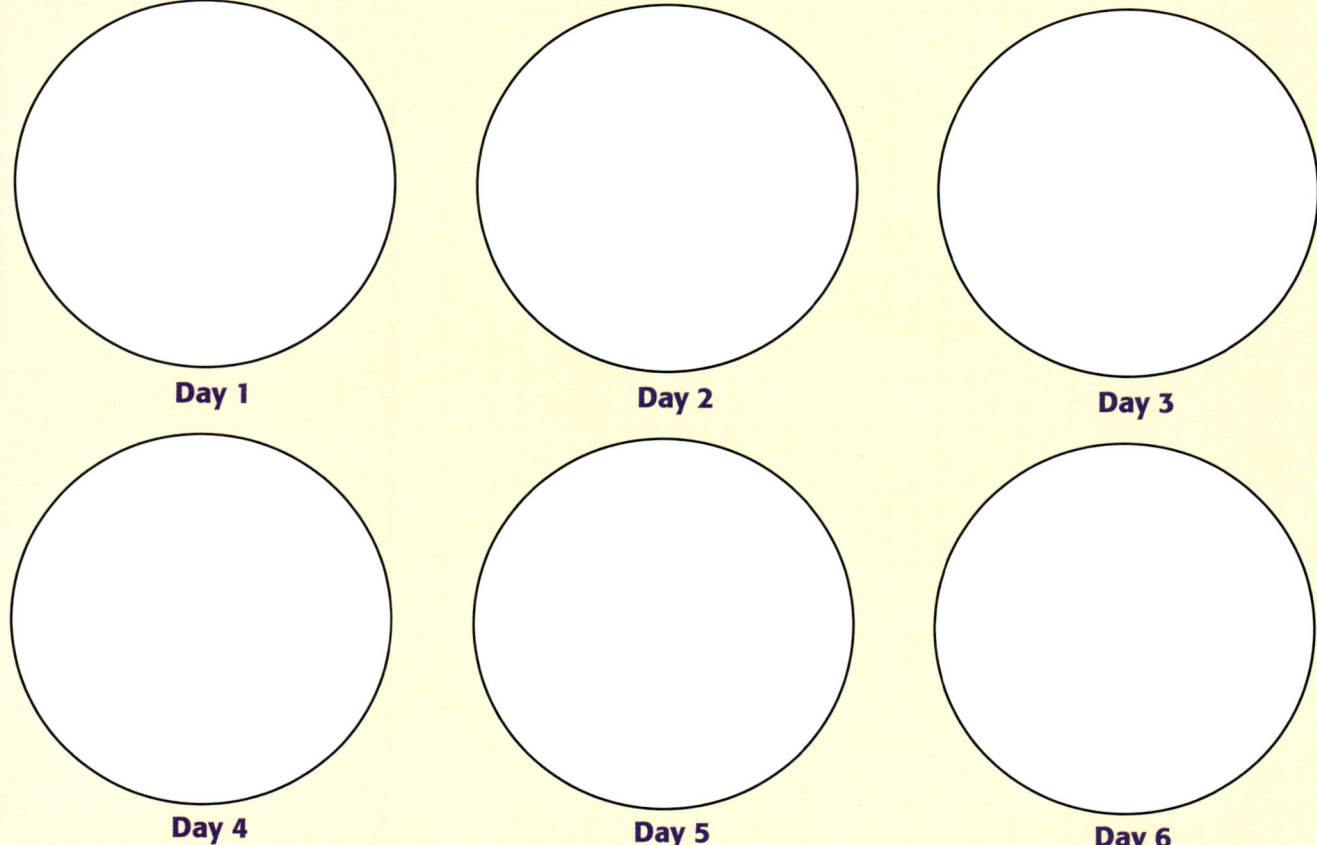

Day 1   Day 2   Day 3

Day 4   Day 5   Day 6

Unit 3 *Night and Day*

Name _____ Section _____ Date _____

Exercise 10.1 Data Sheet

◯ **Day 7**	◯ **Day 8**	◯ **Day 9**
◯ **Day 10**	◯ **Day 11**	◯ **Day 12**
◯ **Day 13**	◯ **Day 14**	◯ **Day 15**
◯ **Day 16**	◯ **Day 17**	◯ **Day 18**

The Moon **CHAPTER 10**

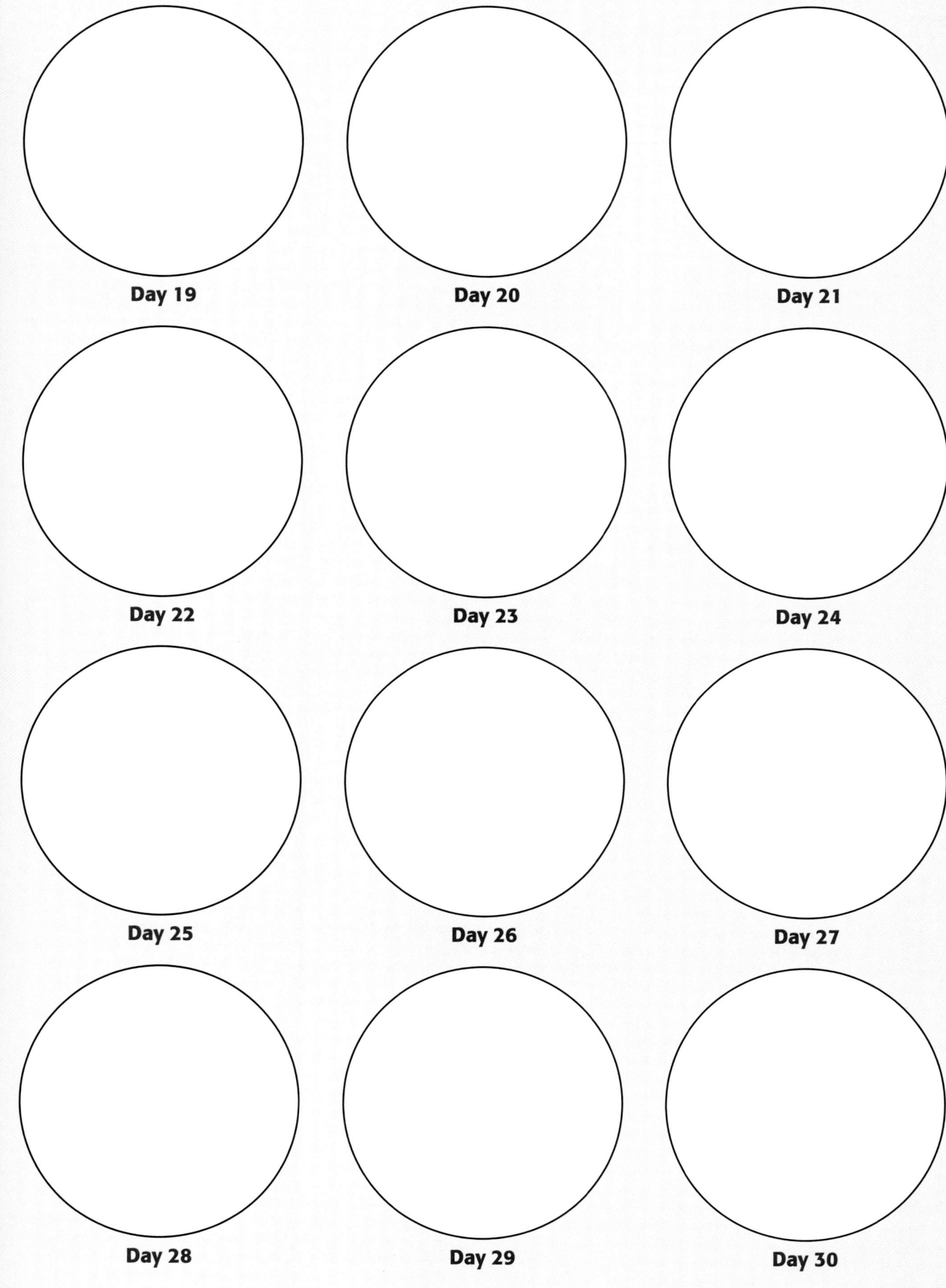

Craters of the Moon

Exercise 10.2

 Procedure

Materials
- Moon chart

1. Using the section on Features of the Moon, pages 186–191, to guide you, find and circle the following features on the Moon charts provided in Figure 10.40.

Albategnius Crater	Messier and Messier A	Schiller Crater
Archimedes Crater	Moretus Crater	Taruntius Crater
Aristarchus Crater	Pitatus Crater	Theophilus and Cyrillus Craters
Clavius Crater	Plato Crater	Torricelli Crater
Copernicus Crater	Posidonius Crater	Triesnecker Crater
Fra Mauro Crater	Ptolemaeus Crater	Tycho Crater
Gassendi Crater	Reiner Crater	
Lacus Mortis	Rupes Recta, the Straight Wall	

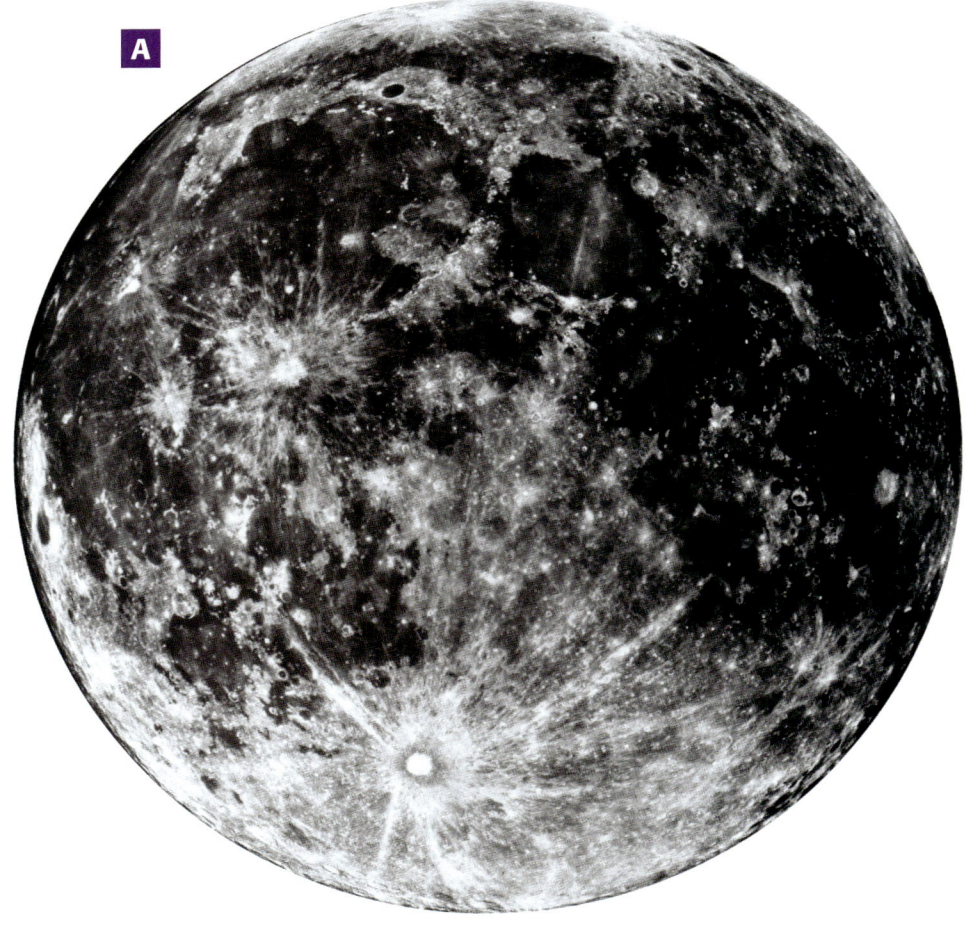

10.40 Moon charts: (**A**) full Moon *(continues)*.

© UC Regents/Lick Observatory

10.40 Moon charts (*continued*): (**B**) last quarter Moon, (**C**) first quarter Moon.

© UC Regents/Lick Observatory

2 Which feature on the list is the largest? Which is the smallest?

3 Which is the Moon's best-known system of rilles?

4 Which feature is known as the Straight Wall?

5 Which is the only double crater on the list?

6 Which crater has part of its wall caved in?

7 Which feature on the list is the darkest?

8 What is another word for "rimae"?

9 Several of the craters listed have visible ray systems. Name one.

10 Name one of the two craters with an unusual, six-sided shape.

Unaided Viewing of the Moon

Exercise 10.3

The goal of this exercise is to introduce you to observing the Moon. This observation is done entirely with the naked eye—that is, with no binoculars or telescope. If you wear eyeglasses for distance (you are nearsighted), wear your glasses when making this observation. This exercise will begin teaching you how to recognize lunar features.

Procedure

Materials
- ❏ Flashlight (if needed, for answering questions and drawing in a dark location)
- ❏ Sketching pencils
- ❏ Eraser
- ❏ Clipboard (to hold your papers)

1. In the circle below, make a sketch of the Moon as you see it with the naked eye. Perfect details are not necessary in this drawing. Later you will go back and identify features you have drawn from the Moon chart. Look for light and dark patches, craters, lines, markings, etc.

2. Use a pencil to make your drawing. You can also use the pencil to shade areas. This drawing is best made around the time of the full Moon.

3 After completing your drawing, answer the following questions:

a How hard was it to differentiate between the Moon's dark features and the light features?

b "Things" in the Moon—such as the man in the Moon, the rabbit, a woman, a witch, etc.—are part of lunar folklore. Did you see any shapes that looked like any of these—or even some other familiar shape? What do you think causes the shapes?

c What colors, if any, did you see?

d How appropriate was Apollo 11 astronaut Buzz Aldrin's comment "magnificent desolation"? Explain your answer.

e Which do you think would be easier to observe, a full Moon or a quarter Moon (50 percent illuminated)? Why?

f What was the hardest thing about observing the Moon for this procedure?

Telescopic Viewing of the Moon

Exercise 10.4

The goal of this lab exercise is threefold: general lunar observing, seeing lunar details, and estimating lunar feature sizes. There will be telescopes (and possibly binoculars) set up for viewing. Each section asks you to make specific observations. Most of these questions and observations are best done at the telescope. You will need the Moon chart provided in your lab manual. Plan to complete the sizes of lunar object calculation after you finish observing, preferably indoors.

Procedure

Materials
- ❏ Telescopes
- ❏ Binoculars
- ❏ Moon chart
- ❏ Sketching pencils
- ❏ Eraser
- ❏ Clipboard (to hold your papers)
- ❏ Flashlight (if needed, for writing answers and drawing in a dark location)

1. Record your observing conditions below.

   Place of Observation: _____

   Overall Sky Conditions: _____

   Start time: _____

   End time: _____

2. Look through all of the telescopes, and complete Table 10.3, page 209, on the data sheet.

3. Which telescope (or binoculars) gave you what you considered the best (or most-impressive) view? What did you like about this view of the Moon?

4. In circle A on page 209, make a sketch of the Moon. Perfect details are not necessary in this drawing. However you will need to identify four (4) features and be able to accurately determine the size of one of the four features from your lunar sketch.

   Telescope: _____   Magnification: _____

5. Identify at least four features on your drawing using the Moon charts (Figure 10.40 on pages 203–204). Try to draw these features accurately, size-wise. How difficult was it to draw the Moon? Specifically, what did you find difficult?

6. In circle B on page 209, make a sketch of one (1) identified lunar feature (we recommend a crater). Try to be as accurate as possible.

   Telescope: _____   Magnification: _____

7. How difficult was it to draw a specific lunar feature? How was this different than drawing the entire visible Moon?

8 Using the Moon chart provided, determine the scale of the photograph. Show your calculations.

Number of millimeters north to south: _____ mm

Moon chart scale: _____ km/mm

9 Measure the diameters (in millimeters) and compute the actual diameters (in kilometers) of the craters listed in Table 10.4 on page 210. There is space provided in the table to show your calculations for determining the diameters of the lunar craters.

10 Choose one of the four features you drew in step 1 (not one of the six listed in Table 10.4). Determine the size of the feature on the Moon chart, and then compare it with the size you drew on your general lunar sketch. How did the size of the crater you drew compare with the same crater on the Moon chart?

Name _____ Section _____ Date _____

Recording Telescopic Observations of the Moon

Exercise 10.4 Data Sheet

TABLE **10.3** Impressions of the Moon

#	Telescope (or Binoculars)			Impression: How the Moon "Looks"
	Diameter	Magnification	Optical Type	
1				
2				
3				
4				
5				

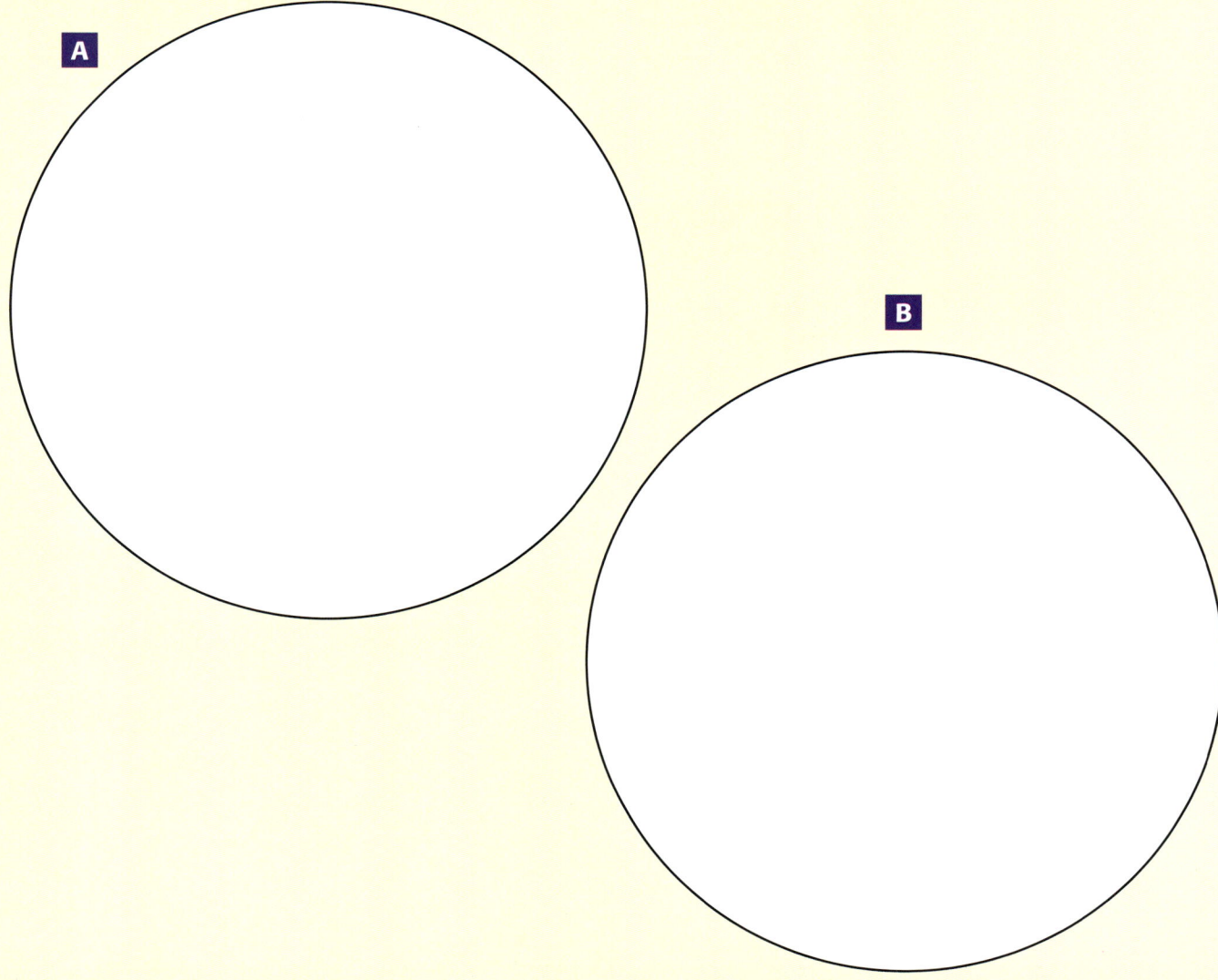

A

B

The Moon   CHAPTER 10   209

Exercise 10.4 Data Sheet

Table 10.4 Calculated Sizes of Lunar Craters

Crater	North to South	East to West	Calculations	Calculated size
Aristarchus	mm	mm		_____ km by _____ km
Copernicus	mm	mm		_____ km by _____ km
Cyrillus	mm	mm		_____ km by _____ km
Plato	mm	mm		_____ km by _____ km
Theophilus	mm	mm		_____ km by _____ km
Tycho	mm	mm		_____ km by _____ km

Unit 3 Night and Day

Night and Day

The Sun

11

LEARNING OBJECTIVES

Upon completion of this chapter, you should be able to:

1. Identify the Sun's main physical features.
2. Explain why sunrise and sunset position varies over the course of a year.
3. Track the Sun's highest altitude and its transit time and explain why this position changes over the course of a year.
4. Prepare a chart of sunrise, sunset, and transit times for your location based on data from the U. S. Naval Observatory's website; compare this to the data you collected
5. Explain why length of the solar day changes over the course of a year.
6. Complete a sketch of a live view of the Sun as seen through a telescope.
7. Identify several types of sunspots, according to the McIntosh classification system, and compute their size.
8. Define all glossary terms.

Sunset over the Rockies.
Courtesy of Vikki Granger, Insight 360 Media

GLOSSARY TERMS

Analemma plot, or graph, of the Sun's position in the sky at a certain time of day (such as noon) at one location measured throughout the year; has the shape of a figure eight

Photosphere surface of a star, including the Sun, which is the layer emitting visible light, about 500 kilometers deep

Granulation mottled (alternating light and dark) appearance of the photosphere of the Sun caused by the convection of gas cells, which are rising and falling

Facula (*pl.* = faculae) bright region of the Sun's photosphere associated with sunspots; seen most easily when near the Sun's limb

Limb darkening phenomenon whereby the edge of the solar disk appears darker than the center due to the light rays from the edge having to move through more of the solar atmosphere to reach us than light rays near the center of the disk

Chromosphere region of the atmosphere of a star (such as the Sun) between the star's photosphere and its corona

CAUTION !

DO NOT look at the Sun through your telescope without a proper solar filter! An example of the danger: Ron Lambert of El Paso, Texas, had his Tele Vue eyepiece capped when he accidentally moved the Sun into the field of view of his telescope for less than one second. The heat caused the eyecap to melt. Note also that sunglasses, even with UV protection, are NOT adequate eye protection for viewing the Sun.

The center of our solar system, and our closest star, is the Sun. It is about 300,000 times closer than the next closest star, Proxima Centauri. Because it is the sky's brightest object, the Sun also is the easiest to observe. We can clearly see its surface features, examine its characteristics in a variety of wavelengths, and glean insight into the nuclear mechanisms of not only the Sun, but also of other stars.

During this lab, you will construct a safe pinhole solar viewer and use it to sketch the positions of sunspots during a clear day. You also will track the Sun in a variety of ways, and construct and use a sundial to track the Sun's movement and tell time.

By tracking the Sun over a period of time, we can better understand the orbital relationship between it and Earth, as well as some of Earth's particular mechanics. Observing the Sun's motion on a daily basis, as well as over the course of a year, will demonstrate Earth's rotational motion and its revolution around the Sun. The location of the observer and the time of year will change where the observer sees the Sun in the local sky. For example, Figure 11.1 captures the apparent track the Sun makes in the local sky from the same place looking in the same direction at the same time every day, usually around noon. The figure-eight visible in the exposures is known as an **analemma**.

Observing the Sun

Solar Features

With proper eye protection, you can see a multitude of features on our Sun (Fig. 11.2). The **photosphere** is the Sun's visible surface and is the lowest observable layer of solar atmosphere. If the seeing (atmospheric steadiness) is good you will spot **granulation** on the photosphere, which observers describe as mottled (having alternating light and dark areas). Vast gas bubbles, whose centers are rising and whose edges are sinking, create the granules.

Faculae, Latin for "little torch," are bright areas visible on the photosphere. Faculae appear all over the solar disk, but observers most often see them near the solar limb where the contrast between the faculae and the darkened limb is highest.

Finally, look for a phenomenon called **limb darkening**. We observe limb darkening because the Sun is a sphere. Near what is seen as the edge of the solar disk, the light must travel farther through the solar atmosphere. This causes the limb to be dimmer than the rest of the disk.

The **chromosphere** (Fig. 11.3), or "sphere of color," lies just above the photosphere. Here, hydrogen atoms emit energy called hydrogen-alpha (Hα) radiation. Hα is reddish-colored light with a wavelength of 656.28 nanometers (nm). Although your eyes can detect the red light of the chromosphere, it

11.1 These 41 solar exposures, plus one foreground image showing the Parthenon in Athens, Greece, were taken at the same time of day starting January 12 and ending December 21, 2002.
Courtesy of Anthony Ayiomamitis

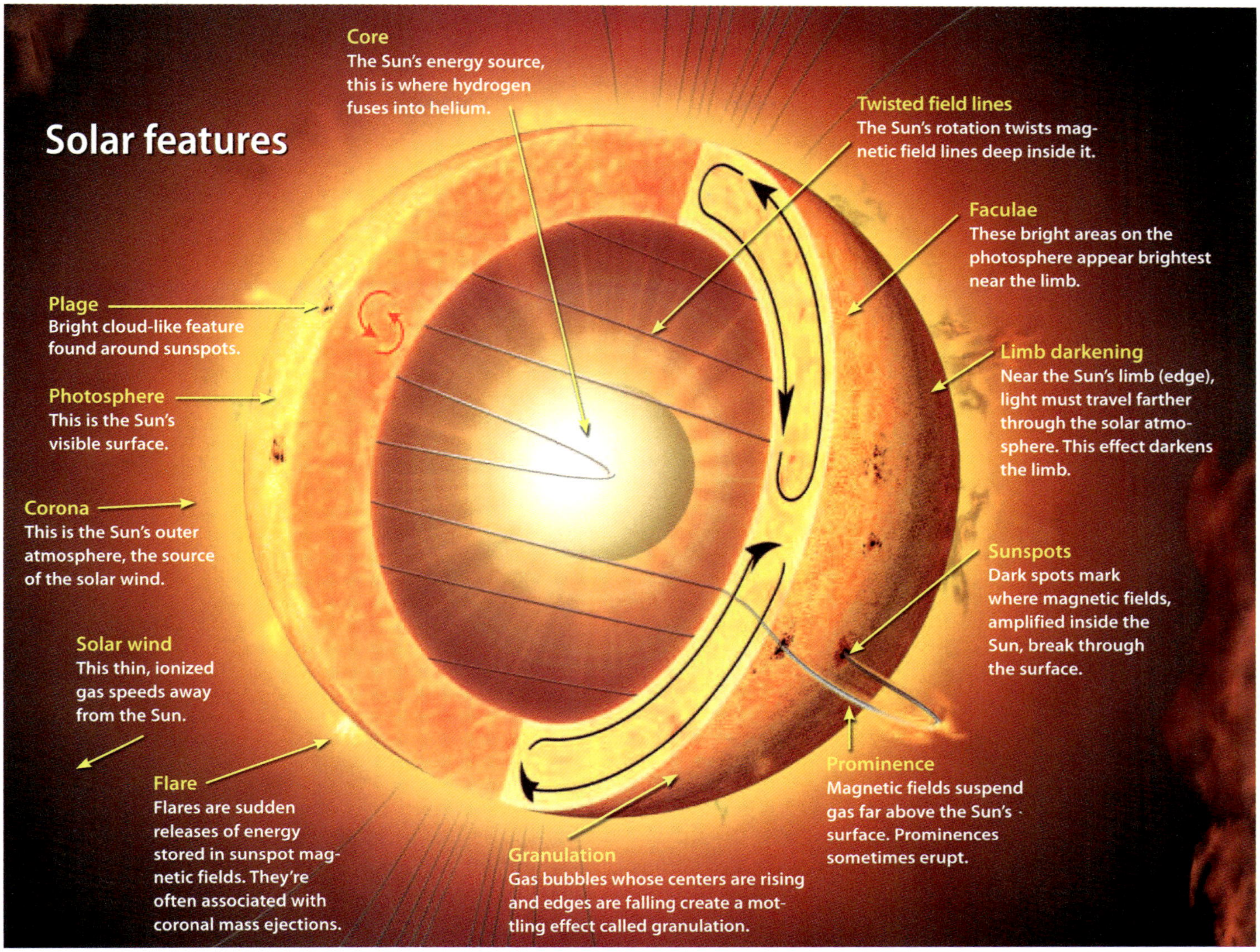

Solar features

Core
The Sun's energy source, this is where hydrogen fuses into helium.

Twisted field lines
The Sun's rotation twists magnetic field lines deep inside it.

Faculae
These bright areas on the photosphere appear brightest near the limb.

Plage
Bright cloud-like feature found around sunspots.

Photosphere
This is the Sun's visible surface.

Limb darkening
Near the Sun's limb (edge), light must travel farther through the solar atmosphere. This effect darkens the limb.

Corona
This is the Sun's outer atmosphere, the source of the solar wind.

Sunspots
Dark spots mark where magnetic fields, amplified inside the Sun, break through the surface.

Solar wind
This thin, ionized gas speeds away from the Sun.

Flare
Flares are sudden releases of energy stored in sunspot magnetic fields. They're often associated with coronal mass ejections.

Prominence
Magnetic fields suspend gas far above the Sun's surface. Prominences sometimes erupt.

Granulation
Gas bubbles whose centers are rising and edges are falling create a mottling effect called granulation.

11.2 Features of the Sun.

Courtesy of Roen Kelly, *Astronomy* magazine

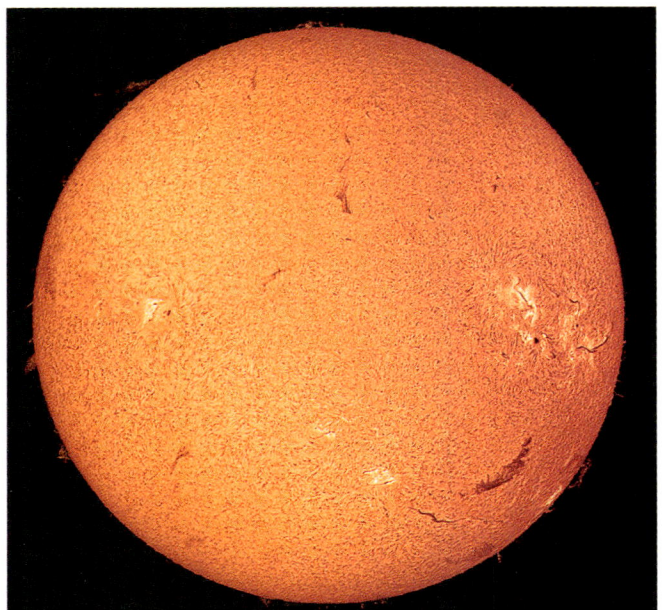

11.3 Chromosphere.
Courtesy of Craig and Tammy Temple

The Sun CHAPTER 11 213

GLOSSARY TERMS

Prominence large-scale, gaseous formation above the surface of the Sun (or theoretically, a star); usually occur over regions of solar activity such as sunspot groups

Filament cloud of gas in the Sun's chromosphere visible as a dark feature against the brighter disk; a straight-on view of a prominence

Plage bright areas in the chromosphere of the Sun that are at a higher temperature than the surrounding areas; also known as bright flocculi

Sunspot temporarily cooler region on the photosphere of the Sun caused by magnetic field variations

Flare bright, short-lived area of the Sun's chromosphere, best viewed through a hydrogen alpha filter; a sudden burst of light particles (protons, electrons, etc.) and electromagnetic energy from the Sun's photosphere; in space they contribute to the solar wind, and on Earth they manifest as an increased low-energy cosmic ray and aurora activity

Umbra for this chapter, the dark inner region of a sunspot

Penumbra for this chapter, the less dark outer region of a sunspot

Sunspot cycle period of approximately 11 years, during which the number of sunspots observed is at a maximum

Solar cycle variation in the activity of the Sun during an 11-year period; most noticeable because of its effect on the number of sunspots visible on the photosphere

mixes with the rest of the Sun's light, which is much brighter. The result is that you cannot see the chromospheres without a Hα filter.

Prominences (Fig. 11.4) are bright gas clouds ejected from the Sun and shaped by its magnetic field. Prominences appear as spikes, loops, "trees," detached regions, and more when you view them at the Sun's edge. When you see them straight on, however, prominences look like dark lines, known as **filaments**, silhouetted against the solar disk. They appear dark because the gas is slightly cooler than the surface beneath it. Another feature that can be seen through a Hα filter is **plages**, which appear as bright areas around **sunspots**.

Solar **flares** occur when the Sun's atmosphere suddenly releases built-up magnetic energy. Solar flares emit radiation storms and are the solar system's largest explosions. Astronomers classify flares by how much of the Sun's area they cover at the time of maximum brightness. Flares range from subflares (smaller than 2 square degrees) to Importance 4 flares, which cover more than 24.8 square degrees. On the Sun, 1 square degree equals roughly 57 million square miles (150 million square kilometers). This is slightly more area than Earth would cover (50.2 million square miles or 130 million square kilometers) if placed on the Sun's surface.

Sunspots (Fig. 11.5A), which are features of the photosphere, come in many shapes and sizes, according to the whim of the Sun's magnetic field. The field traps gas, slowing its motion, and making it cooler than the surrounding area.

Usually, sunspots have two parts: a dark central region called the **umbra** surrounded by a lighter region known as the **penumbra** (Fig. 11.5B). The penumbra's temperature is typically 1,800°F (1,000°C) below that of the photosphere, and the temperature of the umbra is between 2,700°F (1,500°C) and 3,600°F (2,000°C) cooler than the photosphere. The lower temperature of sunspots is why they look darker than the Sun's surface around them.

Approximately every 11 years, solar activity peaks, resulting in greater numbers of sunspots and flares. German astronomer Heinrich Schwabe (1789–1875) discovered the **sunspot cycle** in 1843. This "11-year cycle" varies from as few as 9.5 years to as many as 12.5 years. Solar physicists define the start of any given **solar cycle** as the minimum of solar activity. Usually, they don't realize a new cycle has started until the sunspot activity picks up, which can take a year or more.

11.4 Loop prominence that covers a region much larger than Earth.

Courtesy of NASA/Solar and HelioSpheric Observatory

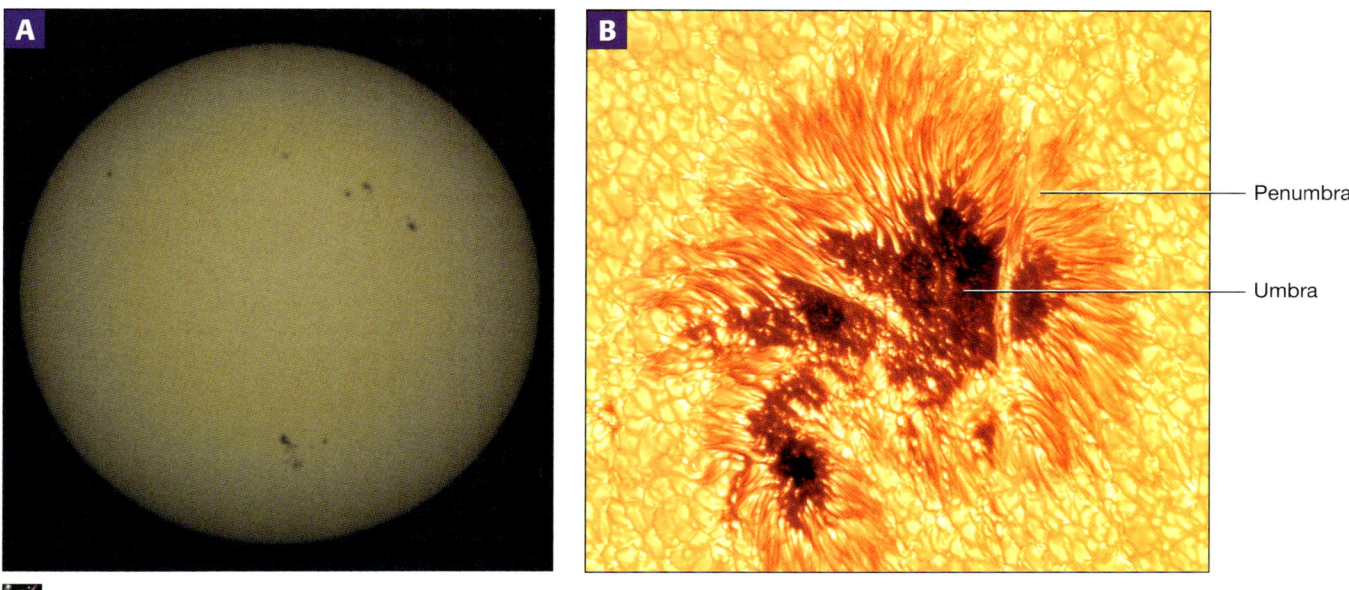

11.5 **(A)** Sunspots, and **(B)** close-up view of a sunspot showing the umbra and penumbra.

(B) Courtesy of Big Bear Solar Observatory/New Jersey Institute of Technology

Visual Solar Filters

A good solar filter (Fig. 11.6) is safe because it does not transmit ultraviolet or infrared radiation, both of which are much more harmful to your eyes than light. It also drops the Sun's brightness to a comfortable level.

Visible-light filters are either specially coated glass or optical-quality Mylar (a thin silver film). The solar image through Mylar looks pale blue; through glass filters it can appear white, yellow, or orange. Glass filters are more expensive but more durable.

All solar filters fit over a telescope's objective (front) end (Fig. 11.7). Some cover the entire objective (full-aperture filters), while others have smaller openings offset from center (off-axis filters). Off-axis filters eliminate secondary-mirror obstructions in reflecting telescopes. All solar filters should have round openings. Other shapes introduce distracting patterns.

NEVER use a solar filter that fits into an eyepiece. Because of how the Sun's focused heat builds up, some of these filters have cracked with devastating results.

The most common solar filters transmit 0.001 percent of the Sun's light. This drops the Sun's brightness by roughly 12.5 magnitudes. Photographic-only solar filters with densities of 3 (7.5-magnitude drop) and 3.5 (8.75-magnitude drop) are also available.

CAUTION! Photographic filters are NOT safe to use to look at the Sun.

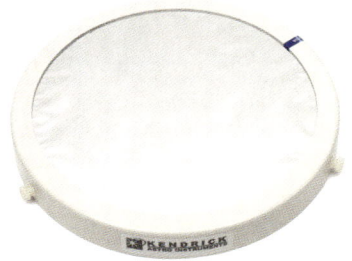

11.6 A visual solar filter allows you to view sunspots and a few other features on the Sun's disk.

11.7 A visual solar filter ALWAYS fits on the front end of a telescope.

Courtesy of *Astronomy* magazine

The Sun **CHAPTER 11**

CAUTION

Use #14 welder's glass only! No other number is safe. Note that stacking welder's glass with lower number ratings (e.g., two #7 welder's glass) to reach #14 is NOT safe.

GLOSSARY TERM

Angstrom a unit of length used to measure light equal to one ten-millionth (0.0000001) of a millimeter

#14 Welder's Glass

Observing the photosphere is easy through visible-light solar filters. A #14 welder's glass (Fig. 11.8) is one safe filter you can use. Small, inexpensive ones are available from welding supply stores. Just hold the glass in front of your eye and then find the Sun through it. Such filters are good only for naked-eye use. You will need a different kind of filter if you want to use a telescope.

Hydrogen-alpha (Hα) Filters

Observing the Sun at the wavelength of Hα light is one of amateur astronomy's fastest-growing segments. Solar flares and prominences are best seen through a Hα filter. All Hα filters center on a wavelength of 656.28 nanometers and allow through only a tiny part of the red light the Sun produces, blocking all other colors. Each filter, however, passes more or less light depending on what astronomers call its bandpass width. Filters have different bandpass widths. The widest of these can be nearly 2 Angstroms (Å) and the narrowest 0.3 Å. One **Angstrom** equals 0.0000001 millimeter. The narrower the bandpass of your filter, the greater detail you will see. Unfortunately, you will also pay more as the bandpass narrows.

Through a 1 Å-bandpass Hα filter, prominences look great but chromospheric detail is low. Through a filter with a bandpass of 0.5 Å, you will see a lot of chromospheric detail but few prominences. Some Hα filters are tunable; you can shift the bandpass' central wavelength slightly to either side.

There are also solar telescopes with built-in solar filters, such as Meade Instruments' Personal Solar Telescope, or PST (Fig. 11.9). This is a hydrogen-alpha telescope that allows you to observe prominences, flares, and the chromosphere.

11.8 Shielding your eyes with a #14 welder's glass is one safe way to view the Sun.

11.9 Meade Instruments' Personal Solar Telescope.
Courtesy of *Astronomy* magazine/Kalmbach Publishing

Check Your Understanding

1.1 Name three features you can see through a visible-light (visual) solar filter.

1.2 Explain the difference between a prominence and a filament.

1.3 A certain sunspot covers 6 square degrees of the Sun's surface. How many Earths would it take to equal that area?

1.4 Why are sunspots dark?

1.5 On average, how often does solar activity peak?

1.6 Why should you never use a solar filter that is screwed into an eyepiece?

1.7 What is the advantage of a narrow bandpass hydrogen-alpha filter?

Exercise 11.1   Tracking the Sun Daily

Procedure

1. You can see one of the Sun's **diurnal** (daily) changes by observing its rising and setting points from your location. First find a spot that gives you a good view of the horizon. Because of trees, buildings, and other obstructions, you might not have a clear line to the horizon. Do not worry, because observing from a regular position over a period of time is more important.

2. There are a couple of options for recording the location of daily sunrise and sunset. Choose one of the following methods:

   a. Sketch your eastern and western horizons on separate pages (Fig. 11.10A), and on a daily basis mark the location of sunrise, along with the date and time on one page, and the sunset details on the other.

   b. Use a camera to take a photo of the location, have multiple prints made, and sketch in the sunrise or sunset location (Fig. 11.10B).

   c. Finally, use a digital device/tablet or cell phone equipped with a camera and each day photograph the sunrise and the sunset.

3. After making your observations, record the photo or sketch numbers, and the date and time of sunrise and sunset in Table 11.1 in the data sheet, pages 219–220. A simple compass to determine east for sunrise and west for sunset will be helpful.

4. Perform this observation every day for a month.

Materials

(optional choices, depending on the method selected to record your observations)
- ❏ Sketching pencils
- ❏ Sketching pad
- ❏ Camera (digital or standard with film)
- ❏ Cell phone with camera
- ❏ Digital tablet with camera
- ❏ Compass (many are available as free, downloadable apps for cell phones)

GLOSSARY TERM

Diurnal activity or event that occurs daily

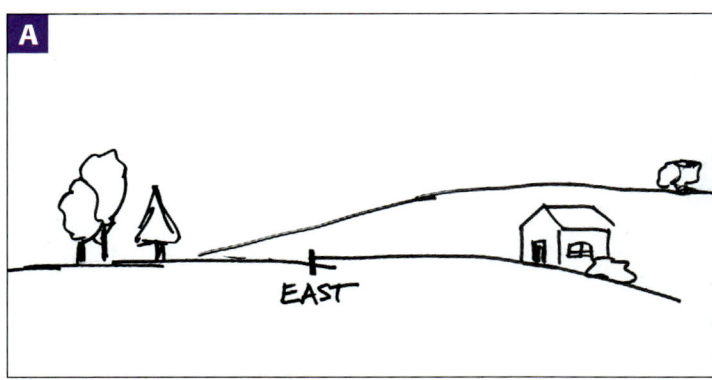

11.10 (A) Sketch or (B) photograph a horizon to help plot the location of sunrise or sunset.

Courtesy of Holley Y. Bakich

Name _____ Section _____ Date _____

Tracking the Sun Daily

Location: _____ Number of daily observations assigned: _____

Observations recorded via: _____ Sketch _____ Sketch on Photo _____ Photograph

_____ Other (alternate method used: _____)

TABLE **11.1** Sunrise and Sunset Data

Date	Time of Sunrise/Sunset	Notes
/ /		
/ /		
/ /		
/ /		
/ /		
/ /		
/ /		
/ /		
/ /		
/ /		
/ /		
/ /		
/ /		

(Continues)

Exercise 11.1 Data Sheet

Date	Time of Sunrise/Sunset	Notes
/ /		
/ /		
/ /		
/ /		
/ /		
/ /		
/ /		
/ /		
/ /		
/ /		
/ /		
/ /		
/ /		
/ /		
/ /		
/ /		

Estimating the Sun's Maximum Altitude

Exercise 11.2

Each day, the Sun reaches a maximum altitude at your location. This is the time the Sun reaches its highest altitude and passes through the **meridian**. Astronomers call the moment when a celestial object climbs highest in the sky a **transit**. Some people also refer to it as "local noon," although the clock time may not read exactly 12:00 p.m. For example, a time of "12:00 p.m." one day might be "1:00 p.m." the next if daylight saving time has gone into effect. The Sun's transit point can be directly overhead only at latitudes between 23.5° north and 23.5° south. The Sun's maximum altitude at any location will vary by 47° throughout the year.

Procedure

Materials
(items needed if you choose to use method 1a)
- ❏ Protractor
- ❏ String
- ❏ Weight
- ❏ Straw

GLOSSARY TERM

Meridian imaginary line that passes from north to south through the overhead point, or zenith; divides the sky into two halves: eastern and western

Transit moment when a celestial object is highest in the sky; also known as "local noon" when the object is the Sun

1. There are several ways to estimate the Sun's maximum altitude. Choose one of the following methods:

   a Build a simple, altitude-finding device using a protractor, a string, a straw, and a weight (Fig. 11.11). This will allow you to "shoot" the Sun when it transits. Use caution doing this, and never look directly at the Sun.

   b Observe over a period of time near midday to estimate when the Sun reaches its highest elevation.

   c Visit a website that provides such data for your location. This will at least let you know when to observe. One of the best is the site provided by the United States Naval Observatory (USNO), the *Sun and Moon Data for One Day*: http://aa.usno.navy.mil/data/docs/RS_OneDay.php. This site has you enter the city, state, and date desired. If you are located in a small city, you will need to enter a larger city close to your location. The USNO site will give you the times of sunrise, Sun transit, sunset, moonrise, Moon transit, and moonset, and the current phase of the Moon for your location based on the information you enter. See the sidebar for an example.

2. Choose your method of determining the maximum altitude of the Sun and make your observations every day for a month. Record your data in Table 11.2 in the data sheet, pages 223–224.

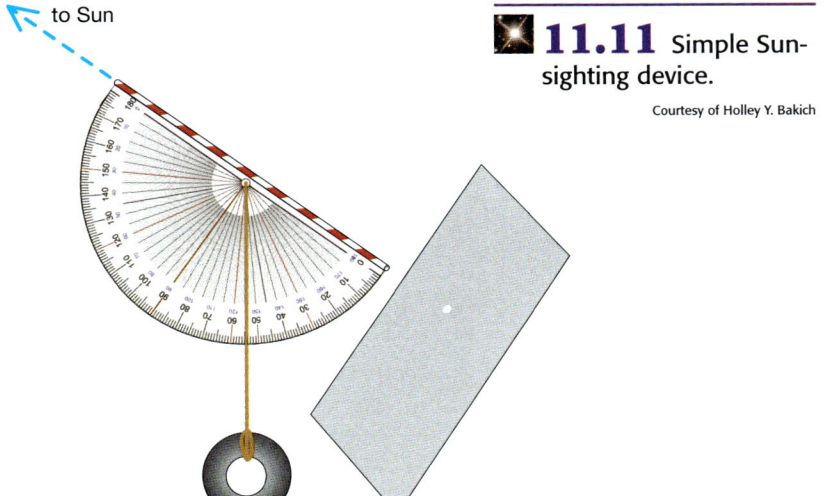

11.11 Simple Sun-sighting device.

Courtesy of Holley Y. Bakich

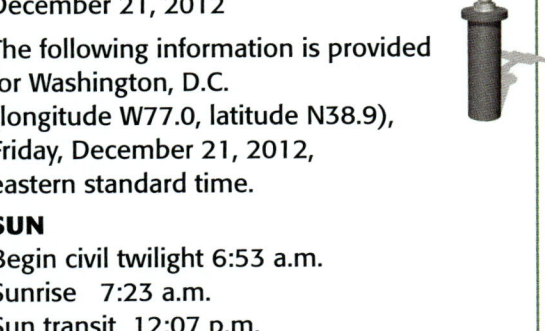

Looking Up

Sample dataset, Washington, D.C., December 21, 2012

The following information is provided for Washington, D.C. (longitude W77.0, latitude N38.9), Friday, December 21, 2012, eastern standard time.

SUN
Begin civil twilight 6:53 a.m.
Sunrise 7:23 a.m.
Sun transit 12:07 p.m.
Sunset 4:50 p.m.
End civil twilight 5:20 p.m.

Exercise 11.3 Determining the Length of the Solar Day

Procedure

1 You can make these observations in several ways. Choose one of the following methods:

   a Determine them from the times you recorded for sunrise and sunset.

   b Most television meteorologists and local newspapers give sunrise and sunset times.

   c Visit the USNO website (http://aa.usno.navy.mil/data/docs/RS_OneDay.php) for the data for your location.

2 Complete Table 11.3 in the data sheet on pages 225–226, indicating the length of the day for your location. You will be able to make some generalizations from your observations, whether taken directly, from a news source, or a reliable online site such as the USNO.

Name _____ Section _____ Date _____

Sun's Maximum Altitude Log

Exercise 11.2 Data Sheet

Location: _____ Number of daily observations assigned: _____

Observations recorded via: _____ Sketch _____ Sketch on Photo _____ Photograph

_____ Other (alternate method used: _____)

TABLE **11.2** Maximum Solar Altitude Data

Date	Time of Maximum Solar Altitude	Altitude	Notes
/ /		°	
/ /		°	
/ /		°	
/ /		°	
/ /		°	
/ /		°	
/ /		°	
/ /		°	
/ /		°	
/ /		°	
/ /		°	
/ /		°	
/ /		°	
/ /		°	

(Continues)

Exercise 11.2 Data Sheet

Date	Time of Maximum Solar Altitude	Altitude	Notes
/ /			
/ /		°	
/ /		°	
/ /		°	
/ /		°	
/ /		°	
/ /		°	
/ /		°	
/ /		°	
/ /		°	
/ /		°	
/ /		°	
/ /		°	
/ /		°	
/ /		°	
/ /		°	

Name _____ Section _____ Date _____

Length of Solar Days

Location: _____ Number of daily observations assigned: _____

Observations recorded via: _____ Sketch _____ Sketch on Photo _____ Photograph
_____ Other (alternate method used: _____)

TABLE **11.3** Length of Solar Day Data

Date	Sunrise time	Sunset time	Length of day	Notes
/ /				
/ /				
/ /				
/ /				
/ /				
/ /				
/ /				
/ /				
/ /				
/ /				
/ /				
/ /				
/ /				
/ /				

(Continues)

Exercise **11.3** Data Sheet

The Sun **CHAPTER 11**   225

Exercise 11.3 Data Sheet

Date	Sunrise time	Sunset time	Length of day	Notes
/ /				
/ /				
/ /				
/ /				
/ /				
/ /				
/ /				
/ /				
/ /				
/ /				
/ /				
/ /				
/ /				
/ /				
/ /				
/ /				

1 As you made your observations over the assigned period of time, what did you notice?

2 What factors cause day lengths to increase or decrease?

Unit 3 *Night and Day*

Classifying Sunspots

Exercise 11.4

Since telescopic solar observing began, astronomers have counted sunspots. They also have divided sunspots into types. Astronomers have used several classification systems for sunspots over the years. One of the easiest to use is the McIntosh classification system (Table 11.4), which debuted in 1966. This three-letter system will classify any type of sunspot. The first letter indicates the overall group structure, the second letter indicates the appearance of the penumbra of the largest spot in the group, and the third letter indicates the distribution of the spots within the group. For example, the McIntosh system would designate the simplest sunspot with no penumbra as Axx.

The dark, inner part of a sunspot is its umbra, and the lighter outer region is the penumbra. Degree measurements relate to the Sun's disk diameter as 180°, so a sunspot 10° across would span one-eighteenth of the Sun's disk.

Remember to always use an approved solar filter to observe the Sun. Don't take a chance where your eyes are concerned.

TABLE 11.4 McIntosh System of Sunspot Classification

First Letter		
A	Individual spot or group of spots without a penumbra or bipolar structure	
B	Group of sunspots without a penumbra in a bipolar arrangement	
C	Bipolar sunspot group where the principal spot is surrounded by a penumbra	
D	Bipolar group where the principal spots have a penumbra; at least one of the two principal spots has a simple structure; length of the group is less than 10°	
E	Large, bipolar group in which the two principal spots, which are surrounded by penumbrae, generally exhibit a complex structure; numerous smaller spots occur between the principal spots; length of the group lies between 10° and 15°	
F	Very large bipolar or complex sunspot group with penumbrae on spots at both ends and whose length is at least 15°	
H	Unipolar (single) group with penumbra	
Second Letter		
x	No penumbra	

(Continues)

r	Rudimentary (incomplete) penumbra, irregular boundaries, width only around 0.2° on the Sun or 3" in the sky; brighter than normal penumbrae, granular-fine structure	
s	Symmetrical, almost circular penumbrae with a typical filament structure that is directed outward, diameter less than 2.5° on the Sun; umbrae of the sunspot form a compact group in the vicinity of the center of the penumbra; included in this class are elliptical penumbrae around a single umbra; spots with s-penumbrae change slowly	
a	Small, asymmetric spot with irregular surrounding penumbra and the umbrae within separated; north-south diameter of 2.5 degrees or less	
h	Symmetrical, penumbrae-like types, but with a diameter of more than 2.5°	
k	Asymmetrical penumbrae like type r but with a diameter of more than 2.5° measured in the north-south direction to avoid elongated leaders that are decaying and inactive; if the diameter exceeds 5°, it is certain both magnetic polarities occur within a penumbra (bipolar group) so the group can be classified as Dkc, Ekc, or Fkc	
Third Letter		
x	Individual spot	
o	Open distribution; the area between the preceding and following principal spots is free of sunspots so that the group clearly comprises two parts.	
c	Compact distribution; area between the main spots is populated with many large spots, of which at least one has a penumbra; in extreme cases the entire sunspot area can be surrounded by one huge penumbra	
i	Intermediate type between o and c; some sunspots without a penumbra can be observed between the principal spots	

Images courtesy of Association of Lunar and Planetary Observers

Procedure

1 Using the McIntosh system outlined in Table 11.4, classify the sunspots in the Data Sheet 11.4, Figures 11.12 through 11.27. Record your answers in Table 11.5, page 229, and Table 11.6, page 230.

Name _____ Section _____ Date _____

Classifying Sunspots

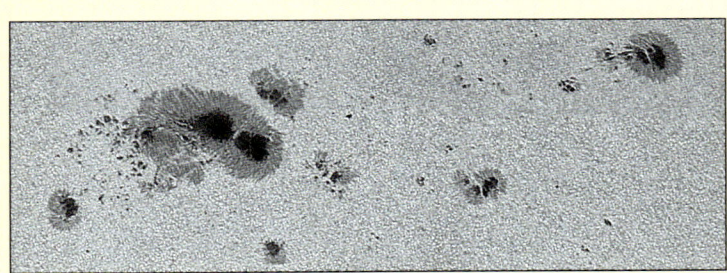
11.12

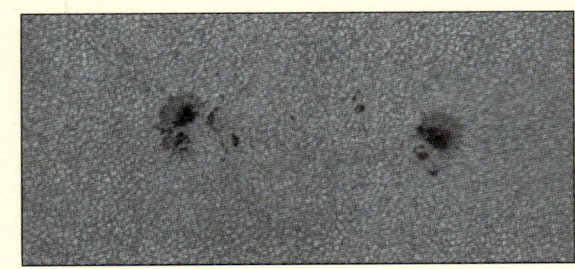
11.13

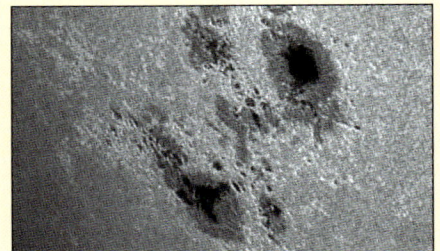
11.14

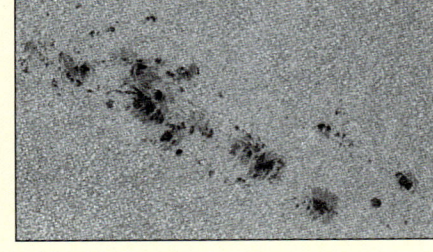
11.15

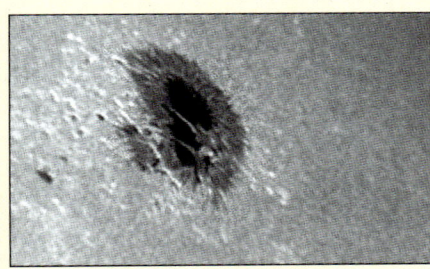
11.16

Images 11.12–11.19 courtesy of David Tyler

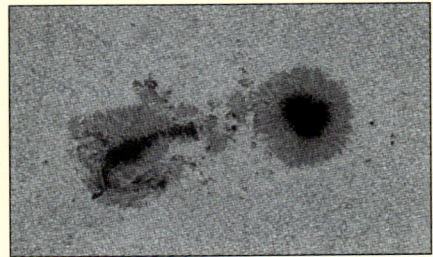
11.17

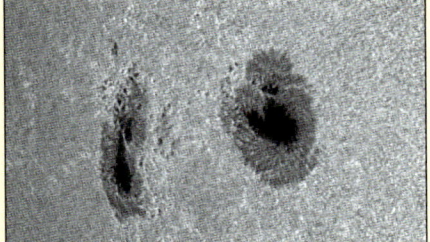
11.18

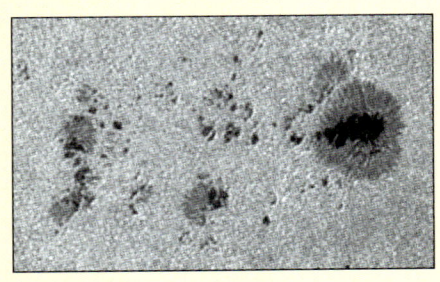
11.19

TABLE **11.5** Sunspot Classifications

Figure Number	First Letter	Second Letter	Third Letter
11.12			
11.13			
11.14			
11.15			
11.16			
11.17			
11.18			
11.19			

Exercise 11.4 Data Sheet

The Sun CHAPTER 11

Exercise 11.4 Data Sheet

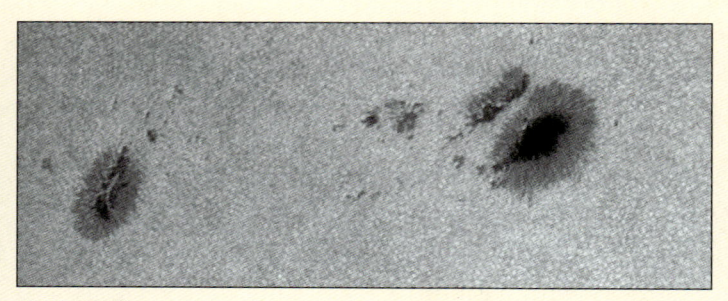
11.20

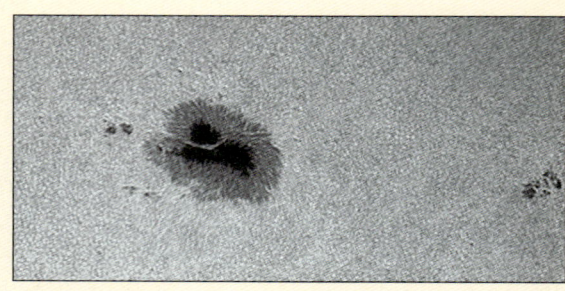
11.21

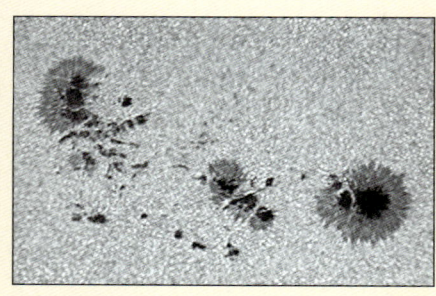
11.22

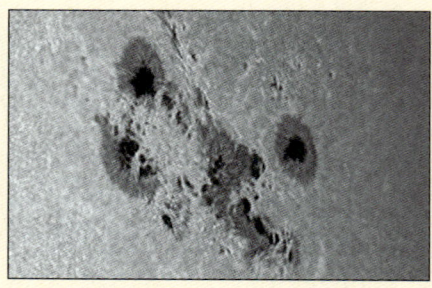
11.23

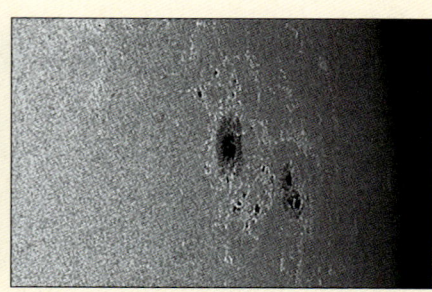
11.24

Images 11.20–11.27 courtesy of David Tyler

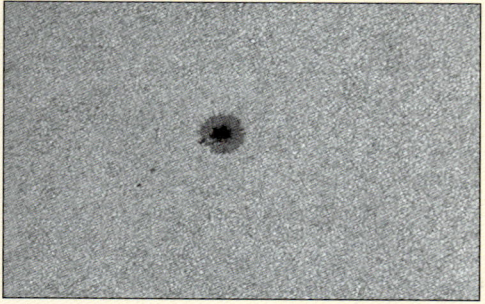
11.25

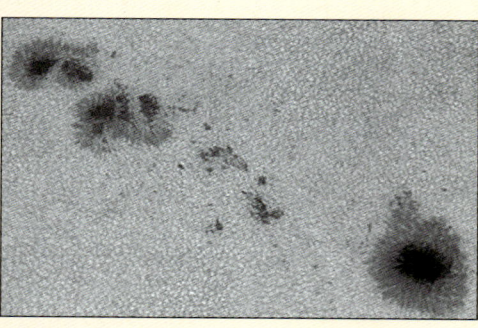
11.26

11.27

TABLE **11.6** Sunspot Classifications

Figure Number	First Letter	Second Letter	Third Letter
11.20			
11.21			
11.22			
11.23			
11.24			
11.25			
11.26			
11.27			

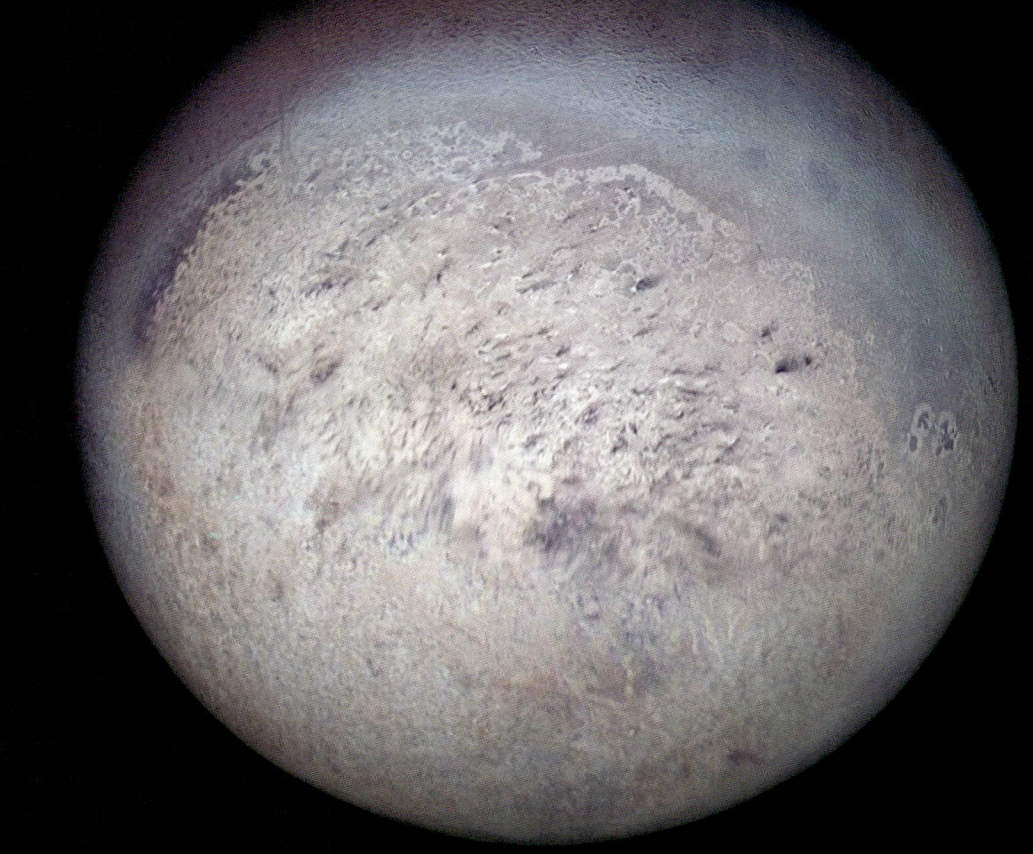

The Solar System
Physical Features

12

LEARNING OBJECTIVES
Upon completion of this chapter, you should be able to:

1. Discuss the scientific assumptions that supported the geocentric solar system model.
2. Explain the significance of the geocentric solar system model in the history of modern astronomy.
3. Differentiate between the geocentric and heliocentric solar system models.
4. Discuss the contributions of Copernicus, Brahe, Kepler, and Galileo to the heliocentric solar system model and reasons for our acceptance of the heliocentric solar system.
5. List the different types of objects found in our solar system.
6. Explain the International Astronomical Union's definition of a planet in our solar system.
7. Compare and contrast the physical characteristics of rocky and gas giant planets.
8. Compare and contrast a solar system and a stellar system.
9. Compare and contrast a planet and an exoplanet.
10. Define all glossary terms.

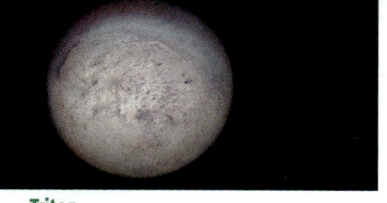

Triton Courtesy of NASA

In this lab, you will conduct an overview of our **solar system** by comparing various objects to one another. You also will simulate the atmospheres of the inner planets Venus and Mars, and of an outer planet such as Jupiter. Finally, you will sketch a planet or planets during class using as a guide a projected high-resolution image obtained by a spacecraft.

Our Solar System

Ever since humankind gazed toward the cosmos, people noticed that some objects seemed to wander independently from the background **stars**. Besides the obvious motions of the Sun and the Moon, observers noted five wanderers: Mercury, Venus, Mars, Jupiter, and Saturn. These naked-eye objects were fairly easy to follow. The fact that these objects appeared to move against the background stars provided the name *planete*, meaning "wanderer" in Greek. From that word evolved the English word **planet**.

The planets, the Sun, and all objects that orbit the Sun are known as the solar system. Astronomers also have discovered planets around other stars in our galaxy. A star and objects—such as planets and/or other stars and other materials—that orbit it are known as a **stellar system**. Planets outside our solar system are known as **exoplanets**.

Geocentric Model

As early astronomers and philosophers pondered the heavens, they considered why these wanderers moved as they did. Three of the planets—Mars, Jupiter, and Saturn—occasionally reversed course and move backward, known as **retrograde motion** (Fig. 12.1). It occurs when an outer planet is near **opposition**, a point in its orbit when it lies opposite the Sun as seen from Earth. At opposition, a planet rises at sunset, is visible all night, and sets at sunrise. At this time, Earth, moving faster, overtakes the planet and the planet appears to move backward.

> **GLOSSARY TERMS**
>
> **Solar system** the Sun and all the objects that orbit it
>
> **Star** self-luminous sphere of gas, many of which produce energy in their cores by nuclear fusion
>
> **Planete** refers to the bright planets Mercury, Venus, Mars, Jupiter, and Saturn; from the Greek, meaning "wanderer"
>
> **Planet** non-stellar, mostly spherical object that orbits a star
>
> **Stellar system** a star and the other objects (such as planets and other stars) that orbit it
>
> **Exoplanets** planets outside our solar system
>
> **Retrograde motion** false (east to west) apparent motion of a planet or other celestial object, farther from the Sun than Earth, on the celestial sphere as seen from Earth
>
> **Opposition** position of two celestial objects when their longitude (as seen from Earth) differs by 180°; when one of the objects is the Sun, opposition means the other object is opposite the Sun in the sky, and therefore visible all night long

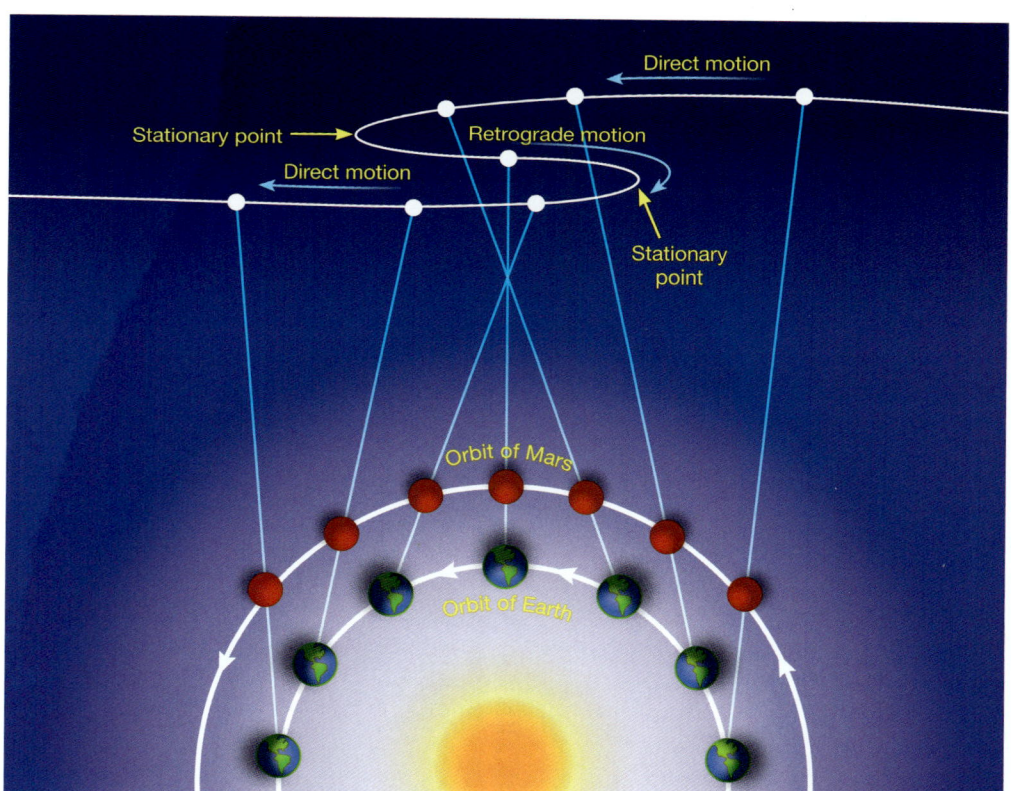

12.1 Example of retrograde motion

Courtesy of Holley Y. Bakich

The preferred solar system model during the second century was the **geocentric** model of Claudius Ptolemy (100 CE–170 CE), a Greek philosopher, astronomer, geographer, and mathematician who lived in Alexandria, Egypt. The geocentric solar system presumed that the nonmoving Earth was the center of the solar system, and the universe in general. In this model, the Moon, the Sun, and the planets moved around Earth, with the background stars much farther away on what early scientists described as a crystalline sphere.

Among the challenges Ptolemy and others faced with the geocentric solar system model was the fact that some planets went backward from time to time. After a period of this backward motion, the planet would again travel in its normal direction. Ptolemy developed a solution he called **epicycles**, which was a loop in a planet's orbit. In Greek, *epicycle* means "on the circle." Ptolemy thought that as a planet moved on a circular epicycle, it usually would be seen traveling in one direction. But there were times, however, that it seemed to reverse course and travel in the other direction. Although Ptolemy gets credit for this system, his Greek predecessors Hipparchus (190 BCE–120 BCE) and Apollonius (15 CE–100 CE) also developed and used it to explain planetary motions. The Ptolemaic solar system was the accepted model for 1,500 years.

> **GLOSSARY TERMS**
> **Geocentric** Earth-centered
> **Epicycle** loop in a planet's orbit; from the Greek, meaning "on the circle"

The geocentric model of our solar system places Earth near the center of the orbit of the Sun, Moon, and planets. As each body revolves around Earth tracing a large circle called the deferent, it also moves in a small circle called an epicycle. The eccentric (+) is the center of the epicycle. To keep the epicycle moving at the same speed, Ptolemy created another point he called the equant (E) (Fig. 12.2).

Today we understand why this background motion of the planets is seen. Each planet orbits the Sun at a different speed; the closer the planet is to the Sun, the faster it orbits. Earth catches up then passes planets farther from the Sun, giving the illusion that the planet is moving backward. The retrograde is due to the two bodies' orbital motions: Earth and the planet we observe from Earth.

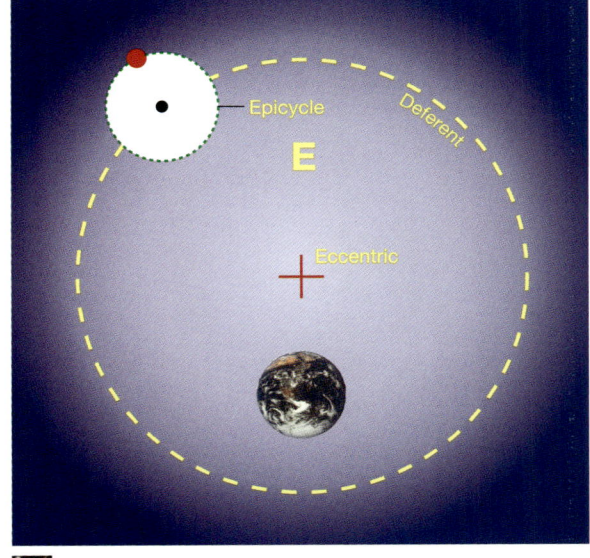

12.2 Geocentric model of our solar system with epicycles.
Courtesy of Holley Y. Bakich

Heliocentric Model

Nicolas Copernicus (1473–1543) was an astronomer and mathematician. Before 1514, Copernicus became interested in the geocentric solar system, but he thought that epicycles made the model too complex. He wrote a treatise on what he considered a better (simpler) model with the Sun at the center, a **heliocentric** solar system. Earth and planets revolve around the Sun, and the Moon orbits Earth. Copernicus continued to collect data and refine his model.

> **GLOSSARY TERM**
> **Heliocentric** Sun-centered

Although the heliocentric model had errors, such as the concept that all the planets moved in circular orbits, the theory was the basis for further work, data collection, and research by men such as Tycho Brahe (1546–1601), Johannes Kepler (1571–1630), Galileo Galilei (1564–1642), and Isaac Newton (1643–1727). Brahe made numerous measurements of celestial objects accurate to better than 0.01° until his death. Kepler attempted to get Brahe's data to fit the heliocentric model, but it didn't work because Copernicus developed his model with circular planetary orbits, and Brahe's data was of the actual observations of elliptical orbits.

Galileo provided the observational evidence confirming the heliocentric model. He made observations of Venus and its phases and discovered four moons in orbit around Jupiter.

> **Looking Up**
> Around 1532 CE Copernicus wrote his definitive work *De revolutionibus orbium coelestium* (*On the Revolutions of the Celestial Spheres*). Because he feared public scorn and criticism, Copernicus held off final publication and distribution of his work until just prior to his death. The publication of *De revolutionibus* marked the beginning of what scientists call the Copernican Revolution and is considered the start of a new beginning in astronomy.

Check Your Understanding

1.1 What is the difference between the geocentric and heliocentric solar system models?

1.2 Who are some of the individuals who contributed to the geocentric model of the solar system, and what did they contribute?

1.3 Who are some of the individuals who contributed to the heliocentric model of the solar system, and what did they contribute?

1.4 Why do we observe retrograde motion?

Classification of Objects in our Solar System

As you examine the solar system and its members, it becomes obvious that it contains several types of bodies. You can start with the center of our solar system, the Sun, and move outward from that vantage point.

Major and Minor Bodies

Next to the Sun, planets are the largest objects in the solar system. Planets are visible because they reflect light.

The International Astronomical Union (IAU) ruled in August 2006 that to be a planet an object must meet three criteria:

1. It must have enough mass for gravity to give it a spherical shape
2. It must orbit its parent star
3. It must reign supreme in its own orbit, having "cleared the neighborhood" of other competing bodies, which means no other planets or asteroids remain in its orbit

These rules apply only to objects in the solar system.

Because of the IAU's acceptance of this definition of a planet, a new category of objects came into being. These objects are called **dwarf planets.** Dwarf planets have enough mass to have a spherical shape and they also orbit the Sun, but they haven't cleared their orbital paths of other objects. Since the inauguration of IAU's new definition, Pluto has been reclassified as a dwarf planet.

Also present in our solar system are minor bodies such as **asteroids** and **comets**. Asteroids are relatively small, rocky or metallic objects. Comets are relatively small, icy objects. Finally transneptunian objects orbit the Sun at a distance greater than Neptune. These include Kuiper Belt objects and bodies in the Oort Cloud, a sphere of comets with compositions much like the Kuiper Belt objects. Scientists consider the outer edge of the Oort Cloud the boundary of the solar system. It is located one-third of the way to the next nearest star.

Other objects include **satellites**, planetary rings, dust, and gas. The various types and characteristics of planetary moons is a study in itself. And not all objects fit neatly in these classifications; for example there are objects that share characteristics of both asteroids and comets.

> **GLOSSARY TERMS**
>
> **Dwarf planet** object massive enough for gravity to control its shape and that directly orbits the Sun; however, it has not cleared its orbital path of other objects
>
> **Asteroid** relatively small, rocky or metallic object *usually* orbiting a star
>
> **Comet** relatively small, icy object *usually* orbiting a star
>
> **Satellite** another name for "moon"
>
> **Rocky planet** one with a solid surface and few to no moons; also referred to as the terrestrial (or earthlike) planets in our solar system
>
> **Gas giant** planet primarily made up of gas with a small metallic core; also referred to as a Jovian (or Jupiter-like) planet

Planetary Types

When examining the planets in our solar system, we find a variety of characteristics. Scientists divide planets into two basic categories: **rocky planet** and **gas giant** (Table 12.1). Some include dwarf planets as a third classification.

Rocky planets, sometimes referred to as the terrestrial (or earthlike) planets, have a solid surface and smaller size when compared to the gas giants, and few to no moons. The gas giants, also referred to as the Jovian (or Jupiter-like) planets, have no solid planetary body (but they may contain a small metallic core), rings, and many moons.

TABLE **12.1** Planet Characteristics

Type	Planets	Diameters	Surface	Craters?	Rings?	Rotation	Period of Revolution	Satellites?
Rocky	Mercury, Venus, Earth, Mars	Small, from 4,880 km to 12,742 km	Solid	Yes	No	1 to 243 days	Short	None to 2
Gas giants	Jupiter, Saturn, Uranus, Neptune	Large, from 49,528 km to 142,984 km	Gaseous, with small cores	No	Yes	9.925 to 17.24 hours	Long	Many satellites

> **GLOSSARY TERMS**
>
> **Rotation** spinning of an object on its axis, often referred to as a "day"; Earth's complete rotation takes 24 hours
>
> **Revolution** orbit of one object around another, often referred to as a "year"; Earth's complete orbit around the Sun takes 365 days
>
> **Albedo** measure of reflected sunlight; albedo is Latin for "white"

All the gas giant planets rotate faster than any rocky planet. To understand why, think of spinning gas versus moving a solid. And only Earth and Mars show a somewhat-regular **rotation** period. Mercury's rotation is long, and Venus rotates backward, perhaps due to a tremendous impact from some other object long ago.

The rocky planets are all closer to the Sun than the gas giants. So, the rocky planets are warmer and their periods of **revolution** (that is, their years) are shorter than those of the gas giants.

Some of the following features and characteristics are visible when observing the planets through a telescope:

- Mercury appears to go through phases like the Moon (because Mercury is closer to the Sun than Earth).
- Venus also appears to go through phases; occasionally markings are visible.
- Features visible on Mars are polar caps, dust storms, gross surface features, and the effects of rotation.
- Things to look for when observing Jupiter are cloud tops, fast rotation, and its four large satellites (Io, Callisto, Europa, and Ganymede).
- Saturn also has visible cloud tops, several satellites, and fast rotation.
- Uranus is characterized by its blue-green color, and Neptune is characterized by its blue color.

How the planets appear to observers on Earth depends on several factors, such as the distance to the planet, its size, and how much light it reflects. **Albedo** is a measure of reflected sunlight. The word *albedo* is Latin for "white." The albedo range is 0 for a non-reflective material to 1.0 for an object that reflects all of the light that strikes it. For some comparisons, charcoal has an albedo of approximately 0.04, whereas snow has an albedo of approximately 0.9. One of the brightest objects in the solar system is Saturn's moon Enceladus, with an albedo of 0.99. Jupiter's satellite Europa exhibits an albedo of 0.67. See Table 12.2 for the albedo of major bodies in the solar system.

TABLE **12.2** Albedo for Planets, the Moon, and the Dwarf Planet Pluto

Object	Albedo	Object	Albedo
Mercury	0.106	Jupiter	0.52
Venus	0.65	Saturn	0.47
Earth	0.367	Uranus	0.51
The Moon	0.12	Neptune	0.41
Mars	0.25	Pluto	0.49 to 0.66

Check Your Understanding

2.1 What are the different types of objects found in our solar system? Give a brief definition of each.

2.2 What is the difference between a star and a planet?

2.3 Explain the International Astronomical Union's rule on what is a planet in our solar system.

2.4 In 2006, the International Astronomical Union (a collection of professional astronomers) decided that Pluto should no longer be a planet. Do you think this was a good decision? Why or why not?

2.5 Several planets have been found revolving around the star Gliese 581. Why might astronomers not call Gliese 581 and its planets a solar system?

2.6 What are the two types of planets found in our solar system?

2.7 List some of the characteristics of each type of planet.

2.8 If a planet has no visible surface, 12 moons, and three broken rings, how would you classify it? Why?

2.9 Why do the gas giants have longer revolutions than the rocky planets?

2.10 What are the observational characteristics of the planets that you might see through a small telescope?

2.11 What is albedo and how does it have an effect on our Earth-based view of objects in the solar system?

2.12 Which planet has the highest albedo? Which has the lowest?

2.13 The Moon has a low albedo, but it appears bright. Why?

Comparative Planetology

Exercise 12.1

 Procedure

Materials
- Figures 12.3 through 12.34 or photographic slides of the same provided by your instructor

1. Examine Figures 12.3 through 12.15 or photographic slides projected in class by your instructor. They show the major bodies of the solar system and a few minor bodies:

- The Sun (Fig. 12.3)
- Mercury (Fig. 12.4)
- Venus (Fig. 12.5)
- Earth (Fig. 12.6)
- The Moon (Fig. 12.7)
- Mars (Fig. 12.8)
- Jupiter (Fig. 12.9)
- Saturn (Fig. 12.10)
- Uranus (Fig. 12.11)
- Neptune (Fig. 12.12)
- Pluto (Fig. 12.13)
- Comet (Fig. 12.14)
- Asteroid (Fig. 12.15)

12.3 Sun. *Courtesy of NASA*

12.4 Mercury. *Courtesy of NASA*

12.5 Venus. *Courtesy of NASA*

12.6 Earth. *Courtesy of NASA*

12.7 Moon. *Courtesy of NASA*

12.8 Mars. *Courtesy of NASA*

Physical Features CHAPTER 12 239

12.9 Jupiter. *Courtesy of NASA*

12.10 Saturn. *Courtesy of NASA*

12.11 Uranus. *Courtesy of NASA*

12.12 Neptune. *Courtesy of NASA*

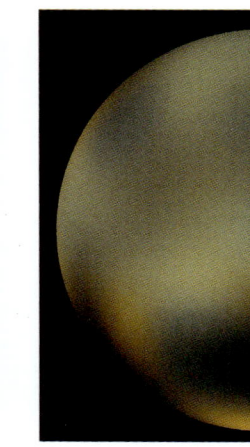
12.13 Pluto. *Courtesy of NASA*

12.14 Comet. *Courtesy of NASA*

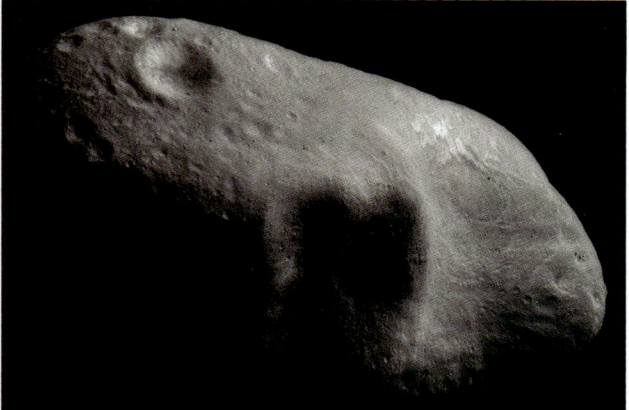
12.15 Asteroid. *Courtesy of NASA*

2 Complete Table 12.3 with characteristics you find for each object. In the second column of the table, list the type of each object shown (in terms of its composition, either gaseous or rocky). In the third column, list any distinctive characteristics you see such as (but not limited to) overall cloud or surface features, coloration, the presence of ice, impact craters, rings, etc.

TABLE 12.3 Characteristics of Planets and Other Objects

Object	Object Type	Characteristic(s) Found in the Image
The Sun		
Mercury		
Venus		
Earth		
The Moon		
Mars		
Jupiter		
Saturn		
Uranus		
Neptune		
Pluto		
Comet		
Asteroid		

Physical Features CHAPTER 12

3 Which objects had similar characteristics? List the object and the characteristic in Table 12.4.

TABLE **12.4** Similar Characteristic Comparison

Object	Characteristic

4 In examining the images of Earth (Fig. 12.6) and Mars (Fig. 12.8), note the polar caps of each planet. Describe the visual differences and similarities between each planet's polar caps.

_____

_____

5 The gas giants each have some sort of ring or ring system. Using Figures 12.16 through 12.19, compare and contrast the rings of Jupiter (Fig. 12.16), Saturn (Fig. 12.17), Uranus (Fig. 12.18), and Neptune (Fig. 12.19). Complete Table 12.5 with ring characteristics for the gas giant planets.

12.16 Rings of Jupiter taken by the Voyager 2 spacecraft after it passed the giant planet and was looking back at the backlit rings and planet's edge.
Courtesy of NASA

12.17 Rings of Saturn taken by the Voyager 2 spacecraft.
Courtesy of NASA

12.18 Rings of Uranus, which are thin and do not reflect much sunlight. These rings are best seen from a position on Earth.
Courtesy of NASA

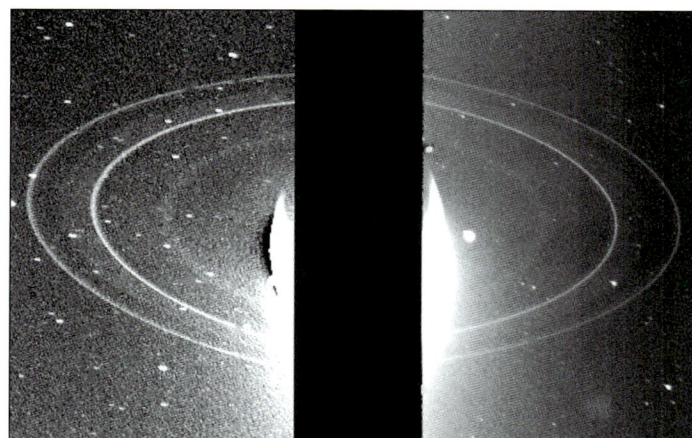

12.19 Rings of Neptune, which are thin and do not reflect much sunlight. These rings are best seen from a position on Earth.
Courtesy of NASA

TABLE **12.5** Ring Description Chart

Planet	Ring Description
Jupiter	
Saturn	
Uranus	
Neptune	

a From your examination of the gas giant rings, what do you find similar?

b What do you find different for each planet's ring structures?

Physical Features CHAPTER 12   243

6 Examine Figures 12.20 through 12.29, which show the following satellites:

- The Moon (Fig. 12.20)
- Deimos (Fig. 12.21)
- Phobos (Fig. 12.22)
- Ganymede (Fig. 12.23)
- Callisto (Fig. 12.24)
- Dione (Fig. 12.25)
- Mimas (Fig. 12.26)
- Rhea (Fig. 12.27)
- Titania (Fig. 12.28)
- Triton (Fig. 12.29)

12.20 Moon. *Courtesy of NASA*

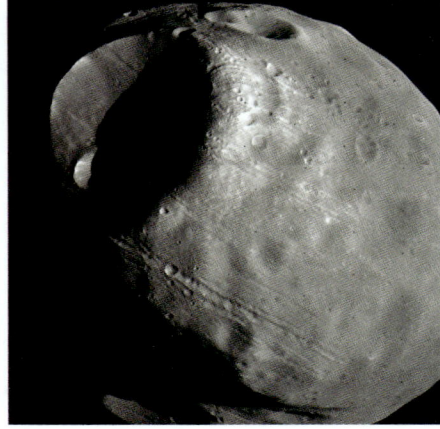
12.21 Deimos. *Courtesy of NASA*

12.22 Phobos. *Courtesy of NASA*

12.23 Ganymede. *Courtesy of NASA*

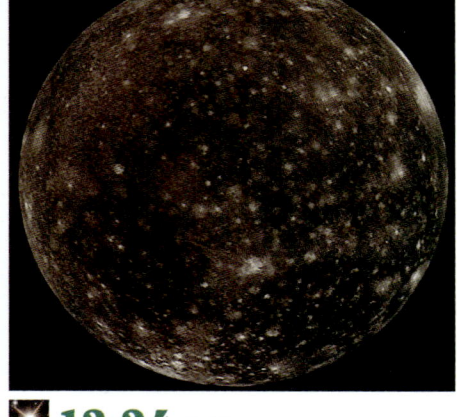
12.24 Callisto. *Courtesy of NASA*

12.25 Dione. *Courtesy of NASA*

12.26 Mimas. *Courtesy of NASA*

12.27 Rhea. *Courtesy of NASA*

Unit 4 *The Solar System*

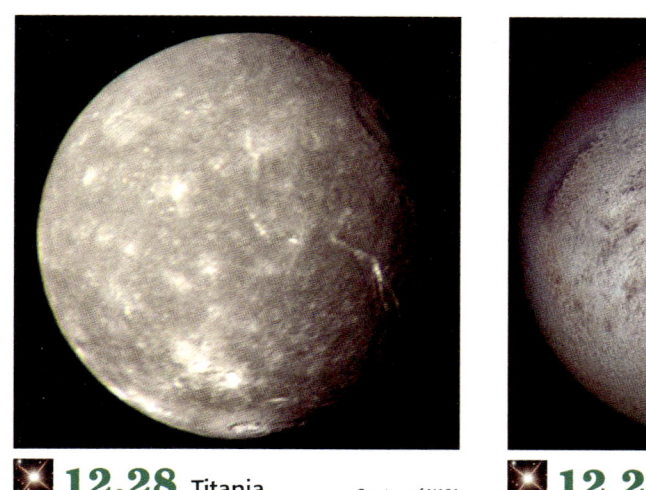
12.28 Titania. Courtesy of NASA

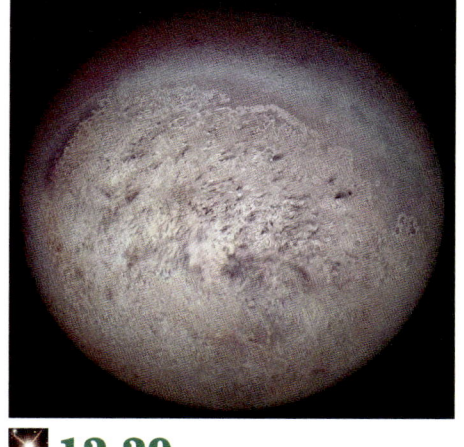
12.29 Triton. Courtesy of NASA

7 Complete Table 12.6 with characteristics for each satellite.

TABLE 12.6 Satellite Comparison, Part I

Object	Planet	Shape	Other Characteristics
The Moon			
Deimos			
Phobos			
Ganymede			
Callisto			
Dione			
Mimas			
Rhea			
Titania			
Triton			

a From your examination, what similarities did you find?

b What differences?

c Why do you think several of these satellites are not spherical in shape?

8 Examine Figures 12.30 through 12.34, which show images of other satellites:

- Io (Fig. 12.30)
- Europa (Fig. 12.31)
- Enceladus (Fig. 12.32)
- Titan (Fig. 12.33)
- Miranda (Fig. 12.34)

12.30 Io. *Courtesy of NASA*

12.31 Europa. *Courtesy of NASA*

12.32 Enceladus. *Courtesy of NASA*

12.33 Titan. *Courtesy of NASA*

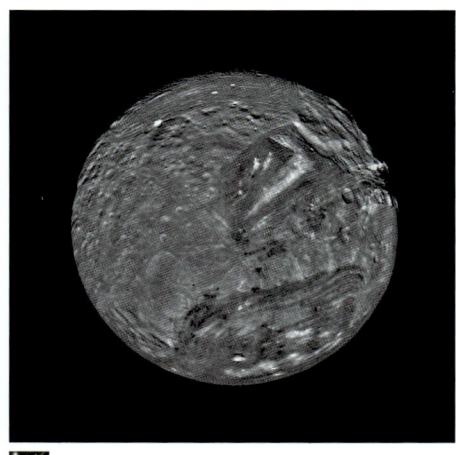

12.34 Miranda. *Courtesy of NASA*

9 Complete Table 12.7 with characteristics for each satellite.

TABLE **12.7** Satellite Comparison, Part II

Object	Planet	Shape	Other Characteristics
Io			
Europa			
Enceladus			
Titan			
Miranda			

a These satellites have some characteristics that are different from other satellites, and from each other. From your examination, what similarities did you find?

b What differences?

Exercise 12.2 Simulating Planetary Atmospheres

Planetary atmospheres are a major characteristic of each planet. Mercury has an almost nonexistent atmosphere, whereas the gas giants are just that: giant balls of gas (some of which exists in a liquid or solid state deep within the planet because of the tremendous pressures at such depths). In this procedure, you will use in-class materials to "create" a planet's atmosphere—first of a small planet such as Venus or Mars and then of a giant planet like Jupiter.

Procedure

Research

1. Research the atmospheres and complete the information for each of the planets and Saturn's satellite Titan, listed in Table 12.8.

Materials
- ❏ 10-gallon aquarium
- ❏ Dry ice
- ❏ Protective gloves
- ❏ Goggles
- ❏ Rubber mallet
- ❏ 500 ml boiling flask or small stoppered bottles, such as a clear gas generating jar, for individual or group work
- ❏ Rubber stopper, solid
- ❏ Rheoscopic fluid (available under the name Pearl Swirl Fluid)
- ❏ Food coloring (blue works best)
- ❏ Candle, match, or other flame source

Table **12.8** Atmosphere Comparison

Planet	Planet Type	Atmosphere Composition
Mercury		
Venus		
Earth		
Mars		
Jupiter		
Saturn		
Titan		
Uranus		
Neptune		

a How do the atmospheres of rocky planets compare with the gas giant planets?

b The satellite Titan has an interesting atmospheric composition. How does it compare with rocky and gas giant planets? Does that surprise you, considering its location in the solar system? Explain why.

Rocky Planet Atmospheres

1. To simulate the carbon dioxide (CO_2) atmospheres of Venus and Mars, you will use the 10-gallon aquarium, rubber mallet, protective gloves, goggles, and dry ice.

2. Crush the dry ice using the rubber mallet. This works best if you place the dry ice in a brown paper bag on a firm surface, and then strike it.

3. Carefully place the crushed dry ice in the bottom of the aquarium.

4. As the dry ice **sublimates**—going from a solid to a gas without passing through the liquid state—it forms a fog-like layer at the bottom of the aquarium. Observe the CO_2 layer, and answer the following questions:

    a Why does the carbon dioxide gas stay at the bottom of the tank?

    b What does the nature of carbon dioxide tell you about the atmosphere of Venus in particular?

5. Using a match, candle, or other flame source as directed by your instructor, place the flame into the aquarium's carbon dioxide atmosphere.

    a Record what happens.

    b Why do you think this reaction occurs with the flame?

CAUTION

USE CAUTION when handling dry ice; it is solid carbon dioxide at a temperature of about −70°C. In the gas state, it is odorless and can be deadly in large quantities.

GLOSSARY TERM

Sublimate to go from a solid state of matter to a gas without passing through the liquid state

Gas Giant Atmospheres

1 Obtain the boiling flask or bottle, stopper, rheoscopic fluid, and food coloring as directed by your instructor.

2 Add water to the flask until 90 percent full.

3 Add 10 ml of rheoscopic fluid and two or three drops of food coloring. Securely place the stopper on the flask.

4 Shake the flask (an easy to moderate agitation is all that is required) to mix the water, rheoscopic fluid, and food coloring.

5 Next, swirling the flask, note the swirls that develop within the flask.

    a Describe what the liquid within the flask tends to do.

    b What happens as the rotating liquid slows?

    c Looking at photos of the gas giants, Jupiter in particular, how does the rotating liquid in the stoppered boiling flask or bottle model the planet's cloud tops as the planet rotates?

    d You can draw parallels between the flask-liquid model and gas giant planets. How does a gas giant's rotation speed affect the atmosphere?

Exercise 12.3 In-Lab Planet Sketching

 Procedure

After completing Exercises 12.1 and 12.2 and noting any additional instructions, you will sketch the Mars image provided by your instructor and perform the activities on the data sheet, pages 251–252.

Materials

❏ Sketching pencils
❏ Notes from Exercises 12.1 and 12.2
❏ Planet image (projected in class)

Name _____ Section _____ Date _____

Planet Sketching

1. In the first circle on the observation form, page 252, use a pencil to draw the projected planet, trying to be as accurate as possible. Use the pencil's point for details that appear sharp; dull or round the point to show shadings.

2. For intensity estimates, roughly draw in the second circle the "outlines" of the planet's prominent features.

3. Next estimate the "intensities" by using a 1–10 system, with 1 being the blackest of black and 10 being the whitest of white. Make 10 intensity estimates.

> **Note**
> Most planet observation forms include two circles for your observations. One is for your sketch, showing details you observed. The second is for your intensity estimates.

4. Most observation forms request information about the time, date, telescope, place where the observation was made, and observing conditions, among other data. For this in-class exercise, however, only your name (observer), the date, and the location will be necessary.

5. After completing the planetary drawing and intensity estimates, answer the following questions as a team:

   a. Compare your planetary sketch with the sketches of others in your team. What are similarities? What are differences?

> **Note**
> Most observers copy from their sketch, rather than re-drawing at the telescope.

   b. Decide among your team which drawing is the best. What was different about that member's work compared to others on the team? What made it the best?

   c. Using the proportionally scaled measurements of a martian feature provided by your instructor, compare the data to each team member's drawing. Quantify your results.

6. Answer the following questions individually.

   a. Think about observing these objects, or any object, at the telescope. What would be some of the challenges and difficulties?

Exercise 12.3 Data Sheet

Physical Features CHAPTER 12   251

b Which object—a planet or a deep sky object (star cluster, nebula, or galaxy)—would be the most difficult for you to draw? Why?

_____

_____

_____

c How can you apply what you learned about sketching to your future observing with a telescope?

_____

_____

_____

In-Lab Sketch

Object _____ Intensity Estimates _____
[1 = Black; 10 = Brilliant White]

Observer _____ Date ____ / ____ /20 ____

Site _____ Start time _____ End time _____

The Solar System
Kepler's Laws of Motion

13

LEARNING OBJECTIVES
Upon completion of this chapter, you should be able to:

1. State Kepler's laws in your own words.
2. Apply Kepler's laws to describe the motion of objects orbiting the Sun.
3. Calculate the eccentricity of any elliptical orbit given the semimajor and semiminor axes.
4. Compute an objects perihelion and aphelion given its orbital eccentricity and semimajor axis.
5. Determine an object's orbital speed given its period, semimajor axis, and the distance from the body it is orbiting.
6. Deduce Kepler's constant for an object orbiting the Sun.
7. Draw ellipses of various eccentricities using different foci.
8. Define all glossary terms.

Kepler's Nova
Courtesy of NASA/JPL-Caltech/ O. Krause (Steward Observatory)

Johannes Kepler's Astronomical Legacy

By studying the three "laws" of the seventeenth-century German astronomer Johannes Kepler (1571–1630), you will learn the shape of the planets' **orbits**, why planets nearer the Sun must move faster, and the relationship between a planet's distance and its orbital speed.

Johannes Kepler (Fig. 13.1) was one of the key figures of the Scientific Revolution. Although Kepler is not as well-known as his contemporary, Galileo, or the reclusive genius, Isaac Newton, who built on Kepler's work, he was a major player in astronomy's history and set the stage for many subsequent important discoveries.

> **GLOSSARY TERM**
>
> **Orbit** path of a celestial object through space as influenced by the gravity of some primary body or bodies

Great Discoveries

By describing the process of human vision, Kepler wrote the first modern book of optics, *Astronomiae Pars Optica*. He discovered the inverse square law of light, the foundation for all modern stellar and galactic distances. His thoughts on the shapes of snowflakes resulted in the first book on crystallography, *De nive sexangula*.

Following Galileo's telescopic discoveries in 1610, Kepler described how the revolutionary new instrument worked. He introduced a few improvements, including one that increased the field of view. His design came into wide use by the middle of the seventeenth century, though in typical Kepler fashion, his name is rarely associated with it. Instead, the design is simply known as the astronomical telescope.

Today, Kepler is primarily known for his three planetary laws, which he derived after a Herculean battle with the data (and often the person) of the great Danish naked-eye observer, Tycho Brahe. Kepler made the first links between physics and astronomy by establishing the idea that the planets had to move due to unseen forces. He thus founded celestial mechanics, and opened the door for Newton's work on gravitation.

13.1 Johannes Kepler.
Courtesy of Wikimedia Commons

Personal History

Kepler was born December 27, 1571. An early bout with smallpox left him with defective vision, believed to be one reason he eventually pursued theoretical rather than observational astronomy.

Although Kepler was a precocious student, he learned to focus his great ability. His first job was as a math teacher in Graz, Austria. It was here that he embarked on his life-long quest for harmony. In a flight of mathematical mysticism, he tried to link the known planets to Plato's five geometrical solids (the cube, tetrahedron, octahedron, icosahedron, and dodecahedron). By nesting each geometrical solid between the orbits of the planets, he thought he had the basis for the number of the planets. Kepler was so pleased with this idea that he compiled it as a "theory of everything" in a book titled *Cosmographicum Mysterium* (*The Cosmic Mystery*).

The book impressed one of the most important astronomers of the day, Tycho Brahe, and the two men began what quickly became a tempestuous collaboration. Brahe had accumulated 20 years of precise observations and needed someone with Kepler's mathematical prowess to make sense of it. Often biting off more than he could chew, Kepler promised he could solve the orbit of Mars within eight days. Mars was particularly difficult, because the observations varied from Brahe's calculations more than any of the planets. Instead of eight days, the task took nearly eight years and more than 900 pages of calculations before Kepler finished.

To harmonize observations with calculations, Kepler first determined Earth's motion as a planet. Consider the motion of an airplane circling an airport as seen from another airplane, and you have some sense of the enormity of

Looking Up

The Kepler Museum is in Weil der Stadt, a small medieval town west of Stuttgart on the edge of the Black Forest in southern Germany. The museum is located in the house that was the birthplace of the genius mathematician and astronomer and where he spent his preteen years. The museum staff tells visitors about Kepler's life, and numerous artifacts and well-preserved copies of all of Kepler's books are on display.

Courtesy of David J. Eicher/ *Astronomy* magazine

his task. "I have spent so much pains on it," Kepler wrote a friend, "that I could have died 10 times."

By insisting that his theoretical orbits agree within the errors of Brahe's data, Kepler created one of the linchpins of the scientific method. In the process, he gave the world two of his three laws of planetary motion: planets orbit the Sun in **ellipses**, and a line drawn from the Sun to the planet sweeps out equal areas in equal times.

Kepler's Nova

In October 1604 a brilliant new star—what astronomers now call a nova (Fig. 13.2)—blazed forth in the evening sky. Kepler observed the star and wrote about it in his 1606 work, *De Stella Nova*. As he observed the star, he noted it did not show any **parallax**.

Parallax is an apparent shift seen as an object is viewed from two locations. A nearby object shows a large parallax, while a distant object shows a small parallax (or none, if you are observing by naked eye the way Kepler did). He concluded the object lay within the sphere of the fixed stars.

This conclusion put Kepler at odds with the church; how could such a phenomenon occur if the heavens were immutable? Such a "new star" showed the starry sphere does, indeed, change.

13.2 Supernova remnant from the "new star" Kepler observed in 1604.

Courtesy of NASA/JPL-Caltech/O. Krause (Steward Observatory)

Triumph Through Adversity

In 1618 Kepler discovered the third of his planetary laws while revising his book, *The Harmony of the Worlds*. He considered this law—that the cube of a planet's distance is proportional to the square of its **orbital period**—one of his greatest achievements. This law tells astronomers a great deal about how the motions of the planets relate to each other, including which planets orbit quickly and which orbit slowly.

In 1627 he completed *The Rudolphine Tables*, a culmination of his life's work, by combining his planetary laws with Brahe's data. The tables remained astronomy's most accurate resource for decades. Astronomers used them to predict the first observed transits of Mercury and Venus in 1631 and 1639, respectively. Successful observations of these events helped solidify acceptance of the heliocentric model and Kepler's laws.

Kepler was the first scientist to freely advocate sharing data and wanted nothing more than to devote his life to philosophical speculation. But time and again he was forced to flee religious turmoil brought on by the Counter-Reformation or to seek secure employment. He died of a fever November 15, 1630. Though his grave was lost in the Thirty Years War, his self-written epitaph remains:

> *I measured the skies, now the shadows I measure,*
> *Skybound was the mind, earthbound the body rests.*

> **GLOSSARY TERMS**
>
> **Ellipse** type of conic section with an eccentricity less than one (a circle is an ellipse with an eccentricity of zero)
>
> **Parallax** apparent angular displacement of a star or other celestial object that results from the revolution of Earth around the Sun; numerically, this is the angle subtended by one astronomical unit at the distance of the particular object; this differs from the solar parallax, which is the apparent displacement of the Sun as seen from two (generally widely separated) places on Earth
>
> **Orbital period** regarding a celestial object, the time interval between two successive, similar events

Looking Up

Kepler-186f: The First Planet in a Star's Habitable Zone

On March 19, 2014, scientists analyzing data from NASA's Kepler Space Telescope announced that they had discovered the first Earth-size planet orbiting within a star's "habitable zone." Being in a habitable zone does not guarantee that a planet is habitable, however. The temperature on the planet depends strongly on its atmosphere. Kepler-186f orbits a star designated Kepler-186 in a system located about 500 light-years from Earth in the constellation Cygnus the Swan. Kepler-186f orbits its star once every 130 days and receives one-third the energy from its star that Earth gets from the Sun, placing it nearer the outer edge of the habitable zone. On Kepler-186f's surface, the brightness of its star at high noon is only as bright as our Sun appears to us about an hour before sunset.

Kepler's Laws

Kepler's laws of planetary motion can be summarized as follows:

- First law—the planets travel around the Sun in elliptical orbits.
- Second law—as a planet orbits the Sun, it sweeps out equal areas of its ellipse in equal periods of time.
- Third law—a relationship exists between the planet's period and its distance from the Sun.

First Law

Kepler's first law of planetary motion states that one body orbiting a second body, such as Earth orbiting the Sun, travels in an ellipse (Fig. 13.3). This was contrary to the Copernican idea that orbits were perfect circles. An ellipse is a two-dimensional shape for which the sum of the lengths of a line drawn from one focus to a point on the ellipse and another line drawn from the other focus to that same point remains constant for every point along the ellipse. (A circle is one type of ellipse. It has both foci at the same point, that is, the center of the circle.)

When scientists describe an ellipse, they usually refer to the ellipse's **eccentricity** (e). In the simplest sense, eccentricity describes how stretched out (less circular and more flattened) an orbit is. An orbit whose eccentricity equals 0 is circular. Eccentricities can approach, but never reach 1. So, for example, an orbit with an eccentricity of 0.09 is close to circular, while an orbit with an eccentricity of 0.99 is extremely stretched out. Mathematically, we can represent eccentricity as:

$$e = \frac{c}{a} = \frac{a - b}{a}$$

Where:
- e = the eccentricity
- a = the **semimajor axis**
- b = the **semiminor axis**
- c = the distance from the orbit's center to one of the two foci

From this definition of eccentricity the closest orbiting point of, for example, a planet to the Sun, called its **perihelion** (R_p), is:

$$R_p = a - c = a - ae = a(1 - e)$$

GLOSSARY TERMS

Eccentricity measurement (from 0 to 1) that is the amount in which the orbit of any solar system object is not circular; an object in a circular orbit would have an eccentricity of 0; mathematically, this is defined as the distance between the focal points of an ellipse divided by twice the length of the major axis

Semimajor axis one-half the greatest distance across an ellipse; the distance from the center to the edge through one of the foci; for a circle, this distance equals the radius

Semiminor axis distance from the center of an ellipse to the edge perpendicular to the line connecting the foci

Perihelion position of an object in solar orbit when it is closest to the Sun; the instant in a given orbit of a planet (or other body) when it is closest to the Sun

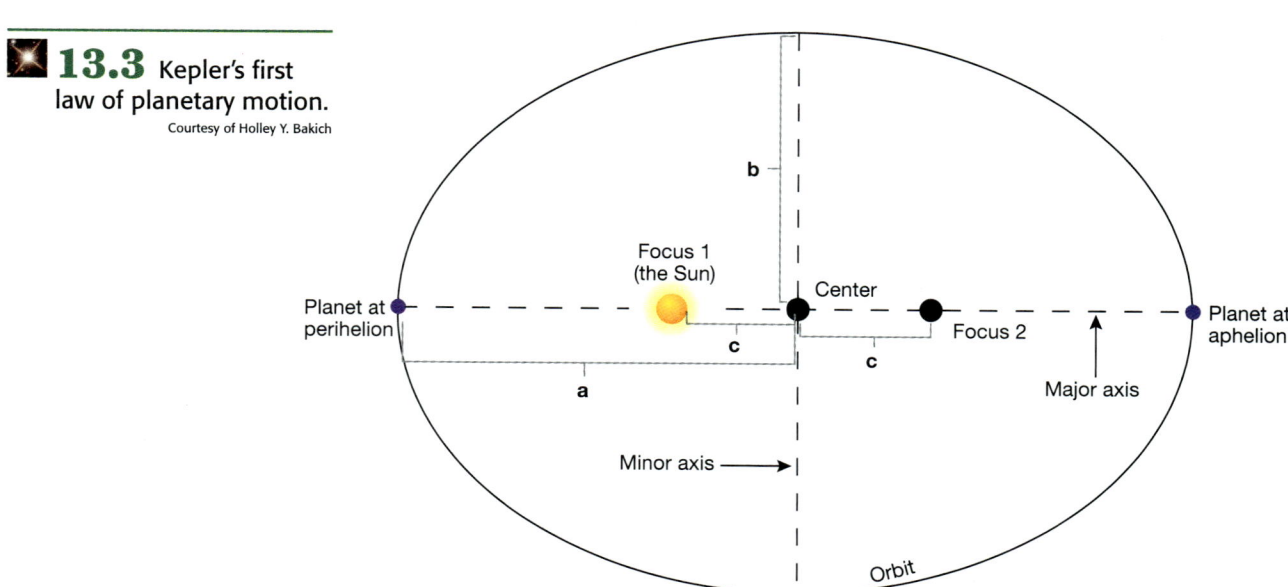

13.3 Kepler's first law of planetary motion.
Courtesy of Holley Y. Bakich

The farthest orbiting point of a planet to the Sun, called its **aphelion** (R_a), is:

$$R_a = a + c = a + ae = a(1 + e)$$

For example: On August 12, 1994, the Moon was at **perigee** (its closest distance to Earth). Its actual distance was 369,453 km. On August 27, 1994, the Moon was at **apogee** (its farthest distance from Earth), when its actual distance was 404,332 km.

The orbital eccentricity of the Moon equals approximately 0.055, and the Moon's semimajor axis is approximately 384,400 kilometers. Use the formulas above to calculate the values of perigee and apogee on those dates.

$$R_p = a(1 - e) = (384{,}400 \text{ km})(1 - 0.055) = 363{,}000 \text{ km}$$
$$R_a = a(1 + e) = (384{,}400 \text{ km})(1 + 0.055) = 406{,}000 \text{ km}$$

Both of these calculated values are within 2 percent of the actual values.

Second Law

Kepler's second law of planetary motion is often referred to as the law of equal areas. It defines the speed of the orbiting body (Fig. 13.4).

Refer to the areas shown in Figure 13.4 as triangles. As the orbiting body moves, it sweeps out a triangle in a specific amount of time (the time between t_1 and t_2) as shown in area 1. The sides are short but the edge is long. In that same period of time (the time between t_3 and t_4) later in its orbit, the body sweeps out area 2. Because the orbiting body is farther from the Sun, the edge will be shorter. But the sides are longer.

Kepler's second law states that these two areas—defined by the same amount of time and referred to as triangles 1 and 2—are equal. For this to be so, the distance traveled by the orbiting body when closer to the Sun must be longer than the distance traveled when farther from the Sun. Because the amount of time is equal, the orbiting body must be traveling faster when closer to the Sun.

If this orbit were circular, the body's speed would be the same at all points in its orbit. Because the orbit is elliptical, the body moves fastest when at perihelion and slowest at aphelion.

Kepler's second law deals with the speed of the orbiting object. You can calculate the speed of an orbiting object at any point in its orbit using the equation:

$$v = \sqrt{\left(\frac{4\pi^2 \cdot a^3}{p^3}\right)\left(\frac{2}{r} - \frac{1}{a}\right)}$$

> **GLOSSARY TERMS**
>
> **Aphelion** position of an object in solar orbit when it is farthest from the Sun; the instant in a given orbit of a planet (or other body) when it is farthest from the Sun
>
> **Perigee** position of the Moon or an object in Earth orbit when it is closest to Earth; the instant in a given orbit of the Moon (or other object) when it is closest to Earth
>
> **Apogee** position of the Moon or an object in Earth orbit when it is farthest from Earth; the instant in a given orbit of the Moon (or other object) when it is farthest from Earth

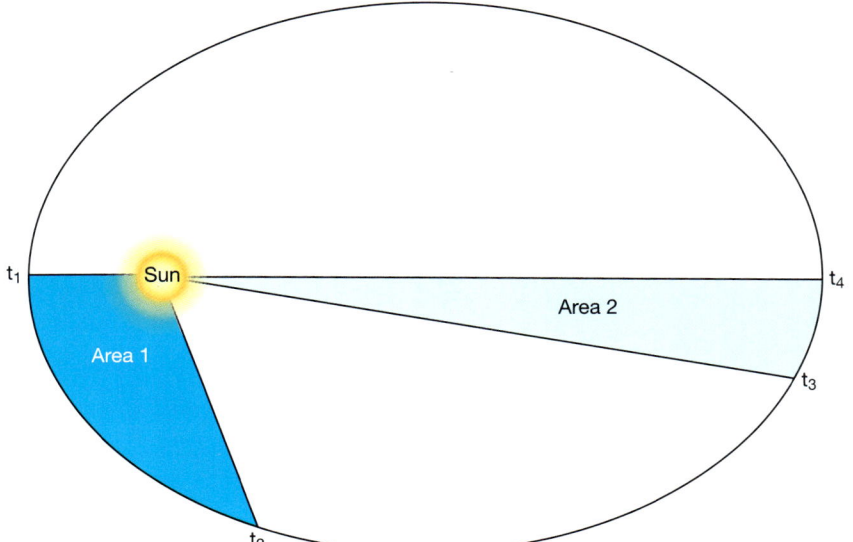

13.4 Kepler's second law of planetary motion.

Courtesy of Holley Y. Bakich

Where:

v = the orbiting object's velocity

a = the semimajor axis of the object's orbit

P = the sidereal period of revolution

r = the distance between the orbiting object and the body being orbited, such as Earth orbiting the Sun, or the Moon orbiting Earth.

Using the Moon's perigee and apogee distances determined in the earlier example, the Moon's maximum and minimum speeds can be determined.

The Moon's maximum speed for a perigee of 363,000 km:

$$v_{max} = \sqrt{\left(\frac{4\pi^2 \cdot (384{,}400 \text{ km})^3}{(2.36 \times 10^6 \text{ sec})^3}\right)\left(\frac{2}{363{,}258 \text{ km}} - \frac{1}{384{,}400 \text{ km}}\right)} = 1.08 \text{ km/s}$$

The Moon's minimum speed for an apogee of 406,000 km:

$$v_{max} = \sqrt{\left(\frac{4\pi^2 \cdot (384{,}400 \text{ km})^3}{(2.36 \times 10^6 \text{ sec})^3}\right)\left(\frac{2}{405{,}542 \text{ km}} - \frac{1}{384{,}400 \text{ km}}\right)} = 0.969 \text{ km/s}$$

Third Law

GLOSSARY TERM

Revolution orbital motion of a planet or other celestial object around the Sun, or of a satellite around a planet

Kepler's third law of planetary motion is a relationship between the orbiting object's semimajor axis and the period of **revolution** (Fig. 13.5). One practical application of Kepler's third law is that the closer a planet is to the Sun, the faster it moves in its orbit. So, Mercury always moves faster than Venus (and the rest of the planets). Venus always moves faster than Earth (and the planets farther from the Sun than Earth). And so on. Recall that the period of revolution is how long it takes the orbiting body to go one time around its star or the object it is orbiting. For Earth it is one year or 365.24 days.

The relationship is stated as follows:

$$a^3 = kP^2$$

Where:

a = the orbiting object's semimajor axis

P = the orbiting object's period to orbit

a = a constant, referred to as Kepler's constant

13.5 Kepler's third law of planetary motion.
Courtesy of Holley Y. Bakich

For example: Determine the Kepler constant for the Sun by using the characteristics of Earth's orbit.

$$a^3 = kP^2$$

Using the average Earth-Sun distance as 1 **astronomical unit** (AU), which rounds off to 150,000,000 kilometers and Earth's period rounded off to 365 days:

$$(150,000,000)^3 = k(365)^2$$

Rearranging to solve for **k**, Kepler's constant:

$$\frac{(150,000,000)^3}{(365)^2} = \frac{3,375,000,000,000,000,000,000,000}{133,225} k$$

Solving for k:

$$k = \frac{133,335}{3,375,000,000,000,000,000,000,000} = 3.9474 \times 10^{-20} \text{ days}^2/\text{km}^3$$

> **GLOSSARY TERM**
>
> **Astronomical unit** unit of distance that is approximately (within about 3/100,000,000) the average distance from Earth to the Sun; this distance is 149,597,870.7 km

The exquisiteness of Kepler's laws of planetary motion is that they do not just apply to Earth (or any planet) orbiting the Sun, but to any object or body orbiting any other, whether it be a satellite orbiting Earth, or Saturn's moon Titan orbiting the ringed planet. Each central object will have a different Kepler constant.

Check Your Understanding

1.1 What about Kepler's first law put him at odds with astronomers who came before him?

1.2 What does Kepler's second law tell us about how fast planets move in their orbits?

1.3 What does Kepler's third law tell us about how fast planets move in their orbits?

1.4 Calculate Kepler's constant for a satellite in geosynchronous orbit around Earth. The semimajor axis of such an orbit is 42,164 kilometers. You can round off Earth's day to 24 hours.

Answer: _____

Exercise 13.1 Applying Kepler's Laws

Procedure

Materials
- ❏ Pins or tacks
- ❏ String
- ❏ Pencil
- ❏ Cardboard
- ❏ Paper

1 Describe the differences and similarities between a circle and an ellipse.

_____

_____

2 On a piece of paper, draw three ellipses with distances between the foci of 1 cm (ellipse X), 2 cm (ellipse Y), and 4 cm (ellipse Z).

  a After marking the three foci distances on your paper, tie the ends of a piece of string that is about 6–8 cm long.

  b Mount your paper on a piece of cardboard into which you can insert pushpins or thumb tacks at the foci.

  c Loop the string around the pushpins or tacks and use your pencil to take up the slack in the string and trace out a complete path with the other two points of contact at the pins (Fig. 13.6).

3 Trace ellipse X, then repeat for ellipse Y and ellipse Z.

4 Using your drawing, complete the data sheet on pages 261–262.

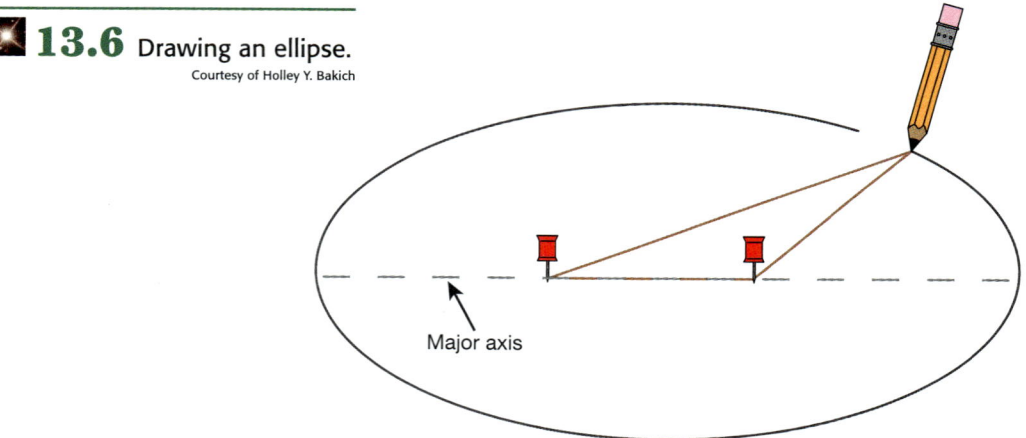

13.6 Drawing an ellipse.
Courtesy of Holley Y. Bakich

Name _____ Section _____ Date _____

Working with Ellipses

Exercise 13.1 Data Sheet

1 Based on your drawings, complete Table 13.1.

TABLE 13.1 Eccentricities of Various Ellipses

Ellipse	Foci Distance (mm)	Length of Major Axis (mm)	Eccentricity: Foci Distance/Length of Major Axis (*show your work and circle the answers*)
X			
Y			
Z			

2 Which ellipse is most eccentric? Which is the least eccentric?

_____

_____

3 Kepler's laws are generally considered in relationship to the planets of the Solar System and its star, our Sun. List other applications and relations of Kepler's laws that occur, such as the Moon orbiting Earth.

_____

_____

4 For the many moons orbiting Jupiter, the closest orbital approach is called perijovian and the farthest orbital distance apojovian. Jovian refers to Jupiter. Calculate perijovian and apojovian distances for the following two moons of Jupiter from the data in Table 13.2 on the following page. Recall:

$$R_p = a(1 - e)$$
$$R_a = a(1 + e)$$

Kepler's Laws of Motion **CHAPTER 13**

TABLE **13.2** Comparison of Io and Sinope

Moon	Semimajor Axis	Eccentricity	Sidereal Period	Diameter
Io	422,000 km	0.000	1.769 days	3,630 km
Sinope	23,700,000 km	0.275	758 days	40 km (irregular)

5 Using the data above and the perijovian and apojovian distances for Io and Sinope, calculate the maximum and minimum speeds for each satellite. Recall:

$$v = \sqrt{\left(\frac{4\pi^2 \cdot a^3}{P^3}\right)\left(\frac{2}{r} - \frac{1}{a}\right)}$$

6 Based on Kepler's first and second laws, why do you think that a planet's distance from the Sun is given in reference sources and books as an average distance?

_____

_____

7 From your calculations above, what can you say about the orbits of Jupiter's moons Io and Sinope?

_____

_____

_____

8 Using Kepler's third law of planetary motion, determine the distance in astronomical units that the planet Jupiter is from the Sun, knowing that Jupiter takes approximately 11.86 years to orbit the Sun once. Recall Kepler's third law:

$$a^3 = kP^2 \quad \text{and} \quad k = 1\frac{AU^2}{y^3}$$

Unit 4 *The Solar System*

The Solar System
Observing the Planets

LEARNING OBJECTIVES

Upon completion of this chapter, you should be able to:

1. Explain why some times are better than others for planetary observing.
2. Identify several features on selected planets.
3. Sketch several planets from images seen through the eyepiece, weather permitting.
4. Use various filters to sketch planets and determine the advantages of each filter.
5. Describe features of the planets as seen through a small telescope.
6. Define all glossary terms.

Mars — Courtesy of NASA

In this chapter, you will take your knowledge of the planets and apply it to observing them in an outdoor lab. Five planets are visible to the naked eye: Mercury and Venus (the **inner planets**), Mars, Jupiter, and Saturn.

Filters

All observers should use color filters (Fig. 14.1) when viewing a planet, because filters bring out faint detail you won't see otherwise. If you have never used color filters, however, please note: Your views won't be colorful through them. Color filters produce **monochromatic** views that enhance the contrast between adjoining planetary features, exaggerating differences in brightness.

Manufacturers label color filters along their circumferences. To use one, screw it into the eyepiece's barrel. All eyepiece filters have threads that match the threaded inside barrels of eyepieces. All you have to do is make sure the eyepiece size fits the filter size. Most eyepieces and filters have a diameter of 1¼", so you should not have any problems.

All color filters work better with larger telescopes. It is a simple rule of light throughput. For example, a #47 Violet filter used with a 4-inch telescope to see **cloud features** on Venus just does not work; the filter transmits only 3 percent of the light hitting it. However, the same filter used with a 12-inch telescope easily reveals features. Table 14.1 lists common color filters used for observing with the percentage of light each one allows through.

> **GLOSSARY TERMS**
>
> **Inner planet** any planet whose orbit around the Sun is closer than that of Earth; Mercury or Venus
>
> **Monochromatic** composed of a single color of light; sometimes used to refer to a black-and-white image
>
> **Cloud features** visible features in the atmosphere of a planet or satellite; may be temporary or permanent

Table 14.1 Common Color Filters and Light Transmission Percentages

Color Filter	Transmission
#8 Light yellow	83 percent
#11 Yellow green	78 percent
#12 Yellow	74 percent
#15 Deep yellow	67 percent
#21 Orange	46 percent
#23A Light red	25 percent
#25A Red	14 percent
#38A Dark blue	17 percent
#47 Violet	3 percent
#56 Light green	53 percent
#58 Green	24 percent
#80A Blue	30 percent
#82A Light blue	73 percent

14.1 Color filter set.

Check Your Understanding

1.1 Explain why you should select lighter filters (those that let more light through) if you use a small telescope.

Mercury

Of the visible planets, Mercury (Fig. 14.2) is the most difficult to observe; many amateur astronomers have never seen Mercury.

Mercury orbits the Sun at an average distance of only 36 million miles (58 million kilometers). Earth is nearly three times as far, so, from our perspective, Mercury always stays near the Sun.

When Mercury is east of the Sun as far as it can get (called greatest eastern elongation), it is seen as an "evening star" low in the western sky. Likewise, greatest western elongation occurs when it is farthest west of the Sun. At such times, it is viewed as a "morning star" in the east before sunrise. Some elongations are better than others because of Earth's tilt and the stretched-out nature of Mercury's orbit. Even at its farthest from the Sun, Mercury appears no more than 28° away (Fig. 14.3). At a "bad" elongation, the planet may be as little as 18° from the Sun.

The numbers above only give angular distance from the Sun. For observers to see a good elongation of Mercury, the Sun's apparent path through our sky—called the **ecliptic**—has to be nearly at a right angle to the horizon. This position puts Mercury higher in the sky than when the ecliptic/horizon angle is small. For evening appearances, Mercury will ride highest when its greatest elongation occurs around mid-March. In the morning sky, Mercury appears highest in mid-September.

Through a telescope, Mercury doesn't look much different than it does with the naked eye. First, its disk appears small—it measures only 7" across at **greatest elongation**. Second, it never appears high in the sky. So, when you observe Mercury, you are looking through the thickest, most distorting part of Earth's atmosphere.

This last problem can be addressed in a way that may sound strange to beginning amateur astronomers: observe Mercury during the day, when it appears highest in the sky. Remember to use extreme caution when attempting this; the planet never strays far from the Sun. For this observation, a go-to (computer-controlled) drive that you previously aligned and left on all day works well. Daytime telescopic viewing of the planet can be improved with the use of a red, orange, or yellow **filter** to reduce the amount of blue light scattered by our atmosphere. Through your telescope, expect to see Mercury go through phases similar to the Moon's.

Most observers detect no surface markings on Mercury. It takes a seasoned observer and excellent atmospheric conditions to see anything at all on the planet, and that is through the largest amateur telescopes. Experienced amateurs, however, have recorded

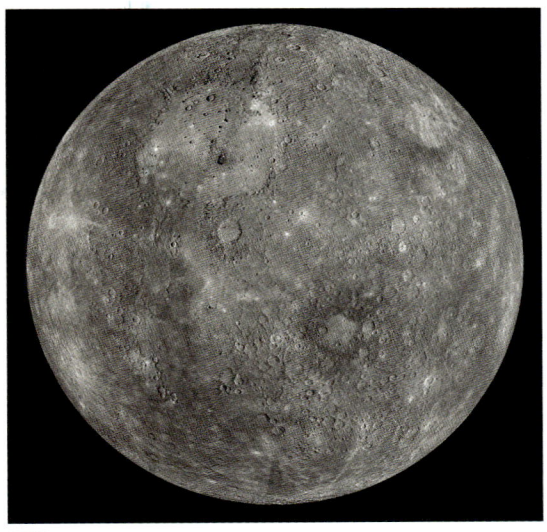

14.2 Mercury. *Courtesy of NASA*

GLOSSARY TERMS

Ecliptic great circle described by the Sun's annual path on the celestial sphere; the mean plane of Earth's orbit around the Sun

Greatest elongation maximum angular distance of Mercury or Venus from the Sun, as seen from Earth

Filter device (usually glass) that transmits light of different wavelengths

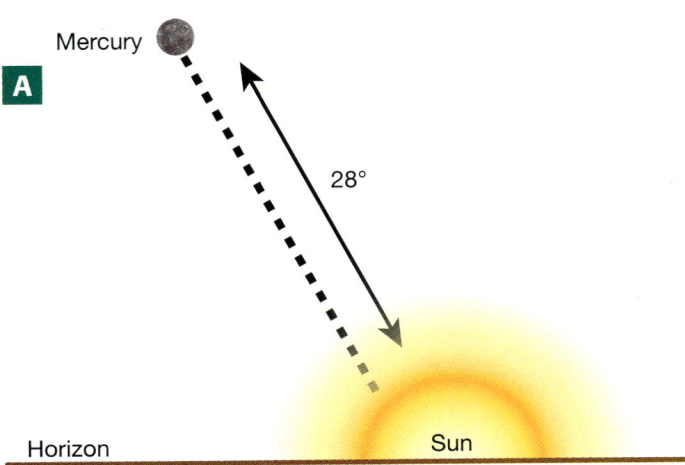
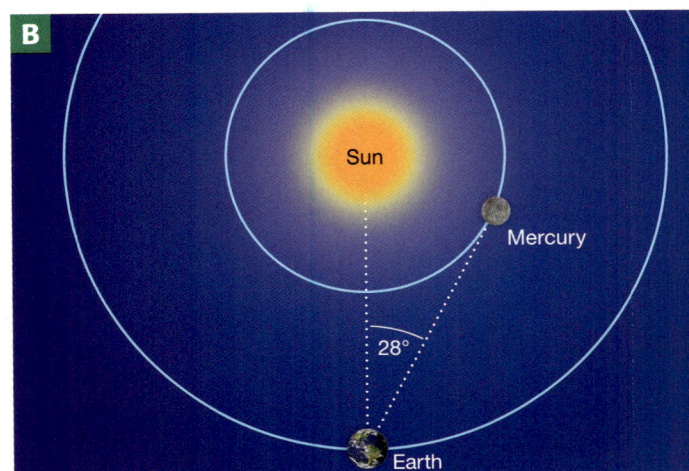

14.3 Example of greatest western elongation of Mercury: **(A)** viewed from Earth, and **(B)** viewed from celestial north (above). *Courtesy of Holley Y. Bakich*

GLOSSARY TERM

Cusp one of the points on a crescent (e.g., a crescent Moon)

dusky markings and occasional bright areas on the planet. The easiest "marking" to observe on Mercury is when the planet's southern **cusp** appears blunted (it doesn't come to a sharp point), but you will need a large telescope to see it.

For most observers, however, just seeing Mercury in the evening sky counts as a successful observation, so set up your telescope, and mark this planet off your "to-see" list.

Check Your Understanding

2.1 Why don't we ever see Mercury across the sky from the Sun?

2.2 Where would you have to look for Mercury when it is at greatest western elongation?

2.3 What is the maximum number of degrees ever between Mercury and the Sun?

2.4 Why don't we ever see details on Mercury?

14

Venus

Venus (Fig. 14.4) is easier to observe than Mercury. Observers routinely follow its phases, along with an aspect easier to observe with Venus than Mercury: size change. While Mercury looks twice as big when between Earth and the Sun than it does when it lies on the far side of the Sun, Venus is more than six times larger when it is nearest Earth than when it lies farthest from our planet. Daytime observations of Venus are also easier than those of Mercury because Venus shines brighter and can lie farther from the Sun (Figs. 14.5 and 14.6).

In December 1610 Italian astronomer Galileo Galilei (1564–1642) became the first to observe the phases of Venus. Galileo wrote that Venus imitated the Moon in appearance. In 1666 French astronomer Giovanni Domenico Cassini (1625–1712) made the first measurements of the rotation rate of Venus. He obtained a value of 23 hours and 21 minutes.

When you view the planet through a telescope, you will see no permanent features discernible in its clouds. The atmosphere is in a

 **14.4** Venus. Courtesy of NASA

266 Unit 4 The Solar System

14.5 Venus setting September 16, 1986, in Tucson, Arizona.

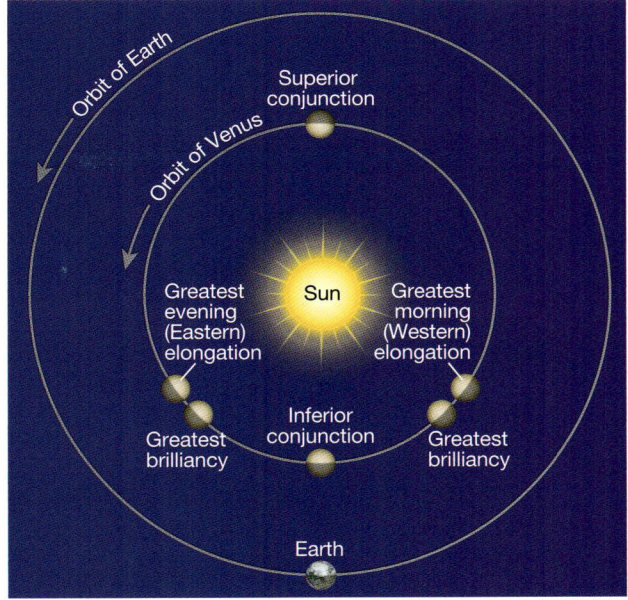

14.6 Highlights of Venus' orbit around the Sun.
Courtesy of Holley Y. Bakich

continuous state of mixing, and any patterns observed quickly dissipate. Features in Venus' atmosphere range from dusty shadings to bright spots. You may be able to spot its most famous feature through a violet filter. This filter doesn't allow in much light, so you will need at least an 8-inch telescope. Look for an immense C or Y-shaped feature centered on the planet's **equator**.

> **GLOSSARY TERMS**
>
> **Equator** imaginary line that divides the northern half of a body from its southern half

Check Your Understanding

3.1 When Venus is at greatest elongation, does it appear closer to the Sun or farther from it than Mercury does when that planet is at greatest elongation?

3.2 How many times larger is Venus when it is nearest Earth compared to when it lies farthest from our planet?

3.3 Who was the first person to observe the phases of Venus?

3.4 Why don't astronomers see long-lasting features as they view Venus?

Mars: The Red Planet

14.7 Mars. *Courtesy of NASA*

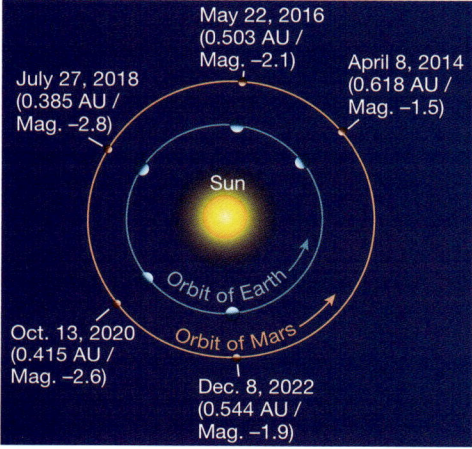

14.8 Oppositions of Mars from 2014 to 2022. *Courtesy of Holley Y. Bakich*

GLOSSARY TERMS

Opposition position of two celestial objects when their longitude (as seen from Earth) differs by 180°; when one of the objects is the Sun, opposition means the other object is opposite the Sun in the sky, therefore visible all night long

Angular size apparent size of a celestial object, measured in degrees, minutes, and seconds, as seen from the Earth; for example, the average angular size of the Sun, as seen from Earth, is 0.53°

Meridian for this chapter, the north-south line that divides a planet in half

Albedo feature bright feature on the surface of Mars or other rocky object, such as the Moon or an asteroid

Observing Mars (Fig. 14.7) is, in one long-time observer's opinion, "Two long years of waiting for four to six weeks of panicked activity." Astronomers' excitement tends to peak during times called **oppositions** (Fig. 14.8). This is when Mars comes closest to Earth, which happens every 780 days (minus 1 hour 26 minutes 24 seconds, to be exact). But each closest approach to Earth by Mars is not really closest, because the orbits of Earth and Mars are not circular. During a distant opposition, Mars can be more than 60 million miles away. Contrast that with a nearby opposition that places Mars less than 35 million miles from Earth, and you might begin to understand what all the fuss is about.

Astronomers use angular measurement to describe how large an object is in the sky. With certain objects, mainly planets, the **angular size** can change a lot. Angular size is measured in degrees (for large objects), minutes (for smaller objects) and seconds (for tiny objects like planets), or a combination of the above. For example, the distance between two stars may be 14 degrees, 29 [arc]minutes, 7 [arc]seconds (denoted 14°29'07"). Mars, on the other hand, varies in size at opposition from a minimum of only 13.8" to a maximum of only 25.1". The maximum size of Mars is the same as that of a quarter seen 650 feet away.

Rotation

Mars' day (which is known as a *sol*) is 37.4 minutes longer than ours. So, if you observe Mars at the same time each night, its markings will move (37.4/1477.4) × 360° = 9.11°/day to the west. In a little more than five weeks, the planet would appear to slowly rotate backward one full spin. All the prominent features of Mars would, at some time during this period, be placed favorably on its **meridian**.

You can also choose to wait for Mars' rotation to bring an object into view, or onto its meridian. Because Mars rotates once every 24.623 hours, in one hour it rotates 360°/24.623 = 14.62°. So, for example, if you are observing a feature on Mars' meridian that has a longitude of 260°, and you want to know when another feature at longitude 296.55° will cross the meridian, divide 36.55° (their separation in longitude) by 14.62°/hour = 2.5 hours. This is the time between meridian crossings of the two features (Fig. 14.9).

Features

The most prominent surface features on Mars are the polar ice caps. With the exception of the polar ice caps, the easiest feature to identify on Mars is a dark, triangular **albedo feature** named Syrtis Major (Fig. 14.10). An albedo feature is a region distinguished by the amount of light it reflects. Albedo features are also areas most subject to seasonal changes. Clouds (Fig. 14.11) can also frequently be found when looking at the features of Mars.

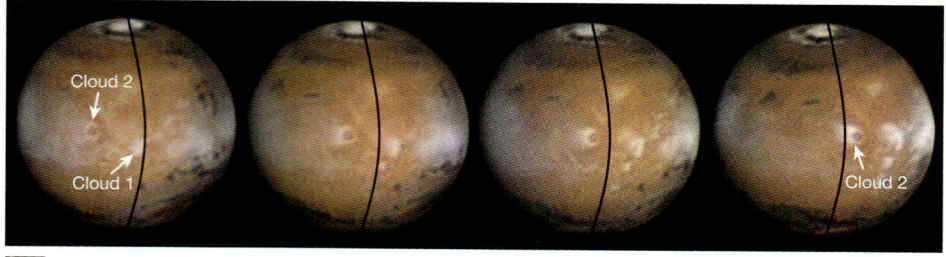

14.9 Meridian crossings of two cloud features separated by 36.55°, which occurred 2.5 hours apart. *Courtesy of Damian Peach*

14.10 Syrtis Major.
_{Courtesy of Steve Lee (University of Colorado), Jim Bell (Cornell University), Mike Wolff (Space Science Institute), NASA}

14.11 Set of color and black-and-white Hubble Space Telescope images showing numerous clouds above Mars' surface.
_{Courtesy of P. James (University of Toledo), T. Clancy (Space Science Institute), S. Lee (University of Colorado), NASA}

Check Your Understanding

4.1 How often does Mars come closest to Earth?

4.2 Why do you think astronomers call Mars' closest approach "opposition"?

4.3 How many degrees does Mars rotate in one hour?

4.4 What is an albedo feature?

4.5 If you are going to sketch Mars, which features should you draw first?

Observing the Planets **CHAPTER 14**

14.12 Jupiter. Courtesy of Damian Peach

Jupiter: The King of the Planets
Features

Jupiter (Fig. 14.12) has an atmosphere that is divided into cloud **bands** of various colors oriented parallel to the planet's equator. The lighter-colored bands are known as **zones**, and the darker-colored bands are called **belts**. Within the zones and belts lie eddies (whirlpools), which may produce temporary spots or streaks within or between Jupiter's cloud bands. Two dark stripes—one above and one below Jupiter's equator—are the North and South Equatorial Belts (Fig. 14.13).

At higher magnification, you will notice Jupiter's **poles** appear flattened, a result of the planet's rapid rotation rate coupled with the fact that it is not a solid body. Jupiter's equatorial diameter surpasses its polar diameter by more than 5,600 miles (9,000 km).

The planet's rotation brings nearly all of its visible area into view in a single night. Individual belts and zones may also become more or less prominent and even may disappear for extended periods.

For more than a century astronomers have been observing the **Great Red Spot** (GRS) and its discovery possibly stretches back to 1664. That year, English astronomer Robert Hooke (1635–1703) probably observed it.

The GRS is an **anticyclone,** a high pressure storm 22° south of Jupiter's equator. The closest analogy is to a terrestrial hurricane. Because the GRS is anticyclonic in Jupiter's southern hemisphere, its rotation is counterclockwise, with a period of about six days.

The actual spot is enormous: It has a north-south width of 8,700 miles (14,000 kilometers) and a variable east-west width of 15,000 to 25,000 miles (24,000 to 40,000 km). Three Earths could fit inside it. In recent years, though, it has been shrinking. The GRS's clouds appear to be about 5 miles (8 km) above neighboring cloud tops. Other features similar to the GRS can be observed, but none are as large. The GRS also drifts in

> **GLOSSARY TERMS**
>
> **Band** planetary feature that stretches all the way around its globe; divided into belts and zones
>
> **Zone** light-colored band in a planet's atmosphere
>
> **Belt** dark band in the atmosphere of a planet
>
> **Poles** two points on a planet, moon, or other celestial object that lie 90° above or below the object's equator
>
> **Great Red Spot** huge storm in the upper atmosphere of Jupiter
>
> **Anticyclone** large-scale circulation of winds around an area of high pressure on all planets with atmosphere; anticyclones rotate clockwise in the northern hemisphere and counterclockwise in the southern hemisphere

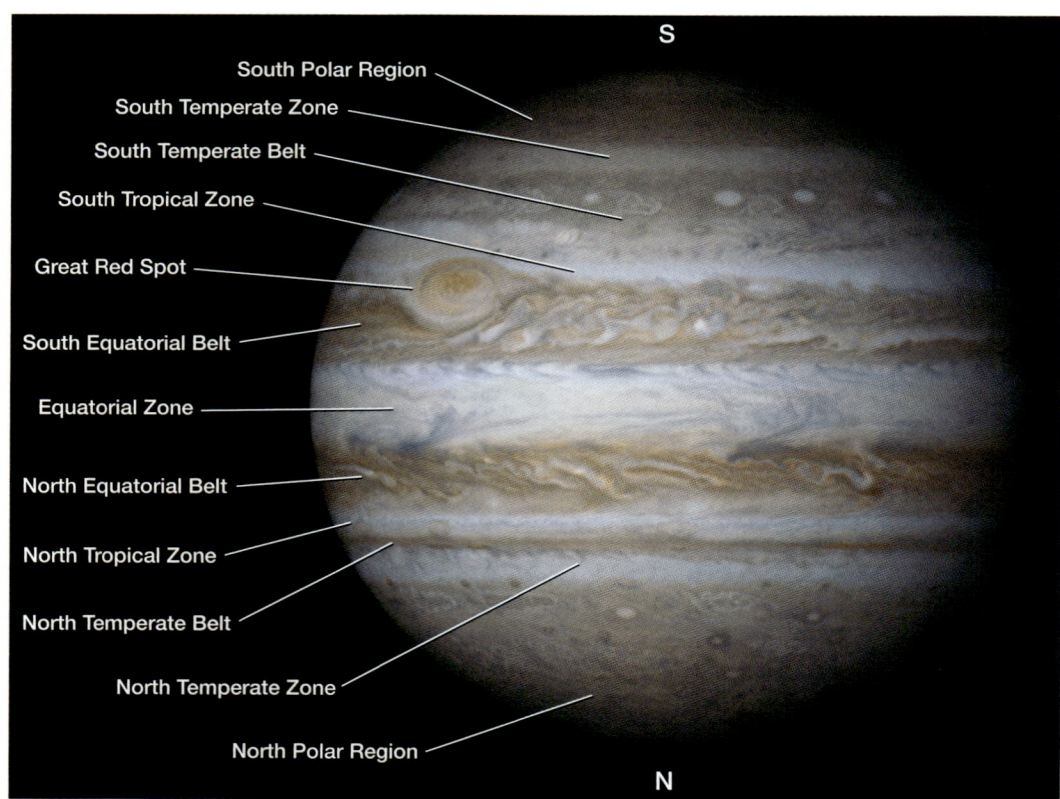

14.13 Jupiter's belts and zones (as viewed through a telescope). Courtesy of NASA

longitudinally through the South Equatorial Belt. The GRS changes color (or fades, in the words of some amateurs) because gases at higher levels and of different compositions condense into clouds above the spot.

Moons

On January 7, 1610, Italian astronomer Galileo Galilei looked at Jupiter through his telescope and saw three stars in a straight line, two on one side of Jupiter and one on the other. The next night, the stars were still there, but their positions had changed. On January 13, 1610, Galileo noticed a fourth star.

After watching these objects for a number of weeks, Galileo came to the conclusion that the four "stars" were actually "planets" revolving around Jupiter in the same way the Moon circles Earth. Galileo's discovery of these four moons made them the first observed solar system objects invisible to the naked eye (Fig. 14.14). When you have observed Jupiter's large moons enough to know which is which and how they move, it will be time to focus on them more closely.

Through a 10-inch or larger telescope, and with good seeing, you can observe details on the **Galilean satellites**. With high magnification (exceeding 350x), you will resolve distinct disks, especially during transits, when the satellites' glare drops because Jupiter provides a lighter background. Try this on Ganymede: look for light shaded frost in its polar regions. Through larger telescopes, you can observe even the colors of Jupiter's satellites.

Four events, called phenomena, involving Jupiter and its four large moons, are possible. A satellite **eclipse** occurs when a satellite moves through, and disappears because of, Jupiter's shadow. A satellite **occultation** occurs when a satellite moves behind Jupiter. Satellites always disappear into occultation at the west side of Jupiter and reappear at the east side. A satellite **transit** occurs when a moon moves in front of Jupiter. A transiting satellite always moves from east to west across Jupiter's face. The satellites themselves often appear as bright dots against Jupiter's dark belts. When a satellite lies in front of the brighter zones, however, it is hard to see unless you have followed it from the time it started to cross the planet's face. A **shadow transit** (Fig. 14.15) occurs when a satellite's

> **GLOSSARY TERMS**
>
> **Galilean satellites** four largest moons (Ganymede, Callisto, Europa, Io) of Jupiter; discovered by Italian scientist Galileo Galilei in 1610
>
> **Eclipse** cutting off light from a celestial body (in this chapter, a moon) as it passes through the shadow of another body
>
> **Occultation** obscuration (total or partial) of any celestial object by another of larger apparent size
>
> **Transit** passage of one object in front of another
>
> **Shadow transit** movement of a moon's shadow across the face of a planet (usually refers to Jupiter)

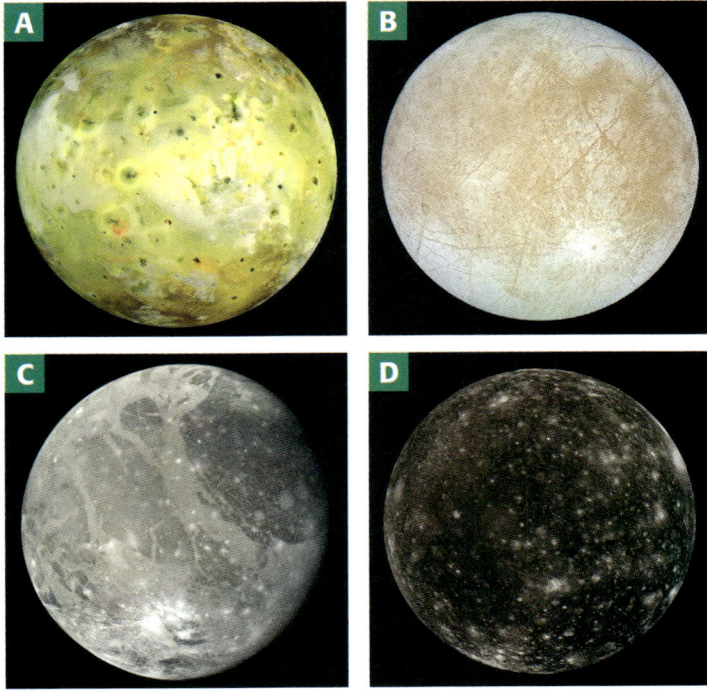

14.14 Spacecraft images of Jupiter's moons: (**A**) Ganymede, (**B**) Callisto, (**C**) Europa, and (**D**) Io.

Courtesy of NASA/JPL

14.15 Double shadow transit of Ganymede (*above*) and Io (*below*) across the surface of Jupiter.

Courtesy of Damian Peach

shadow moves across Jupiter's disk. You will see the shadows as small black dots through any telescope. Transiting shadows also move from east to west across Jupiter. Eclipse events are easier to observe than occultations because eclipses usually occur some distance from Jupiter's disk. Occultations take place at the brilliant planet's edge.

Check Your Understanding

5.1 Why does Jupiter appear slightly flattened when viewed through a telescope?

5.2 What feature on Earth most closely resembles Jupiter's Great Red Spot?

5.3 How did Galileo figure out that the four bright spots near Jupiter were moons?

5.4 Rank the four phenomena of Jupiter's satellites in order of how easy they are to see, with 1 being the easiest. Explain why you think your ranking is correct.

Saturn

When observing Saturn (Fig. 14.16), first look for the Cassini Division, a dark gap between Saturn's two brightest rings, called the A ring and the B ring. After locating the black stripe, note the rings' relative brightnesses. Through 8-inch and larger telescopes, you also may see color differences and textures.

Markings on Saturn's disk are subtle. When you observe it, be sure to look for any bright or dark spots compared to the belt or zone they are in. From night to night, these features may change position. Saturn's zones, shown in Figure 14.17, appear off-white, slate gray, or yellow. Saturn's belts look bluish gray, brown, and red. Such features stand out well through red, orange, or yellow filters. Occasional bright patches look best through a green filter. Highlight the rings using light green or light blue filters. Remember, large telescopes collect more light than smaller ones, so the smaller your telescope, the lighter shade of filter you should use.

Much easier to observe are the positions and shapes of the globe's shadow on the rings or of the ring's shadow on the globe. Watch the shadow shrink in the weeks before opposition and grow afterward. Saturn's rings always have the same orientation in space,

14.16 Saturn. *Courtesy of NASA*

14.17 Saturn: (**A**) belts and zones; (**B**) rings and divisions (as viewed through a telescope). *Courtesy of Roen Kelly, Astronomy magazine*

Observing the Planets **CHAPTER 14** 273

but depending on where it is in its orbit, the rings are seen to tilt at a variety of angles, including edge-on (Fig. 14.18).

14.18 Saturn seen at a variety of angles. Courtesy of Damian Peach

Check Your Understanding

6.1 What is the easiest of Saturn's features to observe?

6.2 At different times, we see Saturn's rings at different angles. Why? Does their tilt change with respect to the planet?

Ice Giants

Uranus and Neptune are almost impossible to see with the naked eye. With the aid of binoculars or a telescope, however, you can find them.

Uranus

The atmosphere of Uranus is usually a featureless haze (Fig. 14.19). Observers first reported details in 1870. Since then, observers have seen markings and belts. Through a small telescope, greenish Uranus appears as a slightly elliptical disk because of its rapid rotation. But in contrast to its quick spin, Uranus moves slowly in front of the stars in our sky. It takes the planet about 44 days to move the width of the full Moon.

Neptune

For an amateur astronomer with a medium-size telescope, Neptune is no problem to find (Fig. 14.20). At opposition, it displays a small blue disk of about magnitude 7.7. That brightness is enough to allow you to view it through binoculars. Unfortunately, the magnification of any binoculars is so low that Neptune will appear like a star through it.

A telescope won't reveal much detail on Neptune, although you will see its largest moon, Triton. You also should see a trace of the deep-blue color caused by the methane in Neptune's atmosphere. Because it lies farther from the Sun than Uranus, Neptune moves more slowly, taking approximately 85 days to traverse a full Moon's span.

14.19 Uranus. *Courtesy of NASA/JPL*

14.20 Neptune. *Courtesy of NASA/JPL*

Check Your Understanding

7.1 On average, approximately how much faster does Uranus travel through the sky than Neptune?

7.2 Describe what you would expect to see of Uranus if you viewed it through a small telescope.

Exercise 14.1 Observing Planetary Characteristics

Procedure

Materials
- ❑ Telescope with at least a 6-inch aperture
- ❑ Clipboard to hold observing forms
- ❑ Red flashlight

1 Set up the telescope or telescopes and accessories as directed by your instructor. Record your telescope setup information below.

Date of observation: _____

Time of starting observations: _____

Place of observation: _____

Telescope diameter: _____

Telescope focal length: _____

Telescope type: _____

Mount type: _____

Weather and sky conditions: _____

2 Use the data sheet on pages 277-278 to draw the planets you are observing.

Planetary Features through the Telescope

1 Planet observed:

Telescope magnification(s) used:

Describe how the planet appeared through the telescope.

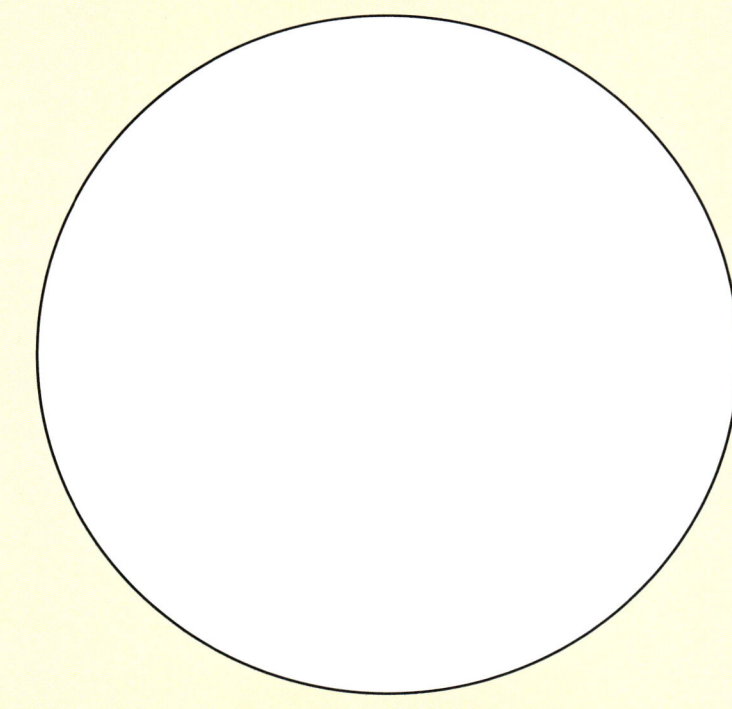

Draw the planet in the circle above. Indicate whether the circle represents the planet or the circular field of view of the telescope. If the latter, be sure to sketch any stars that also appear in the field of view with the planet.

2 Planet observed:

Telescope magnification(s) used:

Describe how the planet appeared through the telescope.

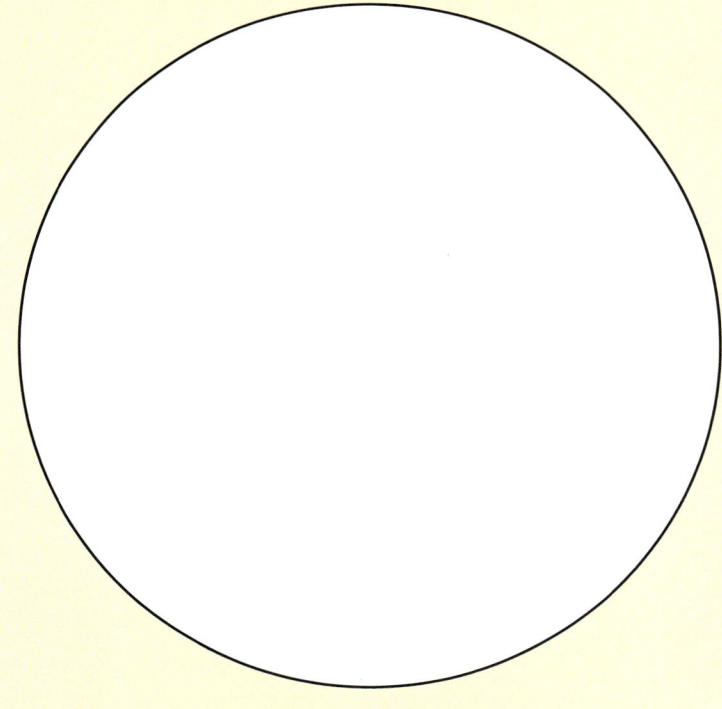

Draw the planet in the circle above. Indicate whether the circle represents the planet or the circular field of view of the telescope. If the latter, be sure to sketch any stars that also appear in the field of view with the planet.

Exercise 14.1 Data Sheet

Exercise 14.1 Data Sheet

3 Planet observed:

_____

Telescope magnification(s) used:

_____

Describe how the planet appeared through the telescope.

_____
_____
_____
_____
_____
_____
_____

Draw the planet in the circle above. Indicate whether the circle represents the planet or the circular field of view of the telescope. If the latter, be sure to sketch any stars that also appear in the field of view with the planet.

---

4 Planet observed:

_____

Telescope magnification(s) used:

_____

Describe how the planet appeared through the telescope.

_____
_____
_____
_____
_____
_____
_____

Draw the planet in the circle above. Indicate whether the circle represents the planet or the circular field of view of the telescope. If the latter, be sure to sketch any stars that also appear in the field of view with the planet.

Unit 4 *The Solar System*

Observing Planetary Motions

Exercise 14.2

Over a period of time, you will observe the motion of a planet as specified by your instructor. You may use the star charts included in the lab manual to plot the position or sketch the bright background stars (only necessary the first time you observe) as well as the planet's location each time you observe. Your instructor also will inform you as to the period of time for observations and the number of observations you will need to make.

 Procedure

Materials
- ❏ Star chart
- ❏ Clipboard to hold observing forms
- ❏ Red flashlight

1. Number each observation on Table 14.2, identifying by the date and time of your observation, place of observation, and overall weather and sky conditions.

2. Use the sketch box on page 280 to plot the position of your chosen planet, as well as other stars in your field of view. Label the first observation "1", the second "2," and so on.

TABLE 14.2 Observations of Planetary Motions

Planet observed:				
Starting observing date:				
Ending observing date:				
Observation	**Date**	**Time**	**Place**	**Sky Conditions**
1				
2				
3				
4				
5				
6				
7				
8				
9				
10				
11				
12				
13				
14				
15				

Continues

Observation	Date	Time	Place	Sky Conditions
16				
17				
18				
19				
20				
21				
22				
23				
24				
25				

3 How difficult was it to notice changes in the location of the planet you observed from night to night?

4 Did the weather or your observing location cause any challenges or problems?

5 How did the time of night you observed change during your period of observations?

Spotlight on Jupiter

Exercise 14.3

During this outdoor lab you will observe Jupiter's moons and then observe the planet through various color filters.

Procedure

Materials
- ❏ Clipboard to hold observing forms
- ❏ Sketching pencils
- ❏ Red flashlight
- ❏ Color filters

1. When you arrive at your destination, your first task is to assess the quality of your observing site. Answer the questions on the data sheet, page 283.

2. In circle 1 on page 284, draw the moons that are visible at their positions relative to Jupiter. Try to be as accurate as possible.

3. Return to the eyepiece at least one hour later and record the moons' positions again in circle 2, page 284.

4. Using the color filters that screw onto the back of the eyepiece or can be carefully held between two fingers, note how various features appear using the filters and record them in Table 14.3, page 284.

5. Answer the following questions:
   a. What filter or filters made Jupiter more visible?

   b. What filter or filters made Jupiter less visible?

   c. List any general impressions regarding using filters to observe Jupiter.

   d. How do you think color filters would work for objects like star clusters, nebulae, and galaxies?

CAUTION
Do not DROP the filter, and do not put your fingers on the colored glass.

Name _____   Section _____   Date _____

Observing Jupiter

1 How dark does the site seem to you? Explain how you came to your conclusion.

_____
_____
_____

2 How far away from the nearest large city are you? How far from the nearest small city?

_____
_____
_____

3 Are there any "light domes" visible? These bright areas of the sky (brightest near the horizon) originate with artificial lights.

_____
_____
_____

4 Are there any clouds? Approximately how much of the sky do they cover?

_____
_____
_____

5 Is the Moon visible? It is best to choose a moonless night to observe, but maybe the nights have been cloudy. What is the Moon's phase?

_____
_____
_____

6 What is your estimate of your site's seeing? Explain your answer.

_____
_____
_____

7 What is your estimate of your site's transparency? Explain your answer.

_____
_____
_____

Exercise 14.3 Data Sheet

Exercise 14.3 Data Sheet

Circle 1

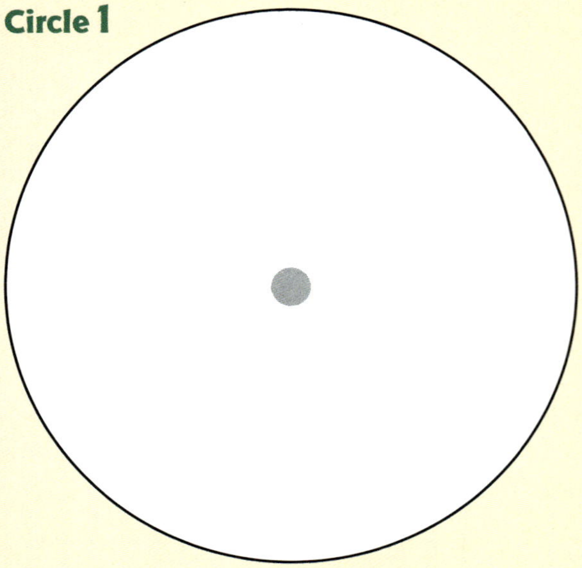

Circle 2

Telescope _____

Eyepiece _____

Magnification _____

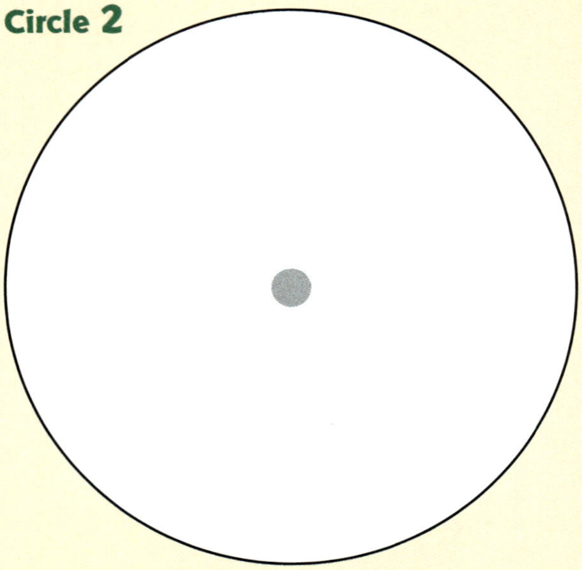

Telescope _____

Eyepiece _____

Magnification _____

TABLE 14.3 Observations of Jupiter through Various Color Filters

Filter	Feature	Notes
No Filter		
Red		
Yellow		
Green		
Blue		

Unit 4 The Solar System

The Solar System

Minor Bodies

LEARNING OBJECTIVES

Upon completion of this chapter, you should be able to:

1. Identify the main parts of a comet.
2. Explain why comets have two types of tails.
3. Describe how comets become visible.
4. Differentiate comets by how concentrated their main bodies appear.
5. Estimate a comet's brightness and calculate the effect of Earth's atmosphere on it.
6. Compare and contrast the different groups of asteroids.
7. Locate the stable points in the orbits of various asteroid groups.
8. Calculate the designation of a newly discovered asteroid.
9. Explain why professional observatory surveys discover most asteroids.
10. Define all glossary terms.

Comet 17P/Holmes Courtesy of Chris Schur

In this chapter, by studying projected images, you will learn to differentiate among comets, asteroids, and meteors. You also will label the major parts of a comet, and compare and image comets and asteroids.

Comets

Many observers rate comets as their favorite observing targets. Most comets appear unannounced, seemingly out of nowhere; some shine brightly; and a few rise to a spectacular level. Unfortunately, bright comets don't appear on a schedule.

Comets are small, irregularly shaped bodies made up of dust grains and frozen gases. Many travel along **elliptical** orbits that occasionally bring them close to the Sun and then take them deep into space, far beyond Pluto's orbit.

Comets develop a surrounding cloud of thinly distributed material called a **coma** that grows as the comet approaches the Sun (Fig. 15.1). A small **nucleus**, usually less than 6 miles (10 kilometers) across, remains hidden within the coma. On March 13, 1986, the Giotto spacecraft approached to within 370 miles (596 kilometers) of the nucleus of Halley's comet. This was the first time a spacecraft had made such close observations of a comet. Giotto collected data that helped astronomers create an enhanced image of the nucleus (Fig. 15.2).

In our solar system's earliest days, planets and comets formed at the same time. The faraway Oort Cloud and the **Kuiper Belt** likely provided much of comets' original material (Fig. 15.3). When far from the Sun, the gas in the comet's nucleus is frozen solid, as seen in the nucleus of comet 103/P Hartley 2 in Figure 15.4A. During the time it is far

GLOSSARY TERMS

Elliptical shape of the path of a celestial body orbiting the Sun; a circle is a special type of elliptical orbit

Coma gaseous, usually spherical, area that surrounds and hides a comet's true nucleus; composed of gas and dust, the coma is evaporated off the comet's nucleus

Nucleus main icy body that is a comet; when close to the Sun a cloud of dust and gas (the coma) surrounds the nucleus

Kuiper Belt disk-shaped region of icy bodies orbiting the Sun that has its inner boundary just beyond Pluto's orbit

15.1 Comet 29P/Schwassmann-Wachmann 1 displays a growing coma as it approaches the Sun from June 16 (bottom left) to July 28 (upper right), 2013.
Courtesy of Damian Peach

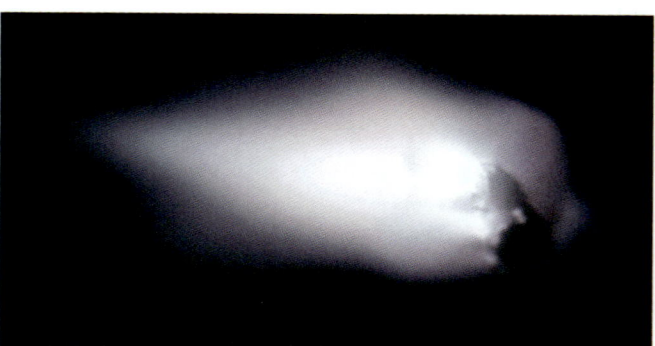

15.2 Nucleus of Halley's comet. © ESA, MPAe, Lindau

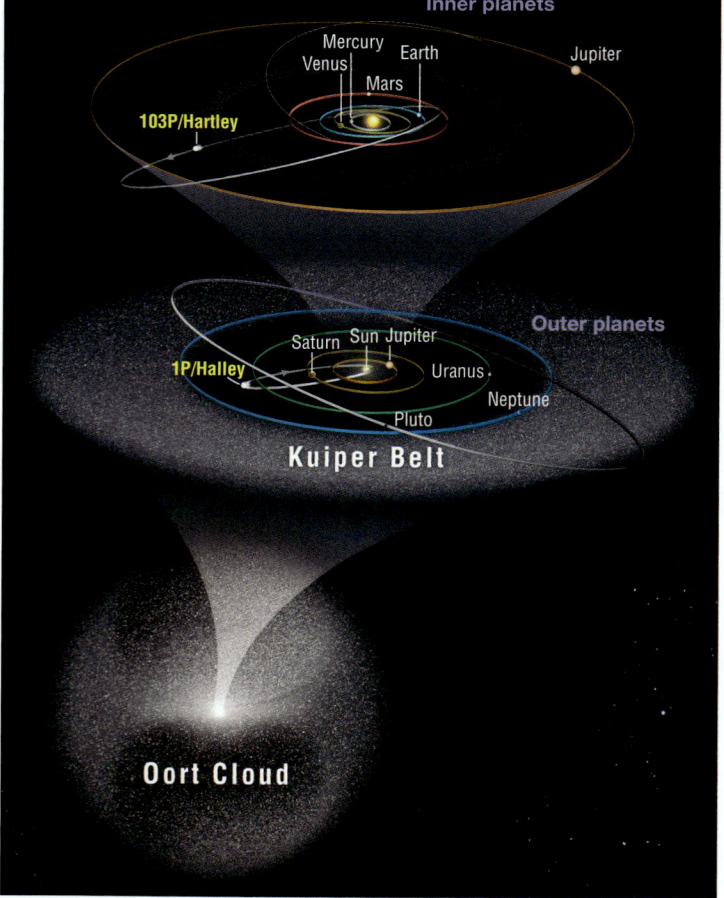

15.3 Oort Cloud and the Kuiper Belt.
Courtesy of Roen Kelly, *Astronomy* magazine

286 Unit 4 The Solar System

from the Sun, the comet appears faint because only sunlight reflected off of the nucleus is seen. When a coma develops, released dust reflects more sunlight, and gas in the coma absorbs ultraviolet radiation and begins to fluoresce (Fig. 15.4B). At about 5 astronomical units (remember that 1 astronomical unit = the average Earth-Sun distance) from the Sun, **fluorescence** contributes more to the comet's brightness than reflected light.

As comets approach the Sun, they develop tails of luminous material that extend from the heads. Some comet tails are enormous, stretching for tens of millions of miles. The **solar wind** accelerates materials away from a comet at different velocities, depending on the size and mass of the material. This can create two types of tails: dust and ion. A **dust tail** contains some mass, so it accelerates slowly and tends to curve to follow the comet's path. An **ion tail** (composed of charged gases) is much less massive, so it accelerates in a nearly straight line, extending directly away from the Sun (Fig. 15.5).

Observing Bright Comets

Comets bright enough to see without optical aid are a treat. One such occurrence took place when Comet 17P/Holmes approached Earth (Fig. 15.6). This comet was discovered in 1892 by Edwin Holmes (1838–1919), and although it is normally a faint comet, it blazed brightly in October 2007.

Comets reward patient observers. If you can see a comet without optical aid, the following questions will help you make a detailed observation:

- What is the comet's altitude? (That is, how high above the horizon is it?)
- In which constellation(s) is it located?
- If you are observing from a dark site, does the target lie against the background of the Milky Way?

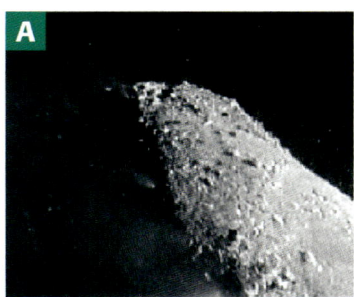
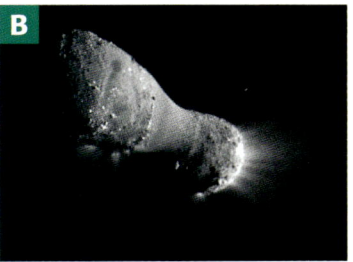

15.4 Comet 103P/Hartley 2 studied up close by the EPOXI mission showing: **(A)** its nucleus, and **(B)** fluorescence.
Courtesy of NASA/JPL-Caltech, UMD

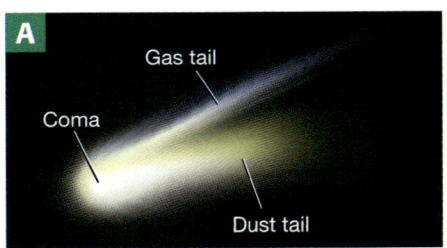
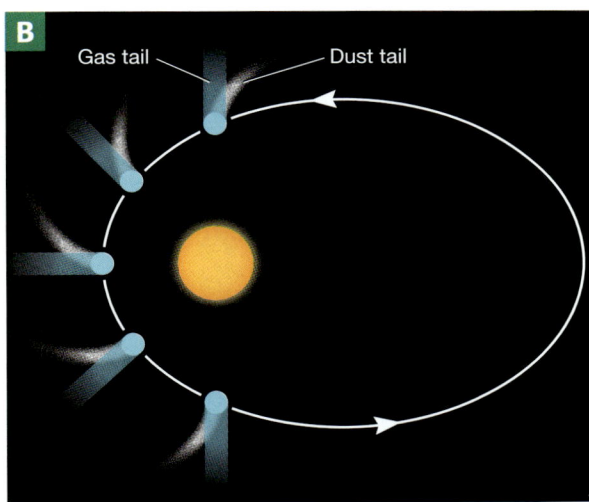

15.5 **(A)** Comet tails, and **(B)** the direction of comet tails in relation to their path and the Sun. Courtesy of Roen Kelly, *Astronomy* magazine

15.6 Comet 17P/Holmes.
Courtesy of Chris Schur

GLOSSARY TERMS

Fluorescence emission of light by a material that has absorbed light or another form of radiation

Solar wind energetic charged particles that flow outward from the Sun

Dust tail one of two types of comet tails, composed of dust and shining by reflected sunlight; dust tails are more curved than ion tails because they follow the path of the comet's orbit around the Sun

Ion tail also called the plasma tail, one of two types of comet tails, composed of ionized molecules; ion tails are generally straight (because the Sun's radiation pushes them in a direction directly opposite where our daytime star is), bluer than dust tails, and can reach lengths of tens of millions of km

- What is the coma's apparent size?
- Can you determine the full extent of the tail and its width both near the coma and at the end?
- Can you see both a dust tail and an ion tail? How do they differ?
- What color is each part of the comet? Look especially at the tail(s). If you choose to sketch the comet, be sure to note the date, time, and the direction of north on your drawing.

For most bright comets, binoculars will provide the best views. Binoculars offer some magnification, darken the sky background a little, and provide a wide field of view. Through binoculars, observe how much farther from the comet's head you can see the last wisps of the tail. If both types of tail are visible, note any increased definition in either and any colors (compared to your naked-eye view). Also note the separation and shapes of the two tails.

Degree of Condensation

One detail to note as you observe a comet is its degree of condensation, or DC (Fig. 15.7). This indicates how much the surface brightness of the coma increases toward its center. As the DC increases, the coma size usually decreases and it appears more defined.

Astronomers assign a DC of 0 to a totally diffuse comet with no brightening toward the center. Between a DC of 3 and 5, the coma's center appears distinctly brighter. By DC 7, the comet exhibits a steep overall increase in brightness toward the center. At DC 8, the coma appears small and dense with well-defined boundaries. At DC 9, the comet looks like a fuzzy star or a planet viewed close to the horizon or in bad seeing.

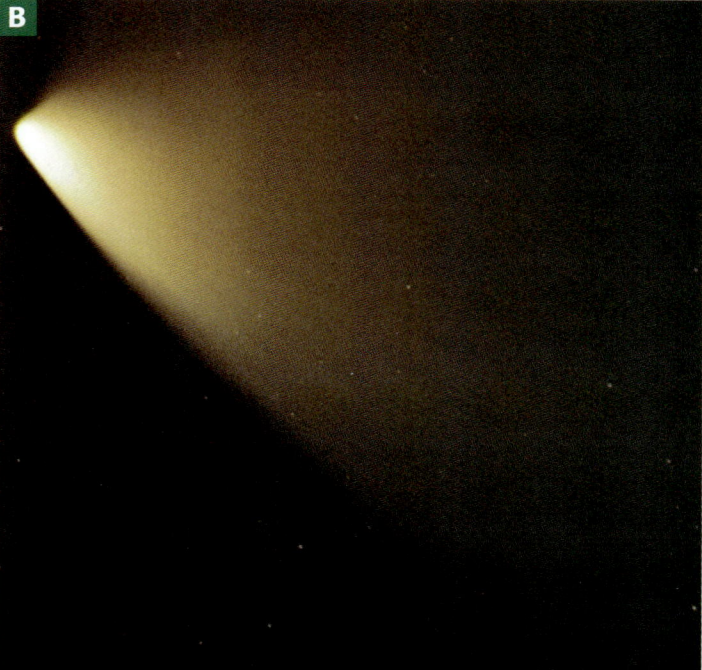

15.7 Comets with: **(A)** degree of condensation of 8, and **(B)** degree of condensation of 2 or 3.

(A) Courtesy of Chris Schur, (B) Courtesy of Gerard Rhemann

15.8 **(A)** Comet 17P/Holmes; **(B)** Comet 103P/Hartley near the Double Cluster in the constellation Perseus in October 2010; and **(C)** Comet C/2012 V2 (LINEAR) against a rich star field in September 2013.

(A) and (B) Courtesy of Chris Schur, (C) Courtesy of Damian Peach

Estimating a Comet's Magnitude

Underestimating the coma's total brightness is one mistake some amateur astronomers make. They concentrate too much on the central portion, which may look like a fuzzy star. Also, from an urban setting the outer coma is often lost because of light pollution.

To estimate a comet's **magnitude** through binoculars (or a telescope for smaller and fainter comets), defocus the images of nearby stars to the same size as the normal in-focus view of the comet's coma, then compare the brightnesses. This technique works better for diffuse comets (Fig. 15.8A). A comet with a high DC usually has a large difference in brightness between the outer coma and the central region (Figs. 15.8B and 15.8C), which makes comparing the comet with out-of-focus stars difficult because the stars' images appear more uniform.

> **GLOSSARY TERM**
> **Magnitude** measure of the amount of light (and in some cases other types of radiation such as infrared) received from a luminous celestial object

Observing Faint Comets

You will be able to find and follow most comets through a 6-inch telescope from a dark observing site away from lights. Because comets are extended objects (not point-sources) light pollution hampers their visibility to a much greater degree than planets or star clusters. If you must observe a comet from near a city, a 10-inch telescope is about the smallest instrument you should use. Most amateur astronomers' telescopes can detect comets when they shine between 10th and 11th magnitude or brighter. Comet C/2011 L4 (PANSTARRS) was bright enough in early 2013 to be seen by many observers without optical aid, even from moderately light-polluted sites (Fig. 15.9).

It is hoped that the comet you are observing will peak in brightness when the Moon is out of the sky. Even that natural form of light pollution hides the outer coma and makes low-contrast features difficult to see. The week on either side of a full Moon usually belongs to imagers rather than to visual observers, because with careful processing photographers can electronically subtract the extra sky light from the image.

Even from a rural location away from city lights, the background sky can be bright enough at low power to hide the fuzzy glow of a faint comet. Adjust the power up to 100x or more to increase the contrast between the comet and the background sky. Try using a dark cloth or towel to cover your head and the eyepiece. By eliminating as much stray light as possible, you can see faint details more easily. On cooler evenings, try not to fog the eyepiece with your breath.

A telescopic observation will allow you to examine the coma in detail. Use a variety of magnifications, and take your time. Although high magnification will darken the field of view, your eye is better at detecting faint structure when it is bigger. Slowly move your eye around, and focus on high-contrast zones.

15.9 Comet C/2011 L4 (PANSTARRS).

Courtesy of Damian Peach

Does the coma look elliptical (and to what degree) at all magnifications? Note any irregularity in its shape or brightness. On many occasions, even observers using medium-size telescopes have seen fragmenting within the coma. If you are lucky enough to see this, the comet will appear to have several bright central regions. When observing at high power, also look for jets. These features will appear as lines, or angular rays, generally in the Sun's direction because **outgassing**, due to solar heating on the comet's sunward side, creates them.

> **GLOSSARY TERM**
>
> **Outgassing** release of a comet's frozen gas due to heating by the Sun

Looking Up

Tips on Photographing Comets

If a bright comet is visible, get out your camera and shoot! Even if you are a beginning photographer, you may be surprised by your results. And in today's digital age, once you have a camera it costs nothing to shoot a lot of images. Try all your camera's settings, save the good pictures, and discard the ones you don't like. Here are a few more tips:

- Mount your camera on a tripod for maximum stability.
- Always use a cable release. If you don't touch your camera, you will limit vibrations that can ruin an image.
- If your camera has a manual setting, use it. Choose a fast ISO setting, and bracket your exposures. Bracketing is a technique in which photographers choose a range of lens openings (f stops) or shutter speeds during a sequence of exposures.
- Set the lens aperture to wide open. If you have more than one lens, use them all. If any have a zoom capability, try changing that setting from one exposure to the next in a series of shots.
- Take the surroundings into account. Some of the most pleasing shots of comets show foreground objects such as buildings, an observer with binoculars or a telescope, or even a body of water that reflects the comet's image.
- For piggyback shots in which you connect a camera to a motor-driven telescope, try a slower ISO setting, this will produce less electronic noise and may allow you to record more detail.
- Don't be afraid to experiment with colored filters that screw onto your camera lens. A light blue filter will accentuate the bluish ion tail. A light or dark yellow or even an orange filter (if the comet is bright) may bring out unseen detail in the dust tail.
- Always plan to record more of the comet's tail than you can see with your naked eye because the camera will capture more of it than your eyes will pick up. Frame the comet in the shot appropriately as seen in the image below.

Courtesy of Damian Peach

Check Your Understanding

1.1 Give a short description of a comet that describes its nucleus, coma, and tail.

1.2 How is a comet's dust tail different from its ion tail?

1.3 Give several reasons why binoculars are best for observing bright comets.

1.4 Which comet would be more visible in the sky, one with a degree of condensation (DC) of 0 or one with a DC of 9? Explain your answer.

1.5 What is the best way to estimate the brightness of a comet?

1.6 Why is it especially important to observe comets from a dark site?

1.7 How can you tell if a comet has fragmented?

Asteroids

The most exciting celestial objects in the last decade have been asteroids (also called minor planets), such as 433 Eros pictured in Figure 15.10. Most observers would like to discover a comet, but few people will. However, many people have been discovering asteroids, and some people have discovered a lot of asteroids.

Asteroids are small, rocky bodies that orbit the Sun. The first asteroid, Ceres, was discovered by the Italian astronomer Giuseppe Piazzi (1746–1826) on the first day of the nineteenth century, January 1, 1801. Ceres is also the largest of the asteroids. English astronomer Sir John Herschel (1792–1871) bestowed the name *asteroid* on these bodies. As of this writing, individuals and professional asteroid surveys have discovered more than a quarter of a million asteroids.

Asteroid Types

Main Belt asteroids (Fig. 15.11) orbit in a zone between 1.8 and 4.0 AU from the Sun, essentially between Mars and Jupiter. Generally, this area is referred to as the Asteroid Belt. Within the belt are areas where there are very few asteroids, called **Kirkwood gaps**. The gravity of Jupiter is responsible for these gaps.

Asteroids in the Main Belt are divided into families with mythological names. Among them are the Cybeles, Eos, Floras, Hildas, Hungarias, Hygiea, Koronis, Phocaea, Themis, and Veritas. Kirkwood gaps separate asteroid families. One theory is that each family may have been a single body shattered by an impact with another asteroid long ago, leaving the pieces to occupy the same orbit as the original. Astronomers name the families after the main asteroid in them.

Near-Earth asteroids (NEAs or, more usually, **NEOs** when called **near-Earth objects**) are those that come close to Earth's orbit at some point in their own orbits. Each has a perihelion of 1.3 AU or less. There are three subgroups of NEOs.

> **GLOSSARY TERMS**
>
> **Main Belt** area between the orbits of Mars and Jupiter where most asteroids are found
>
> **Kirkwood gap** region in the asteroid belt essentially clear of asteroids caused by Jupiter's gravity
>
> **Near-Earth object (NEO)** asteroid with a statistical chance of colliding with Earth; one whose orbit crosses Earth's orbit

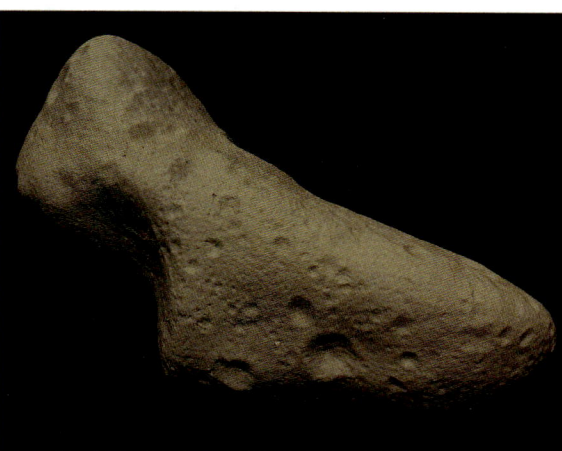

15.10 Visualization of the asteroid 433 Eros made from measurements taken by the Near Earth Asteroid Rendezvous (NEAR) Shoemaker spacecraft in February 2000.
Courtesy of NEAR Project, NLR, NASA/JHUAPL, Goddard SVS

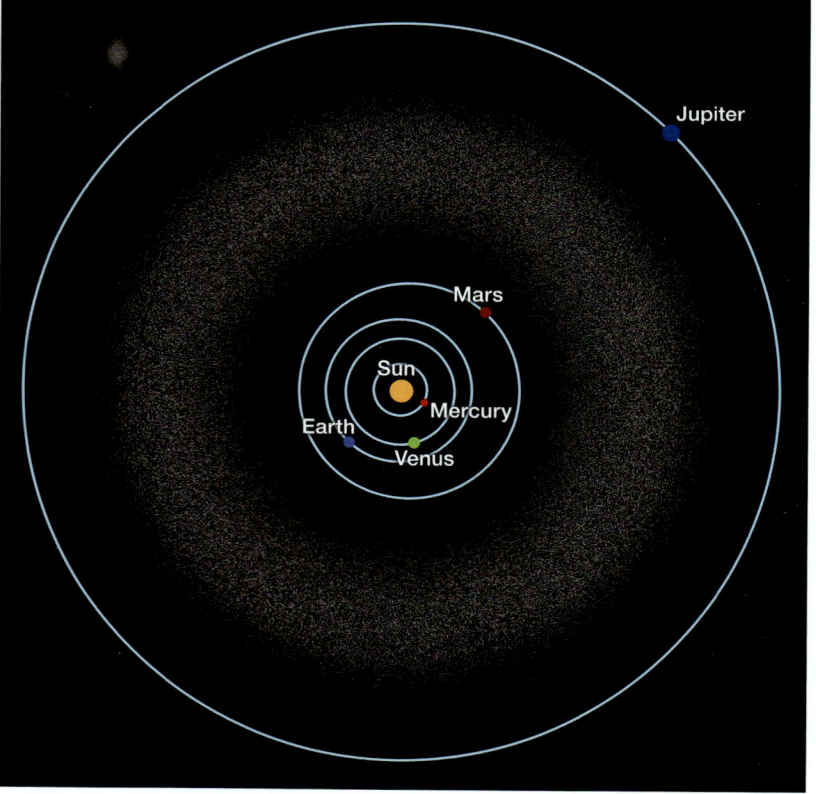

15.11 Main Belt asteroids.
Courtesy of Roen Kelly, *Astronomy* magazine

The Aten and **Apollo asteroids** (Fig. 15.12) cross Earth's orbit. The Amor asteroids do not reach Earth's orbit but do cross that of Mars. Table 15.1 shows each subgroup, its semimajor axis, and its perihelion.

The **Trojan asteroids** (Fig. 15.13) lie along the orbital path of Jupiter, at stable **Lagrangian points**. The French mathematician Joseph Louis Lagrange (1736–1813) suggested that when objects in the same plane form an equilateral triangle with the Sun, they can share the same orbit without catching up to or colliding with each other. In the Earth-Sun system, there are five Lagrangian points, designated L1 through L5. The two largest groups of asteroids lie at the L4 and L5 points on Jupiter's orbit, at an angle of 60° to both the Sun and Jupiter. Their orbits are stable, with L4 leading and L5 trailing Jupiter.

Transneptunian objects (TNOs) are, as their name indicates, solar system bodies located past the orbit of Neptune. Astronomers often group them with asteroids, but they are small, icy bodies more similar to comets. Their range of distances is 30 AU (Neptune) to farther than 50 AU. Observations show that TNOs lie within a thick band around the ecliptic. This region is generally referred to as the Kuiper Belt and objects there are often known as Kuiper Belt objects (KBOs).

Table 15.2 shows the three main types of asteroids by their chemical composition; a dozen or so other rare types of asteroids fall within this classification scheme. Astronomers have assigned them their own letter designations associated with clustering of brightness and color or spectral properties.

> **GLOSSARY TERMS**
>
> **Apollo asteroids** group of near-Earth objects named for 1862 Apollo, the first asteroid of the group to be discovered
>
> **Trojan asteroids** asteroids in Jupiter's orbit that follow it around the Sun at one of the planet's stable Lagrangian points
>
> **Lagrangian points** five locations related to a planet's orbit around the Sun where asteroids can maintain stable orbits
>
> **Transneptunian object** any solar system object that orbits the Sun farther from it than Neptune

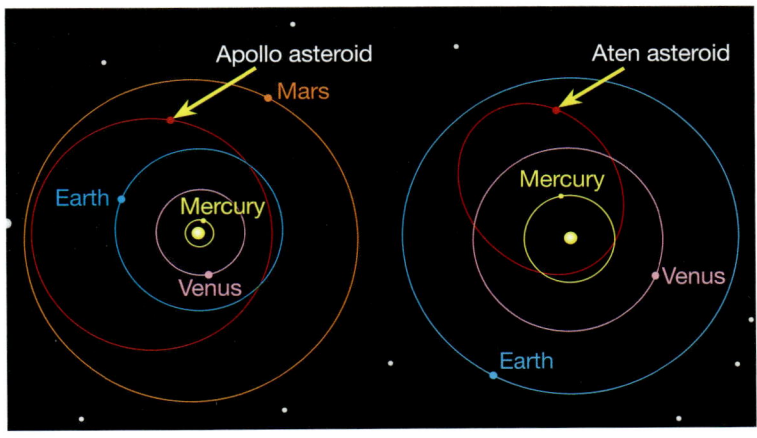

15.12 Apollo and Aten asteroids. *Courtesy of Roen Kelly, Astronomy magazine*

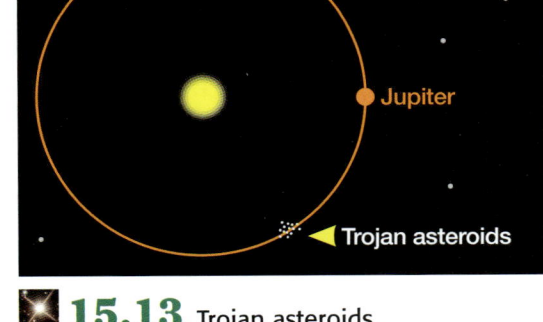

15.13 Trojan asteroids. *Courtesy of Roen Kelly, Astronomy magazine*

TABLE **15.1** Near-Earth Asteroid Subgroups

Subgroup	Semimajor Axis	Perihelion
Aten	less than 1.0 AU	less than 0.983 AU
Apollo	greater than 1.0 AU	less than 1.017 AU
Amor	greater than 1.0 AU	between 1.017 and 1.3 AU

TABLE **15.2** Asteroid Types by Chemical Composition

Type	Percent of Total	Albedo	Color	Properties
C	75	0.03–0.06	Grayish	Rich in carbon
S	17	0.10–0.22	Greenish to Reddish	Metal-rich silicates; brighter than C-type
M	8	0.10–0.18	Reddish	Mainly iron and nickel; the brightest type

Asteroid Designations

Upon discovery, astronomers assign an asteroid a temporary designation consisting of six characters. The first four characters are the year of the discovery, the next character is a letter indicating the half-month in which the discovery occurred (they don't use the letters "I" and "Z"), and the last character is another letter numbering from 1–25 the asteroids discovered in that half-month (this time only "I" is unused). For example, the asteroid 2000 MB would be the second asteroid discovered in the second half of June 2000. If there are more than 25 asteroid discoveries in a half-month then the last character reverts to "A" but with the addition of a number. An example is Asteroid 2012 DA14. This asteroid was the 39th asteroid discovered in the second half of February 2012. It is a 30-meter-wide lump of rock that set a record as the closest pass to Earth ever of an object of this size.

On February 15, 2013, Asteroid 2012 DA14 passed within just 17,200 miles of Earth's surface. The image in Figure 15.14 was captured on the night of closest approach. It shows the asteroid speeding through the field of view. Each streak represents a single 30-second exposure, which revealed the asteroid's rapid motion across the sky.

When astronomers have studied an asteroid long enough to determine its orbit, they assign it a permanent number. This number is simply the next in a sequential list. At this point, the discoverer can suggest a name for the asteroid to the Minor Planet Center (MPC). Once the MPC issues a citation for the asteroid, its name becomes official. Usually both number and name are listed, as in 1 Ceres, 131245 Bakich, or 21 Lutetia (Fig. 15.15).

Observing Asteroids

Although asteroids are essentially point sources and look no different than stars through amateur telescopes, many are bright enough to observe even through 4-inch instruments. The key to recognizing an asteroid visually is to observe it several times.

Making a sketch of the eyepiece's field of view at two different times is one method. A second method is to mark a printed star chart with the position of the possible asteroid and then return to the field a couple hours later.

A third method involves photography through your telescope. Take a picture of what you think is the asteroid's location. Several hours later, photograph the same area again. When you compare the two images, one (or more) of the points of light will be in a different position; that is the asteroid. Whatever method you use, it is best to observe asteroids visually near opposition when they are brightest and when their relative motion is greatest.

15.14 Path of Asteroid 2012 DA14.

Courtesy of Damian Peach

15.15 Asteroid 21 Lutetia.
Courtesy of ESA 2010 MPS for OSIRIS Team/MPS/UPD/LAM/IAA/RSSD/INTA/UPM/DASP/IDA

The Big Asteroid Surveys

Professional sky surveys are finding more objects than ever before, and these discoveries are pushing the magnitude limit of amateur equipment. Figure 15.16 plots nearly 500,000 minor planets (or asteroids) discovered during the past 200 years. The average magnitude of minor planet discoveries has continued to trend toward fainter objects during the last 160 years (Fig. 15.17).

The four professional observatories that participate in the surveys are the Catalina Sky Survey, Lincoln Near Earth Asteroid Research Program (LINEAR), Spacewatch, and Lowell Observatory Near-Earth Object Survey (LONEOS), which is shown in Figure 15.18. Information about these observatories is listed in Table 15.3. The goal of these professional observatories is a complete survey of all asteroids down to a relatively small size. The main reason is that astronomers want to identify potentially life-threatening NEOs. These surveys use telescopes larger than the average amateur astronomer uses as well as sensitive CCD cameras. The telescopes usually scan a large region of the sky several times each night. Software notes objects identified to have moved from one image to the next. The software then determines if the object is a typical Main Belt asteroid, or something different, such as an NEO.

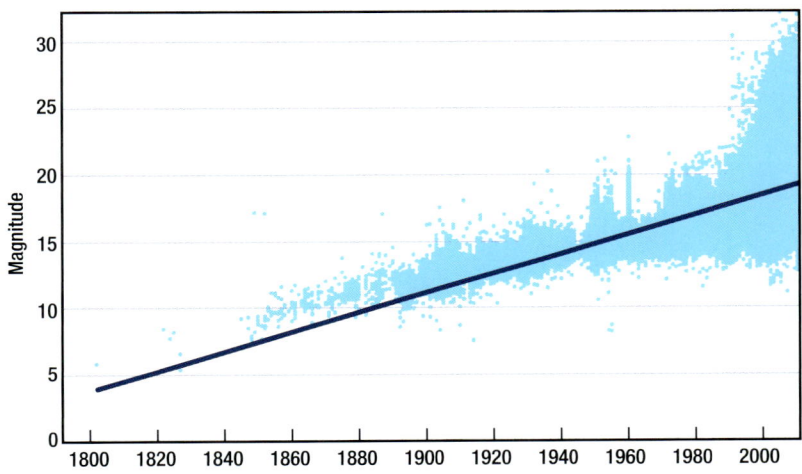

15.16 Number of minor planet discoveries by year.
Courtesy of Roen Kelly after Stephen G. Cullen, Astronomy magazine

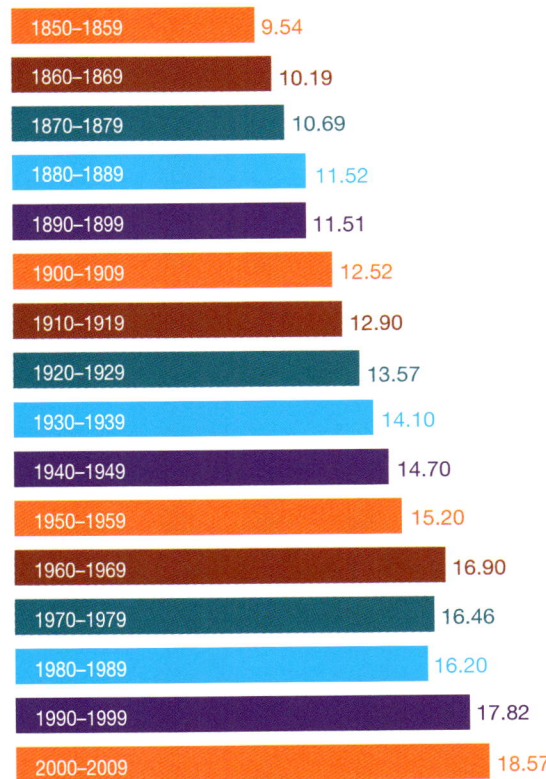

15.17 Average apparent magnitude of minor planet discoveries since 1850.
Courtesy of Roen Kelly after Stephen G. Cullen, Astronomy magazine

15.18 Lowell Observatory Near-Earth Object Search (LONEOS).
Courtesy of Lowell Observatory

Minor Bodies **CHAPTER 15** **295**

There are three reasons these surveys discover far more objects than amateurs discover:

1. Their mountaintop locations give them more clear nights.
2. Computer automation controls them so they maximize their imaging time.
3. Their optics are large, so they collect more light and thus record fainter objects.

TABLE 15.3 Surveys Used to Seek NEOs

Survey	Nightly Area Covered	Times Photographed	Telescope Used	Location
Lowell Observatory Near-Earth Object Survey (LONEOS)	600 square degrees	three	0.6m Schmidt Telescope	Flagstaff, Arizona
Catalina Sky Survey	417 square degrees	three	1.5m, 0.68m, and 0.5m Schmidt Telescopes	Atop Mount Lemmon and Mount Bigelow near Tucson, Arizona, and at Siding Spring Observatory in Australia
Lincoln Near Earth Asteroid Research program (LINEAR)	1,200 square degrees	five	Two 1m wide field telescopes	Near Socorro, New Mexico
Spacewatch*	24.6 square degrees	three	0.9m and 1.8m reflectors	At Kitt Peak National Observatory southwest of Tucson, Arizona

*The Spacewatch survey has a different strategy than the others. Their procedure searches a smaller area but to a much fainter limiting magnitude.

Check Your Understanding

2.1 Which was the first asteroid discovered, and who found it?

2.2 What features separate asteroid families within the asteroid belt?

2.3 Name two asteroid subgroups that cross Earth's orbit.

2.4 What range of distances do transneptunian objects have?

2.5 When do astronomers assign an asteroid a permanent number?

2.6 Approximately how many asteroids have astronomers discovered in the last two centuries?

Making a Comet Model

Exercise 15.1

This may be done individually, as a group, or as a classroom demonstration. The size of the comet—and materials used—may be adjusted, depending on the number of comets or the size of the comet/comets made by students in the laboratory.

Procedure

1. Collect all materials and supplies at your lab station. Put on lab apron, goggles, and gloves.
2. Line the mixing bowl with the garbage bag.
3. To the lined mixing bowl add water, charcoal, sand, ammonia, rubbing alcohol, starch, and pancake syrup. Each of these represents compounds found in comets, with the exception of starch, which is used as a binder in the model.
4. Mix the components in the bowl as completely as possible, using the spatula or spoon.
5. Wearing heavy-duty gloves for protection, obtain the dry ice as directed. If the dry ice is not in a bag or wrapped, obtain some newspaper or other wrapping material as directed and wrap it around the dry ice.
6. Break up the dry ice into tiny pieces, using the mallet.
7. Carefully add the crushed dry ice to the mixture in the mixing bowl. Note the gas released as you add the dry ice to the mixture.
8. Wearing gloves, remove the model comet from the mixing bowl, lifting it with the garbage bag liner.
9. Next remove the comet model from the garbage bag.
10. Observing the comet model, note the following:
    a. What color is the comet model? Is the color consistent throughout?

    b. Does the comet model continue to outgas? What gas is it giving off?

    c. If you pick up (always use protective gloves) and move the comet model (or as an alternate, use a fan or hair dryer to blow air onto it), what happens to the gas being given off by the comet model?

Materials

- ❏ Mixing bowl
- ❏ Spatula or mixing spoon
- ❏ Mallet
- ❏ Goggles
- ❏ Heavy-duty gloves (to protect for low temperatures)
- ❏ Apron
- ❏ Garbage bag, 7- or 13-gallon size
- ❏ Newspaper or other paper material in which to wrap the dry ice
- ❏ Water, 1 L
- ❏ Dry ice, 2½ to 3 pounds
- ❏ Charcoal, ½ cup
- ❏ Sand, ½ cup
- ❏ Pancake syrup, 10 ml
- ❏ Ammonia, 5 ml
- ❏ Rubbing alcohol, 10 ml
- ❏ Starch, 1 tablespoon

CAUTION !
USE SAFETY PROCEDURES as directed by your instructor at all times.

d Over a period of time, what happens to the comet model?

11 Dispose of the comet model as directed by your instructor.

Exercise 15.2 Comet Observing and Imaging

Procedure

Materials
- ❏ Comet ephemeris
- ❏ Telescope, eyepieces, etc.
- ❏ Red flashlight
- ❏ Clipboard (to hold your papers)
- ❏ Sketching pencils

Part I

1 Set up the lab's telescope or go to the observatory as directed.

2 Make comet observations, completing the observing log in Table 15.4 on the data sheet, page 299.

  a Comet designation: Name or official designation, for example, Comet ISON, C/2012 S1, or C/2012 S1 (ISON).

  b Time of observation: Local or universal time as directed by your instructor.

  c Location: Observatory or observing location.

  d Sky conditions: Can be general conditions; include percent of cloud cover, a seeing estimate, limiting magnitude estimate, and other details as directed.

  e Magnification(s): Calculate the magnification through each eyepiece used.

  f Magnitude estimate: Make an estimate of the comet's brightness based on comparison to other stars.

  g Reference star or stars: Which star or stars did you use to make your magnitude estimate?

  h Estimate of coma diameter: Make an estimate of the size of the comet's coma through the telescope by knowing the eyepiece's field of view.

  i Estimate of tail length: Use just your eyes for long tails or make a telescopic estimate by knowing the eyepiece's field of view.

3 Make drawings at the telescope in the space provided on the data sheet as directed by your instructor.

Part II

1 Image the comet as directed. Record the additional information in Table 15.5 on the data sheet, page 300. Include as many details as you can (camera type, system focal ratio, etc.)

2 Attach printed images or submit images as directed by your instructor.

3 Turn in your completed data sheet before leaving the lab.

Name _____ Section _____ Date _____

Observing and Imaging Comets

Exercise 15.2 Data Sheet

Part I

TABLE 15.4 Comet Observation Chart

Comet designation	
Date of observation	
Time of observation	
Location	
Sky conditions	
Telescope used	
Magnification(s)	
Magnitude estimate		Reference star or stars	
Estimate of coma diameter	
Estimate of tail length	
Comments and notes:	

Minor Bodies CHAPTER 15

Part II

TABLE 15.5 Comet Imaging Log

Imaging system	

Imaging Log			
#	Exposure Time	Exposure Length	Notes
1			
2			
3			
4			
5			
6			
7			
8			
9			
10			

Study your images. Describe the visibility and appearance of the coma and tail. Did the comet change (appearance, location) during the imaging session?

Comparing Asteroids

Exercise 15.3

Procedure

Materials
☐ Asteroid images from space probes and Earth

1. Review the images of asteroids shown in lab, making the following in-class observations.

   a. Asteroids as seen from Earth: Characterize how asteroids look through Earth-based telescopes.

   b. Asteroids as seen from space probes: review Figures 15.19–15.23, making notes of specific characteristics in Table 15.6.

TABLE 15.6 Asteroid Comparison Chart

Asteroid	Space Probe	Size	Color(s)	Craters? Numbers? Sizes?	Other Features

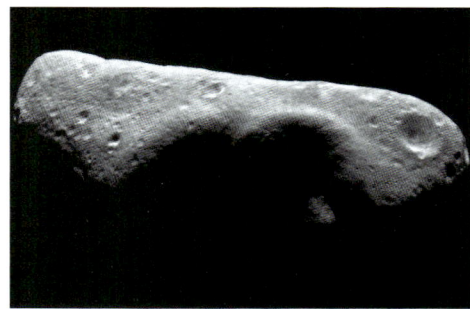
15.19 Eros.

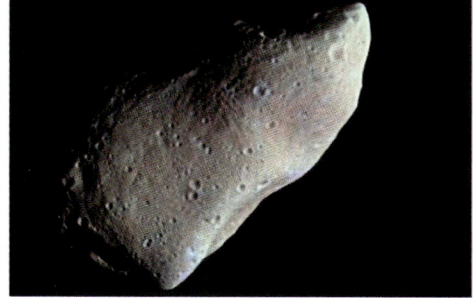
15.20 Gaspra.

15.21 Ida.

15.22 Lutetia.

15.23 Vesta.

Figures 15.19–15.23 courtesy of NASA

Minor Bodies **CHAPTER 15** **301**

Exercise 15.4 Asteroid Observing and Imaging

Procedure

Materials
- ❑ Asteroid ephemeris
- ❑ Telescope, eyepieces, etc.
- ❑ Red flashlight
- ❑ Clipboard
- ❑ Sketching pencils

Part I

1. Set up the lab's telescope or go to the observatory as directed.

2. Make asteroid observations, completing the observing log in Table 15.7 on the data sheet, page 303.

   a. Asteroid: Name or official designation; for example, 4 Eros.

   b. Time of observation: Local or universal time as directed by your instructor.

   c. Location: Observatory or observing location.

   d. Sky conditions: Can be general conditions; include percent of cloud cover, a seeing estimate, limiting magnitude estimate, and other details as directed.

   e. Magnification(s): Calculate the magnification through each eyepiece used.

   f. Magnitude estimate: Make an estimate of the asteroid's brightness based on comparison to other stars.

   g. Reference star or stars: Which star or stars did you use to make your magnitude estimate?

   _____
   _____
   _____

3. Make two drawings at the telescope in the spaces provided on the data sheet, separated by the length of time (or days) as directed by your instructor.

Part II

1. Image the asteroid as directed. Record the additional information in Table 15.8 on the data sheet, page 304. Include as many details as you can (camera type, system focal ratio, etc.)

2. Attach printed images or submit images as directed by your instructor.

3. Turn in your completed data sheet before leaving the lab.

Name _____   Section _____   Date _____

Observing and Imaging Asteroids

Exercise 15.4 Data Sheet

Part I

TABLE **15.7** Asteroid Observation Chart

Asteroid	
Date of observation	
Time of observation	
Location	
Sky conditions	
Telescope used	
Magnification(s)	
Magnitude estimate		Reference star(s)	
Comments and notes:	

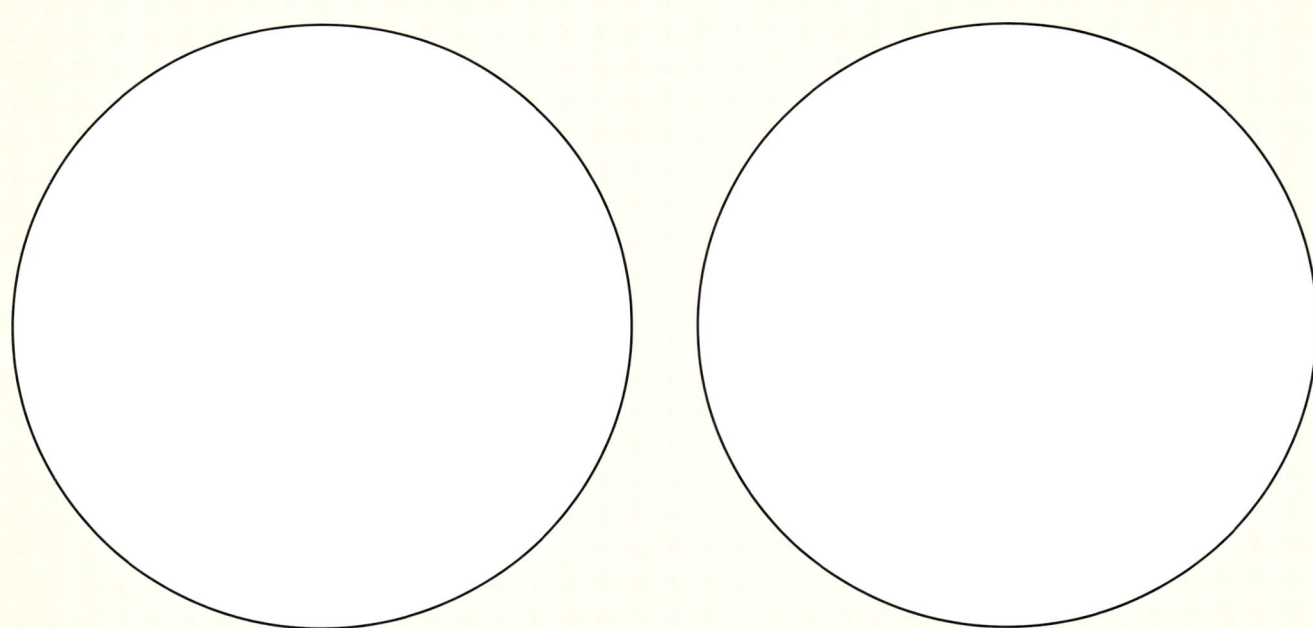

Were you able to detect the asteroid's movement against the background stars during your period of observation?

Minor Bodies CHAPTER 15   303

Part II

TABLE 15.8 Asteroid Imaging Log

Imaging system			
Imaging Log			
#	Exposure Time	Exposure Length	Notes
1			
2			
3			
4			
5			
6			
7			
8			
9			
10			

1 Study your images and answer the following questions.

  a How easy, or difficult, is it to see the asteroid?

  _____

  _____

  b How did the asteroid's position and brightness change in the time from the first to the last image?

  _____

  _____

The Solar System

Meteorites

16

LEARNING OBJECTIVES

Upon completion of this chapter, you should be able to:

1 Describe the importance of certain meteorites to science.
2 Describe how scientists test for certain materials in meteorites.
3 List the main characteristics of meteorites.
4 Explain how to identify a meteorite from a "meteorwrong."
5 Plot the trails of meteors from a known shower on a star chart using the right ascensions and declinations of their start and end points.
6 Determine the radiant of a meteor shower by plotting the starting and ending points of a given number of meteors from the same shower.
7 Determine the hourly rate of a meteor shower given the time observed and the total number of meteors seen.
8 Define all glossary terms.

Geminid Meteor Shower
Courtesy of Babak Tafreshi/Science Source

On any clear, moonless night from a dark site, an observer can see approximately six **meteors** per hour and perhaps more if the observation occurs after midnight. Several dozen times each year, however, meteors appear in greater numbers. Astronomers call those times **meteor showers**.

Individual members of meteor showers don't move randomly across the sky. Rather, all their points diverge from a single point on the sky called the **radiant**. Almost none of the meteors appear to start at the radiant, but all their **trails** lead back to that point.

Sometimes a meteor falls to Earth and becomes a **meteorite**. Over the years, meteorites have grown popular as collectibles. For buyers, the motivation is obvious. First, it is simply special to hold one of these space visitors, the ultimate alien. Second, you can ponder what that meteorite in your hand represents: the formation of our solar system, a most likely cataclysmic collision in space, and a truly unique specimen. Finally, it turns out that meteorites can be a good investment, but that is just a bonus.

A personal experience usually gets someone started collecting space rocks. A person may become interested in beginning a collection after observing a bright **fireball** in the sky, or perhaps after hearing on the news about a meteorite that landed nearby. Whatever the inspiration, the person now wants his or her own piece of the solar system.

This chapter will introduce you to the three major types of meteorites. You also will employ a process of elimination to classify a meteorite in the classroom and determine where various classes of meteorites originate.

Identification and Classification of Meteorites

It is important to understand the basic characteristics that identify a meteorite; begin by asking these four simple questions:

1. Does the sample attract a magnet? Most meteorites do because of the iron in their compositions (Fig. 16.1A).
2. Does it have a fusion crust? Such an indicator is an unusually dark, thin surface layer that melted during the meteorite's fiery plunge through Earth's atmosphere (Fig. 16.1B).

GLOSSARY TERMS

Meteor streak of light in the night sky caused by a meteoroid entering Earth's upper atmosphere and burning due to friction with the atmosphere; sometimes called a "shooting star" or "falling star"

Meteor shower annual result of Earth's passage through a meteoroid stream

Radiant apparent origin of a meteor shower, when the paths of all shower meteors are traced backward in the sky

Trail streak left by a meteor; it can be light or smoke

Meteorite any meteoroid that (after becoming a meteor) lands on Earth's surface

Fireball meteor bright enough to cast a shadow; the magnitude limit (visual) for labeling a meteor a fireball is generally given as −5

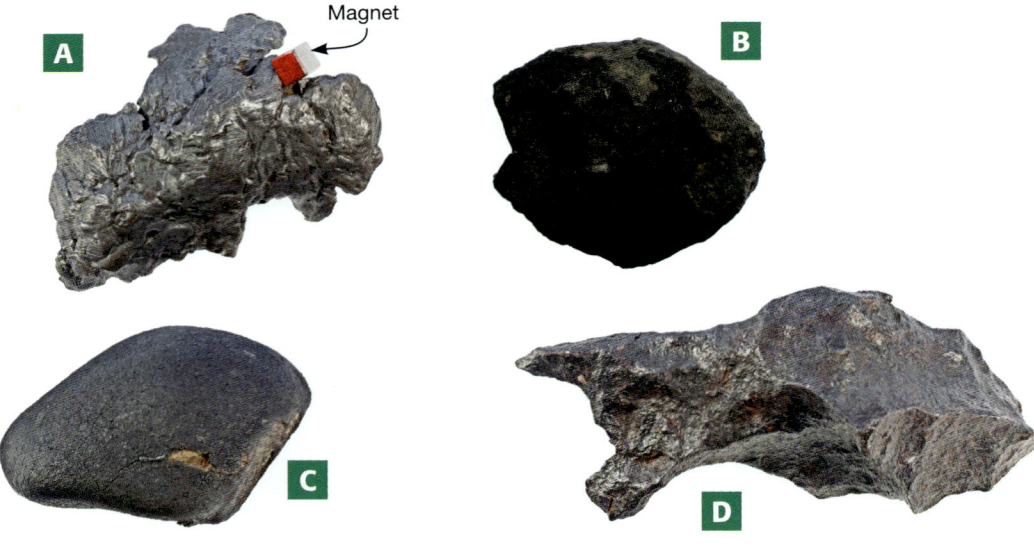

16.1 Clues for recognizing a meteorite: (**A**) magnetism, (**B**) fusion crust, (C) aerodynamic shape, and (**D**) regmaglypts.

3. Does the sample exhibit aerodynamic shaping? If the meteorite isn't tumbling during its high-speed flight through Earth's atmosphere, it will acquire a characteristic conical shape (Fig. 16.1C).

4. Does it have thumbprint-like indentations in its surface? The heat of entry into Earth's atmosphere gouges out these indentations, called regmaglypts, as parts of the specimen vaporize (Fig. 16.1D).

Not every meteorite will have all four characteristics. And, of course, exceptions to these rules exist. If a meteorite has been on Earth some time, the atmospheric flight characteristics of a fusion crust, aerodynamic shaping, and regmaglypts could be erased due to erosion from wind, water, and/or chemicals. Still, these questions are a good starting point.

There are three basic classifications, or types, of meteorites:

1. Stone meteorites
2. Iron meteorites
3. Stony-iron meteorites

Stone Meteorites

Stone meteorites (Fig. 16.2), also called stony meteorites or stones, make up 94 percent of all observed falls. Scientists believe most originate from the crusts and mantles of **asteroids**, while a few might be from **comets**. There are even some that scientists have directly linked to the Moon and Mars based on their chemical composition.

Stone meteorites contain 75 percent to 90 percent silicate materials, such as olivine and pyroxene. These minerals contain silicon, oxygen, and one or more metals. The majority of stone meteorites also have a small percentage of some nickel-iron alloy.

The two major groups of stones are chondrites and achondrites. Chondrites are so named because they contain chondrules, small spherical crystals of minerals embedded in the stony material. Achondrites consist of material similar to terrestrial volcanic and igneous rocks.

Lunar and martian meteorites fall into the achondrite group. Lunar meteorites are impact breccias—rocks formed by the melting and solidification of loose fragments that were shattered during lunar-impact events. Lunar meteorites have a slightly green fusion crust and a gray interior with angular inclusions of often-brighter materials. Scientists believe lunar meteorites were blasted off the Moon as ejecta from impact events.

A small number of recovered meteorites originated on Mars. Astronomers refer to these space rocks as the SNCs, after Shergotty, Nakhla, and Chassigny, the first three identified martian meteorite subgroups. All share characteristics that together point to a martian origin.

> **GLOSSARY TERMS**
>
> **Asteroid** small bodies composed of rock and metal that orbit the Sun; most (95 percent) are located in a belt between the orbits of the planets Mars and Jupiter; also known as minor planet
>
> **Comet** one of many relatively small solar system objects that orbit the Sun, composed of frozen gases and dust; the parent bodies of most meteor showers

16.2 Types of stone meteorites: (**A**) achondrite (northwest Africa 1929), and (**B**) chondrite (Cocklebiddy).

Iron Meteorites

Iron meteorites (Fig 16.3), or irons, are actually a combination of iron, nickel, and a trace of cobalt. Iron is the predominant metal, with between 5 percent and 50 percent nickel (as an alloy with the iron). Iron meteorites divide into three basic subgroups: octahedrites, hexahedrites, and ataxites. These groups have different nickel-iron ratios.

Octahedrites are the most common structural group of the iron meteorites, containing between 9 percent and 18 percent nickel. They get their name from the fact that the crystal structure is octahedral in shape. Hexahedrites have much-lower nickel concentrations than octahedrites, always below 5.8 percent. They do not show the same pattern as octahedrites, instead showing parallel lines when polished and etched. Ataxites are composed of more than 18 percent nickel and show no pattern when polished and etched.

Stony-Iron Meteorites

Stony-iron meteorites (Fig. 16.4), or stony-irons, are mixtures of silicates and nickel-iron in roughly equal proportions. Only about 1 percent of all falls have been stony-irons. Scientists divide them into two main groups: pallasites and mesosiderites. Pallasites contain plentiful olivine within their nickel-iron matrices. Mesosiderites have approximately equal amounts of stony and metal components.

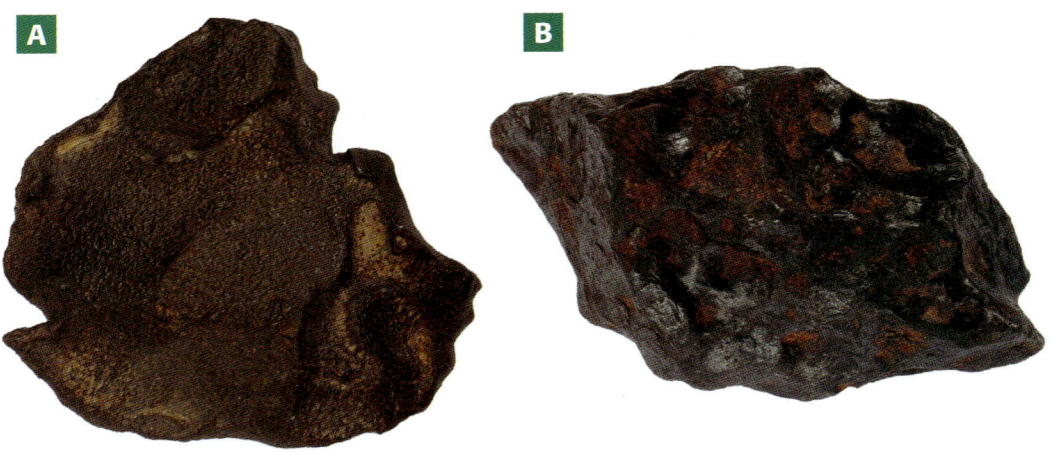

16.3 Types of iron meteorites: (**A**) ungrouped (Gebil Kamil), and (**B**) octahedrite (Odessa).

16.4 Types of stony-iron meteorites: (**A**) mesosiderite (Vaca Muerta), and (**B**) pallasite (Fukang).

Check Your Understanding

1.1 Why do meteorites have dark crusts?

1.2 How do regmaglypts form?

1.3 Which is the most common type of meteorite?

1.4 What type is a stone meteorite if it came from Mars?

1.5 What is the second most common metal in iron meteorites?

Starting a Collection

Collectors usually purchase meteorites at a per-gram price, which can range from cents to thousands of dollars, depending on the rarity of the specimen. The physical characteristics of the purchased meteorite also can vary; not all will be an unbroken or uncut specimen (Fig. 16.5A).

A popular type is a slice or end piece (Fig. 16.5B). The exterior of a meteorite tells only part of the story. By slicing and polishing the meteorite, you can see the internal matrix and composition of the sample. Think of cutting the specimen like slicing a loaf of bread: You end up with slices and two end pieces. The end pieces show the exterior of the meteorite on one side and the interior on the opposite side. The individual slices show the interior on both sides as well as the edge around the slice, like the crust on a piece of bread.

You can also buy fragments (Fig. 16.5C), pieces broken off a meteorite (either in flight or when it strikes the ground). These may exhibit a fusion crust but will not have a cut and polished side.

Due to the high cost of many meteorites, a number of collectors obtain very small samples, called micro-mounts, of many different meteorites. Weighing up to a few grams each, they can be whole specimens, slices, end pieces, or fragments.

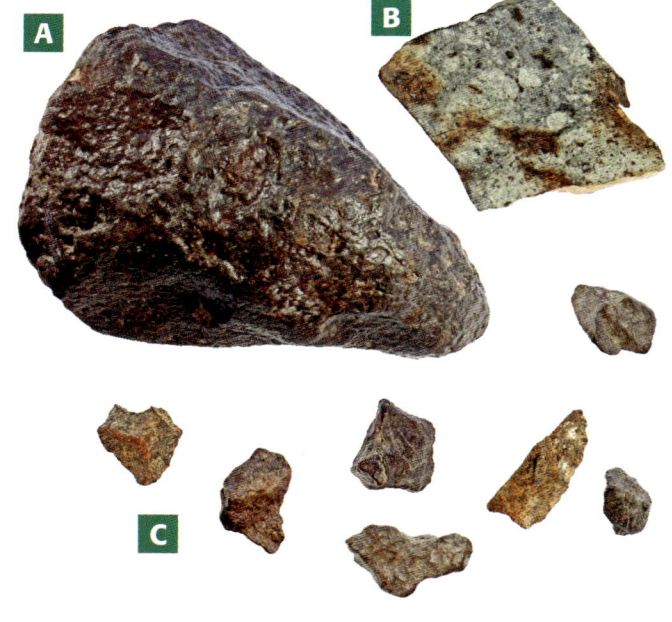

16.5 Examples of meteorite types: (**A**) uncut, (**B**) end piece showing internal matrix and composition, and (**C**) fragments.

GLOSSARY TERM

Widmanstätten pattern crystal patterns seen in any iron meteorite after someone has cut, polished, and etched a face of it with a mild acid solution

A meteorite's name, like Sikhote-Alin from Russia, usually refers to the closest post office or identified geographic region where the meteorite fell or was found. When many meteorites are discovered in one location, a number may follow the name (as is the case for northwest Africa finds).

Choosing what type of physical characteristics you desire in your meteorite collection depends on several things, from personal preference to the size of your collection. For example, if you want only a few samples, consider whole meteorites with matching slices/end pieces. Any quality basic collection includes whole meteorites and slices of all three major classifications. Figure 16.6 shows a slice of a whole Gibeon iron that has been polished and then etched to show the **Widmanstätten pattern**. The Widmanstätten pattern (also called the Thomson structure) gives the collector a look at the exterior of the meteorite as well as the inside story. It occurs in iron meteorites because broad kamacite mineral bands sandwich between narrow taenite bands as the meteorite cools.

16.6 Widmanstätten pattern, broad kamacite mineral bands sandwiched between narrow taenite bands.

Beyond the basics, you can collect meteorites in a variety of human-created shapes, from knives to jewelry. Some watches sport an etched iron meteorite face, often at astronomical prices. Another option is collecting HAMmers, meteorites that have struck humans, animals, or man-made objects. Other popular pieces are tektites and shatter cones. These glass rocks and underground geological features are earthly results of a meteorite's violent impact with our planet and, in themselves, can make for a fascinating collection.

As with any collection, there are many ways to present or store your meteorites. You can display micro-mounts in small square plastic boxes or round containers called gem jars (Fig. 16.7A). You can store slices, end pieces, and small whole meteorites in Riker mounts—flat boxes with transparent tops primarily used for biological specimens (Fig. 16.7B). Easel stands for collectible plates are ideal for slices, as long as they are sturdy. You can place whole specimens on a stand of your choosing, a caliper-like mount, or under cover. A variety of glass and plastic presentation boxes and jars are readily available.

No matter what you collect or how you display your meteorites, they can be as addictive as observing all of the Messier objects, the Lunar 100, or Mars on a regular basis. Meteorite collecting can complement your love of the night skies or become a major focus. Either way, holding something with the story of a meteorite is special, as special as the meteorite itself.

16.7 Examples of meteorite display methods: (**A**) micro-mounts featuring small meteorite samples in a gem jar set, and (**B**) Riker mount featuring a Thuathe stone.

Check Your Understanding

2.1 What is the advantage of slicing a meteorite?

2.2 Give another name for a meteorite's Thomson structure.

2.3 How do tektites relate to meteorites?

2.4 Describe two ways a collector might choose to display meteorites.

Looking Up

Buyer Beware!

When buying meteorites, be sure to work with reputable dealers. Established sellers offer authentic samples, both in classification and name. An unscrupulous dealer can take a common meteorite and, by changing its name, charge many times more for the sample. Is it a meteorite? Yes, but it is not what you think you purchased; unfortunately, you have paid for a switched sample. Good dealers also will prepare meteorites correctly, especially specific irons. Some samples are subject to oxidation, such as Nantan, an infamous iron from China. Such iron meteorites can literally turn to a pile of rust on your display shelf if not properly prepared.

You can purchase meteorites at gem and mineral shows, on the Internet from dealers, or on auction sites like eBay. Use caution, especially on auction sites.

Many serious collectors track common rocks being sold as meteorites or even as rare lunar or Martian specimens. Serious collectors refer to these samples as "meteorwrongs"; they are not meteorites!

The International Meteorite Collectors Association (IMCA) is a group of meteorite collectors, sellers, and researchers whose primary purpose is to assist collectors looking for authentic meteorites and help people learn more about meteoritics. IMCA members subscribe to a specific set of ethics; members who sell meteorites can have their membership revoked if they break the IMCA code. Learn more at www.imca.cc.

Exercise 16.1 — Meteorite Sample Study

 **Procedure**

Materials
- ❑ Meteorites
- ❑ Magnet

1 You will find several stations set up in the lab. All but the last station will contain identified meteorite samples. Each lab team of two people will spend five minutes per station. Study each meteorite carefully, noting the name, classification, and characteristics in the data sheets on pages 313–316. Make certain you return each sample back to its original station.

2 The last station will have several unknown samples for you to identify after you complete your observations at the previous stations. You may disagree with your lab partner or other students whether an unknown is a meteorite or "meteorwrong." Regardless of your identification, be sure to justify why you think any of the unknowns are meteorites or "meteorwrongs." Your explanation is the critical part of your answer, not simply giving a correct response.

3 For each case, explain how you came to your conclusion in the data sheet on pages 317–318.

4 Make certain you place the unknown sample back on the correct unknown numbered sample card.

Name _____   Section _____   Date _____

Meteorite Sample Classifications

Station Number _____

Name _____   Classification _____

☐ **Whole Meteorite**   ☐ **Partial Meteorite**   ☐ **Slice** *(Check one)*

Describe the meteorite, noting any specific characteristics. Is it magnetic? Does it have any fusion crust? Can you see regmaglypts? Are there any other distinguishing features? What specifically do you see in this sample that would help you identify an unknown specimen?

_____
_____
_____
_____
_____
_____
_____

Station Number _____

Name _____   Classification _____

☐ **Whole Meteorite**   ☐ **Partial Meteorite**   ☐ **Slice** *(Check one)*

Describe the meteorite, noting any specific characteristics. Is it magnetic? Does it have any fusion crust? Can you see regmaglypts? Are there any other distinguishing features? What specifically do you see in this sample that would help you identify an unknown specimen?

_____
_____
_____
_____
_____
_____
_____

Station Number _____

Name _____   Classification _____

☐ **Whole Meteorite**   ☐ **Partial Meteorite**   ☐ **Slice** *(Check one)*

Describe the meteorite, noting any specific characteristics. Is it magnetic? Does it have any fusion crust? Can you see regmaglypts? Are there any other distinguishing features? What specifically do you see in this sample that would help you identify an unknown specimen?

_____
_____
_____
_____
_____
_____
_____

Exercise 16.1 Data Sheet

Exercise 16.1 Data Sheet

Station Number _____

Name _____ Classification _____

☐ **Whole Meteorite**   ☐ **Partial Meteorite**   ☐ **Slice**   (Check one)

Describe the meteorite, noting any specific characteristics. Is it magnetic? Does it have any fusion crust? Can you see regmaglypts? Are there any other distinguishing features? What specifically do you see in this sample that would help you identify an unknown specimen?

_____
_____
_____
_____
_____
_____
_____

Station Number _____

Name _____ Classification _____

☐ **Whole Meteorite**   ☐ **Partial Meteorite**   ☐ **Slice**   (Check one)

Describe the meteorite, noting any specific characteristics. Is it magnetic? Does it have any fusion crust? Can you see regmaglypts? Are there any other distinguishing features? What specifically do you see in this sample that would help you identify an unknown specimen?

_____
_____
_____
_____
_____
_____
_____

Station Number _____

Name _____ Classification _____

☐ **Whole Meteorite**   ☐ **Partial Meteorite**   ☐ **Slice**   (Check one)

Describe the meteorite, noting any specific characteristics. Is it magnetic? Does it have any fusion crust? Can you see regmaglypts? Are there any other distinguishing features? What specifically do you see in this sample that would help you identify an unknown specimen?

_____
_____
_____
_____
_____
_____
_____

Name _____   Section _____   Date _____

Meteorite Sample Classifications

Station Number _____

Name _____   Classification _____

☐ **Whole Meteorite**   ☐ **Partial Meteorite**   ☐ **Slice**   (Check one)

Describe the meteorite, noting any specific characteristics. Is it magnetic? Does it have any fusion crust? Can you see regmaglypts? Are there any other distinguishing features? What specifically do you see in this sample that would help you identify an unknown specimen?

_____
_____
_____
_____
_____
_____
_____

Station Number _____

Name _____   Classification _____

☐ **Whole Meteorite**   ☐ **Partial Meteorite**   ☐ **Slice**   (Check one)

Describe the meteorite, noting any specific characteristics. Is it magnetic? Does it have any fusion crust? Can you see regmaglypts? Are there any other distinguishing features? What specifically do you see in this sample that would help you identify an unknown specimen?

_____
_____
_____
_____
_____
_____
_____

Station Number _____

Name _____   Classification _____

☐ **Whole Meteorite**   ☐ **Partial Meteorite**   ☐ **Slice**   (Check one)

Describe the meteorite, noting any specific characteristics. Is it magnetic? Does it have any fusion crust? Can you see regmaglypts? Are there any other distinguishing features? What specifically do you see in this sample that would help you identify an unknown specimen?

_____
_____
_____
_____
_____
_____
_____

Exercise 16.1 Data Sheet

Exercise 16.1 Data Sheet

Station Number _____

Name _____ Classification _____

☐ **Whole Meteorite** ☐ **Partial Meteorite** ☐ **Slice** *(Check one)*

Describe the meteorite, noting any specific characteristics. Is it magnetic? Does it have any fusion crust? Can you see regmaglypts? Are there any other distinguishing features? What specifically do you see in this sample that would help you identify an unknown specimen?

_____
_____
_____
_____
_____
_____
_____

Station Number _____

Name _____ Classification _____

☐ **Whole Meteorite** ☐ **Partial Meteorite** ☐ **Slice** *(Check one)*

Describe the meteorite, noting any specific characteristics. Is it magnetic? Does it have any fusion crust? Can you see regmaglypts? Are there any other distinguishing features? What specifically do you see in this sample that would help you identify an unknown specimen?

_____
_____
_____
_____
_____
_____
_____

Station Number _____

Name _____ Classification _____

☐ **Whole Meteorite** ☐ **Partial Meteorite** ☐ **Slice** *(Check one)*

Describe the meteorite, noting any specific characteristics. Is it magnetic? Does it have any fusion crust? Can you see regmaglypts? Are there any other distinguishing features? What specifically do you see in this sample that would help you identify an unknown specimen?

_____
_____
_____
_____
_____
_____
_____

Name _____ Section _____ Date _____

Unknown Sample Classifications

Unknown Sample Number _____

☐ **Meteorite**   ☐ **Meteorwrong**   *(Check one)*

What characteristic(s) and/or comparison(s) did you base your decision on?

_____
_____
_____
_____
_____
_____

Unknown Sample Number _____

☐ **Meteorite**   ☐ **Meteorwrong**   *(Check one)*

What characteristic(s) and/or comparison(s) did you base your decision on?

_____
_____
_____
_____
_____
_____

Unknown Sample Number _____

☐ **Meteorite**   ☐ **Meteorwrong**   *(Check one)*

What characteristic(s) and/or comparison(s) did you base your decision on?

_____
_____
_____
_____
_____
_____

Exercise 16.1 Data Sheet

Exercise 16.1 Data Sheet

Unknown Sample Number _____

☐ **Meteorite**   ☐ **Meteorwrong**   (Check one)

What characteristic(s) and/or comparison(s) did you base your decision on?

_____
_____
_____
_____
_____
_____
_____

Unknown Sample Number _____

☐ **Meteorite**   ☐ **Meteorwrong**   (Check one)

What characteristic(s) and/or comparison(s) did you base your decision on?

_____
_____
_____
_____
_____
_____
_____

Unknown Sample Number _____

☐ **Meteorite**   ☐ **Meteorwrong**   (Check one)

What characteristic(s) and/or comparison(s) did you base your decision on?

_____
_____
_____
_____
_____
_____
_____

Unit 4 *The Solar System*

Finding the Radiant of the Leonid Meteor Shower   Exercise 16.2

A radiant is seen because of perspective. All meteors that belong to a shower are moving at parallel paths when they encounter Earth. The radiant, therefore, is the vanishing point of those trails. Meteors that appear in parts of the sky far from the radiant produce correspondingly long trails, while those coming from the radiant appear as points of light.

The radiant is usually a small area of the sky rather than a point. This is due to errors in observation, to the slightly different directions and apparent speeds of the shower members, and to the fact that the particles in space are spread out rather than concentrated at one point.

A meteor shower gets its name from the constellation in which the radiant lies. For short- to moderate-duration showers, the radiant doesn't move appreciably. For showers of long duration, however, the radiant's position may shift because it depends on the direction of the meteor stream and on the direction of Earth's orbit.

One of the ways observers evaluate meteor showers is by calculating their hourly rates. This number is calculated by taking the total number of meteors seen during a specified time and dividing by the number of observing hours. For example, assume someone watched the Perseid meteor shower from 1:30 a.m. until 5:00 a.m. During that time, they saw 210 meteors. You would find the hourly rate as follows:

Total number of meteors/hours observed = 210/3.5 = 60 meteors per hour.

The star chart (Fig. 16.8) on page 322 shows right ascensions from 7 hours to 13 hours and declinations from 60° north to 30° south. The central portion of the star chart shows the constellation Leo the Lion to scale as it would appear in the sky. You will use this chart to plot the radiant.

Procedure

Materials
- ❏ Pencil
- ❏ Ruler
- ❏ Star chart (Fig. 16.8)

1 On the star chart (Fig. 16.8, page 322), plot the beginning and ending points of the 40 meteors listed in Table 16.1. Then join the two points with straight lines. Use a sharp pencil and make light lines.

2 Indicate with an arrowhead the direction the meteor is traveling. Also place a small number corresponding to the data point near the trail.

3 After you plot all the trails, extend them by drawing dotted lines backward from the direction of flight until they all intersect. Because this exercise is imprecise, the dotted lines will not intersect at a point. The radiant, therefore, is the center of the region formed by the intersection of the lines.

TABLE **16.1** Beginning and Ending Points of 40 Meteors

Meteor Number	Beginning		Ending		Time (local)
	R.A.	Dec.	R.A.	Dec.	
1	11h05m	32.5°	11h28m	35°	1:00 a.m.
2	11h37m	33.1°	12h18m	36.2°	
3	11h55m	21.1°	12h21m	19.6°	
4	11h22m	16.6°	12h01m	20.7°	
5	10h04m	22.2°	10h34m	21°	
6	11h02m	18.6°	11h18m	17.8°	
7	11h28m	17.2°	11h52m	14.1°	
8	10h58m	16°	11h12m	13.2°	
9	12h00m	6.2°	12h28m	1.6°	
10	11h04m	11.2°	11h20m	8°	
11	11h43m	-8°	12h04m	-16°	
12	11h03m	3.5°	11h10m	1.2°	
13	11h10m	-6°	11h17m	-9.5°	1:10 a.m.
14	10h55m	-8°	12h01m	-14.6°	
15	10h15m	3.5°	10h17m	1.5°	
16	10h14m	8.5°	10h19m	5.8°	
17	10h07m	13.2°	10h07m	12°	
18	10h04m	14.5°	10h04m	12.5°	1:20 a.m.
19	10h35m	-6.9°	10h41m	-11.5°	
20	10h28m	-3°	10h31m	-7°	
21	10h21m	-1°	10h26m	-5°	
22	10h10m	-5.5°	10h16m	-10°	
23	10h06m	3.4°	10h07m	0°	
24	10h05m	-2.8	10h08m	-9°	
25	10h02m	-18.2°	10h06m	-22.2°	1:30 a.m.
26	10h01m	6.4°	10h02m	3.4°	
27	9h51m	10.3°	10h10m	10°	
28	9h14m	-16°	9h10m	-20.5°	
29	9h03m	-3°	8h55m	-8.2°	

Continues

TABLE **16.1** Beginning and Ending Points of 40 Meteors *(continued)*

Meteor Number	Beginning		Ending		Time (local)
	R.A.	Dec.	R.A.	Dec.	
30	9h21m	6°	9h10m	1.6°	1:40 a.m.
31	9h54m	20°	9h49m	19°	
32	8h48m	4.2°	8h36m	0.8°	
33	8h03m	-2.5°	7h50m	-7°	1:50 a.m.
34	8h12m	2.5°	7h55m	-0.9°	
35	8h23m	6.1°	8h09m	2.6°	
36	7h48m	3.4°	7h33m	0.9°	
37	9h43m	21.8°	9h25m	18.4°	
38	9h39m	24°	9h20m	24.5°	
39	9h34m	27°	8h50m	31.7°	
40	10h07m	37.1°	10h09m	42.1°	2:00 a.m.

4 Upon completing steps 1 through 3, answer the following questions:

a What is your best estimate of the right ascension and declination of the radiant?

b How do the lengths of the meteor trails near the radiant compare with those far away? Explain the reason for this difference.

c Why are so few meteors within the Sickle of Leo itself? What would a meteor in this region look like?

d Are there any trails that do not appear to come from the radiant? What do you think is the reason for this?

5 What was the hourly rate for this meteor shower during the hour the observer watched?

6 What was the hourly rate for the first half hour?

7 What was the hourly rate for the first 20 minutes?

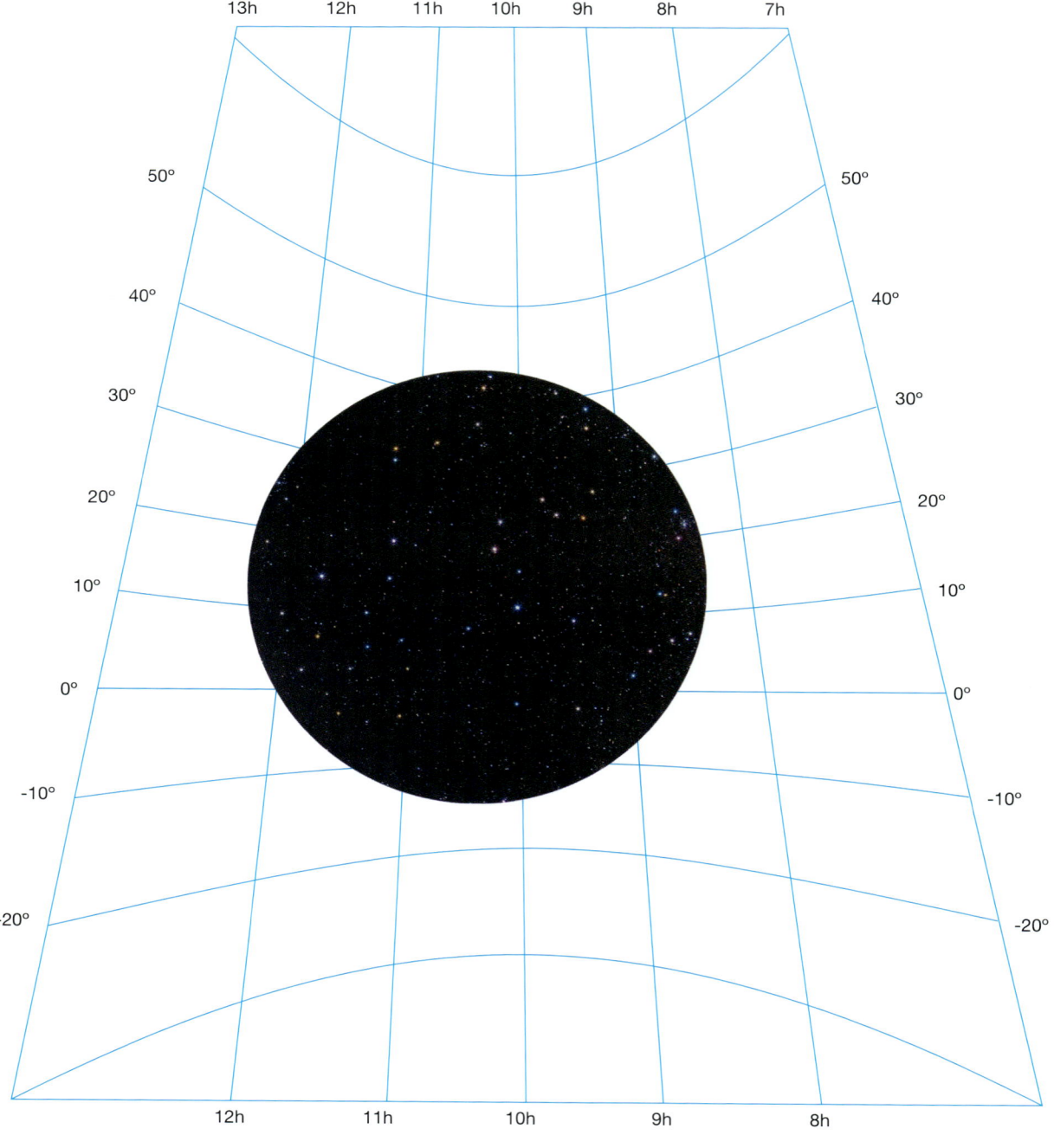

16.8 Star chart.

Courtesy of Holley Y. Bakich

The Solar System

Transits and Eclipses

LEARNING OBJECTIVES

Upon completion of this chapter, you should be able to:

1. Compare and contrast transits of Mercury and Venus.
2. Compare and contrast lunar and solar eclipses.
3. Explain why transits occur only at certain times of the year.
4. Explain the four contacts (key timings) of transits and eclipses.
5. Identify the two main parts of the shadows of Earth and the Moon.
6. Describe why an eclipse doesn't happen every month.
7. State the phase the Moon must be at for various eclipses to occur.
8. Explain how to observe transits and solar eclipses safely.
9. Define all glossary terms.

"Diamond Ring" Courtesy of Mike D. Reynolds

CAUTION

Sunlight—especially focused through a telescope or binoculars—can harm your eyes and even blind you. To observe the transit, you will need an approved **solar filter** that fits your telescope. Several manufacturers sell such filters, which are made of aluminized Mylar film or specially coated glass. Make sure you place the solar filter at the front of your telescope (before the optics), not at the eyepiece, and fasten it securely. Be sure a gust of wind can't dislodge the filter. This warning goes double for binoculars. Each lens requires a solar filter.

GLOSSARY TERMS

Solar filter accessory that lets you view the Sun safely, either with just your eyes or through binoculars or a telescope

Transit as used in this lab, the passage of one celestial body in front of another; alternate definition: the passage of a celestial body across an observer's meridian

Eclipse obscuration of light from a celestial body as it passes through the shadow of another body; such obscuration may be total or partial

Ecliptic apparent path of the Sun through the sky; the plane of Earth's orbit around the Sun

Transits and eclipses are dramatic events loved by amateur astronomers around the world. There was a time, even as recently as a century ago, when scientists could learn a lot from eclipses and transits. Studies of the Sun's outer atmosphere happened during eclipses, and measurements during transits helped to determine the distance scale of the solar system. Today, these events yield no great science, and yet we still look forward to them. Eclipse trips and cruises are big business; public observing sessions of eclipses help astronomy clubs promote their existence and programs; and images and videos of eclipses are among the most numerous produced by amateur astronomers. In this lab, you will learn how to calculate distances using transits, and if you are lucky, perhaps even observe a lunar eclipse! If one is not occurring during your term, your professor may give you the option to observe a video of an eclipse.

Transits

Of the planets visible to the naked eye, Mercury, with its orbit closest to the Sun, is the most difficult to observe. Venus ranges about twice as far from the Sun's brilliant disk as Mercury, but it is not difficult to observe because, except for the Sun and the Moon, Venus is the brightest celestial object we can see. Occasionally, however, you can catch Mercury or Venus in broad daylight as they cross the solar disk. The passage of one celestial body in front of another is called a transit.

Transits of Mercury

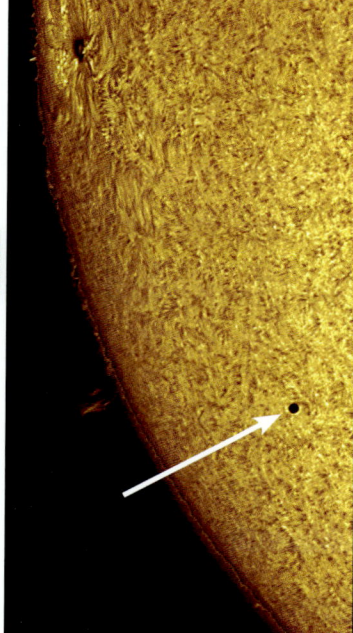

17.1 Mercury crossing the Sun's disk on November 8, 2006, as seen through a Hydrogen-alpha filter. Courtesy of Phillip Jones

Mercury orbits the Sun at an average distance of only 36 million miles (58 million kilometers). Earth is nearly three times as far, so, from our perspective, Mercury always appears near the Sun. Several times each year, the planet appears either farthest east of the Sun (when Mercury is visible in the evening sky) or farthest west (when we see the planet in the morning sky).

Four, and occasionally five, times each year Mercury passes between the Sun and Earth. Astronomers refer to this position as inferior conjunction. Mercury's orbit tilts 7° to the **ecliptic**, so most of the time it lies above or below the ecliptic at inferior conjunction, and there is no lineup with Earth. If inferior conjunction occurs within a day or so of the planet crossing the ecliptic, a transit will result. Figure 17.1 shows a single-exposure shot of a transit of Mercury in 2006 and Figure 17.2 shows a multiple-exposure shot of a transit of Mercury in 2003.

On those rare occasions—13 or 14 times each century—that Mercury lies directly between the Sun and Earth, we see it cross, or transit, the Sun's disk. Transits of Mercury occur near May 8 or November 10. In May, Mercury crosses the ecliptic heading south, and in November, the planet intersects the ecliptic moving north. Table 17.1 shows the projected next seven transits of Mercury.

324 Unit 4 The Solar System

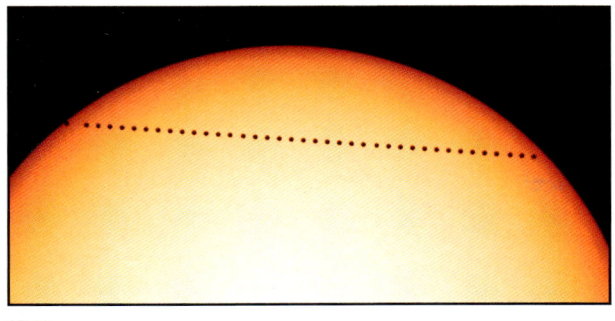

17.2 Entire 2003 transit of Mercury captured by the SOHO spacecraft. *Courtesy of NASA/SOHO*

TABLE 17.1 Future Transits of Mercury

Date	Starting Time (UT)	Duration
May 9, 2016	11h12m	7 hours 30 minutes
November 11, 2019	12h35m	5 hours 29 minutes
November 13, 2032	6h41m	4 hours 26 minutes
November 7, 2039	7h17m	2 hours 58 minutes
May 7, 2049	11h03m	6 hours 41 minutes
November 9, 2052	23h53m	5 hours 13 minutes
May 10, 2062	18h16m	6 hours 41 minutes

Transits of Venus

It takes 584 days for Venus to go from one **inferior conjunction** to the next. Venus' orbit is tilted 3.4° to the ecliptic, so most of the time it lies above or below the ecliptic at inferior conjunction, and there is no lineup with Earth. Twice every 106 or 122 years, Venus lies directly between the Sun and Earth and a transit occurs.

All transits of Venus (Fig. 17.3) fall within several days of June 8 and December 9, which are the dates the orbit of Venus crosses the ecliptic (the plane of Earth's orbit around the Sun). These points are also called **nodes**. The **longitudes** of the nodes of Venus' orbit, measured from the Sun, are approximately 77° and 257°. If Venus passes through inferior conjunction when its longitude is near 77° (around June 8) or 257° (around December 9), a transit occurs. Transits of Venus always occur in pairs eight years apart. A recurring pattern of either 105.5 or 121.5 years between the last transit of one pair and the first transit of the next pair also exists. Table 17.2 shows the projected next four transits of Venus.

Transits of Venus have allowed astronomers of the past to learn two important facts. First, they discovered that Venus has no moon. Some astronomers had thought that if Venus had a moon in a close orbit, the brilliancy of the planet would overwhelm its light, and such a moon would remain undiscovered. It was, therefore, a matter of some importance for astronomers to carefully examine Venus' immediate vicinity during transits. If a satellite of any appreciable size had existed, it would have been detected against the brightly lit background of the Sun.

Another fact first gleaned about Venus as a result of a transit was the existence of its atmosphere. This was first discovered by Russian scientist Mikhail Vasilyevich Lomonosov (1711–1765) during the transit of 1761. If Venus had no atmosphere, that feature would be totally invisible just before the planet started crossing the solar disk. All anyone would see, therefore, would be the black outline of the planet. Likewise the planet would relapse into total invisibility immediately after the transit. Lomonosov's observations proved

> **GLOSSARY TERMS**
>
> **Inferior conjunction** point at which Mercury or Venus lies between Earth and the Sun
>
> **Node** intersection of a planet's orbit and the ecliptic (the plane Earth makes as it orbits the Sun)
>
> **Longitude** degree measurement (from 0° to 360°) that defines where a planet is in its orbit around the Sun

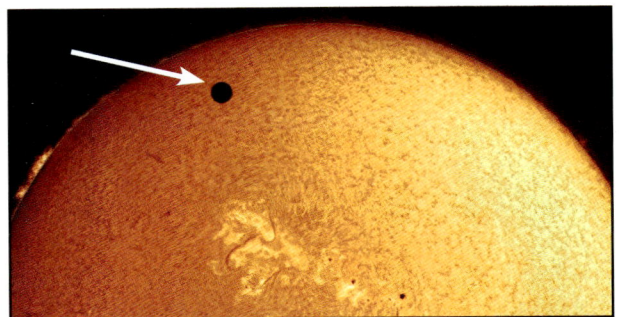

17.3 Venus crossing the solar disk on June 5, 2012. *Courtesy of John Chumack*

TABLE 17.2 Future Transits of Venus

Date	Starting Time (UT)	Duration
December 11, 2117	2h51m	5 hours 41 minutes
December 8, 2125	16h01m	5 hours 31 minutes
June 11, 2247	11h43m	5 hours 44 minutes
June 9, 2255	4h50m	7 hours 0 minutes

Transits and Eclipses **CHAPTER 17** **325**

17.4 Venus has a thick atmosphere made up primarily of carbon dioxide. *Courtesy of NASA*

otherwise, however. The glow before and after the transit showed conclusively that an atmosphere surrounded Venus.

As Venus gradually moved off the Sun, Lomonosov saw the circular edge of the planet extending out into the darkness bounded by an arc of light. Some observers with large telescopes under extremely favorable conditions have been able to follow the planet until it moved entirely away from the brilliant solar background. At that point the globe of Venus, though invisible, had a distinct circle of light surrounding it. The only explanation possible is that an atmosphere surrounds Venus (Fig. 17.4).

Points of Contact

During a transit, astronomers look for four points of contact, which are shown in Figure 17.5. The moment Mercury or Venus encounters the solar disk is called **first contact**. Minutes later, Mercury's or Venus' entire disk appears silhouetted against the Sun. This is **second contact**. For the next three to six hours, the black circle slowly crosses the brilliant solar sphere. Mid-transit occurs when Mercury or Venus lies closest to the Sun's center. At some point, the planet touches the Sun's eastern edge (which is to the west, or right, in our sky), marking **third contact**. Figure 17.6 shows this clearly. Several minutes later, at **fourth contact**, the planet exits the Sun's disk and the transit ends.

Because Mercury spans only 10" (1/180 the Sun's diameter), the best way to follow its trek is through a properly filtered telescope. Choose an eyepiece whose magnification nearly fills the field of view with the Sun's image. (You can select the right eyepiece in advance of the transit by targeting the full Moon, which is the same apparent size as the Sun.)

You can spot the planet against the Sun's disk through filtered binoculars, but Mercury will appear tiny. In fact, at $7\times$ or $10\times$, you will have trouble distinguishing it from a sunspot. Still, it is better to watch the transit through binoculars than not at all. Just wait 10 minutes or so, and you will be able to pick up the planet's motion and direction.

Although it is unlikely that anyone now alive will see a transit of Venus, we can still compare the planet's size to that of Mercury. The diameter of Venus' disk will be approximately 1/32 (3.125 percent) that of the Sun, so careful observers hand-holding proper solar filters will be able to detect the transit naked eye.

GLOSSARY TERMS

First contact beginning of a transit or eclipse; for a transit, the initial appearance of Mercury or Venus against the Sun's disk; for an eclipse, the initial appearance of the Moon's disk against that of the Sun

Second contact for a transit, the moment when all of Mercury or Venus appears silhouetted against the Sun; for an eclipse, the beginning of totality, when the Moon blocks out all of the Sun's bright disk

Third contact for a transit, the last moment that the entire planet appears against the Sun's disk; for an eclipse, the end of totality, when the Sun's bright disk begins to reappear

Fourth contact end of a transit or eclipse; for a transit, the last moment when Mercury or Venus appears against the Sun's disk; for an eclipse, the last moment when the Moon's disk appears against that of the Sun

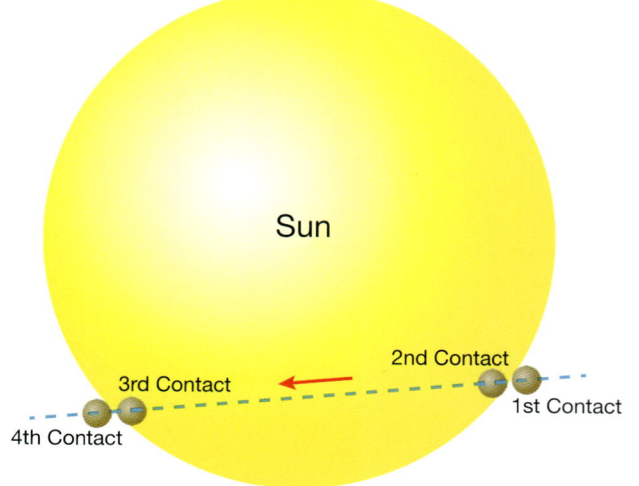

17.5 Telescopic view of four points of contact during a transit. *Courtesy of Holley Y. Bakich*

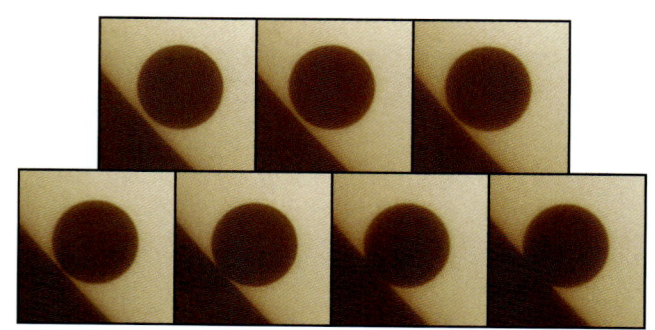

17.6 Third contact during the transit of Venus on June 8, 2004. *Courtesy of Damian Peach and David Tyler*

You also might be able to make an interesting observation through a telescope equipped with a **hydrogen-alpha filter**. If a solar prominence happens to be in the right place, you may be able to see Mercury or Venus silhouetted against it before or after the actual transit begins.

Calculating Distances using Transits

In 1716 British astronomer Sir Edmund Halley (1656–1742) presented a paper to the Royal Society in London in which he developed a method to determine the distance from Earth to the Sun using a transit of Mercury or Venus. Halley's method requires at least two astronomers to make simultaneous transit observations from two locations on Earth. Because parallax causes the transit to occur at slightly different times for two observers at different sites, Venus doesn't enter or leave the Sun's disk at the same moment for both. Also, Venus' position in front of the Sun is not quite the same from the two locations. Halley died in 1742 before he could observe either of the transits of Venus in 1761 and 1769; however, the method he developed can still be used to determine the length of an astronomical unit (AU). Exercise 17.1 explains Halley's method in greater detail and asks you to perform calculations to determine the AU.

An Alternate Definition

If you read enough astronomical literature or listen to enough observers, eventually you will encounter the word *transit* used in a different context. When any celestial body crosses an observer's meridian that event is called a transit. The observer's meridian is the line that joins the north and south horizon points and also passes through the overhead point as seen in Figure 17.7.

Used in this way, two types of transits are possible: upper transit and lower transit. At upper transit (in most discussions the "upper" is dropped), a celestial body is at its highest point in the sky. For example, when the Sun is at upper transit for an observer, it is highest in the sky and lies due south. At that particular location, it is local midday. (This is not necessarily "12:00 noon," although it can be.) Following this example ahead one-half an Earth rotation, the Sun would lie at its lowest possible point—lower transit—and it would be local midnight at the observer's location. Amateur astronomers want to observe objects "when they transit" because that is when they climb highest in the sky and appear through the least amount of Earth's atmosphere.

For the majority of celestial objects (the Sun, Moon, planets, and most stars), lower transit occurs when the object lies below the horizon. Certain lower transits, however, are visible. Depending on your location, some stars around the celestial poles never set. These are called circumpolar stars. When a circumpolar star lies directly beneath the celestial pole, it is both at lower transit and also above the horizon. If it is night and clear when this occurs, that star will be visible at lower transit.

> **GLOSSARY TERM**
> **Hydrogen-alpha filter** type of solar filter that allows through a specific wavelength of light; with such a filter you can see solar prominences, flares, and the Sun's chromosphere

17.7 Observer's meridian (dashed line).

Courtesy of John Chumack and Holley Y. Bakich

Looking Up
Transits from Other Planets?

For a transit to be observable from any location, the transiting planet must be between the observer and the Sun, and such events have happened. The most recent occurred May 11, 1984, when Earth transited the disk of the Sun for a theoretical observer on Mars. Earth took about eight hours to traverse the face of our daytime star. The Moon began its transit about six hours later, so for two hours anyone then standing on Mars would have seen both Earth and the Moon as black dots crossing the Sun.

Check Your Understanding

1.1 What is the position of Mercury or Venus when either lies at inferior conjunction?

1.2 Why do transits of Mercury and Venus occur near certain dates?

1.3 At or between which of the four contact points can you see the complete disk of Mercury or Venus silhouetted against the Sun?

1.4 During a transit, how much larger does the (dark) disk of Venus appear than that of Mercury?

Lunar Eclipses

Eclipses of the Moon occur when the Moon passes into the shadow cast by Earth. Shadows of Earth and the Moon have two parts: the **penumbra** is the faint outer shadow; the **umbra** is the dark inner shadow. The reason there are two shadow densities is that the Sun, which is casting the shadows, is not a point source. So, some light from the top of the solar disk enters the shadow cast by the sunlight originating from the bottom of the solar disk.

Lunar eclipses always take place at full Moon because that is when the Moon lies opposite the Sun as seen from Earth. You can see a lunar eclipse from any part of the Earth where the Moon is in the sky at the time (at night). This means that, over time, far more lunar eclipses than solar eclipses can be seen from any particular location.

Full Moon occurs approximately every 29.5 days. However, we don't experience a lunar eclipse at every full Moon because the Moon's orbit tilts to Earth's orbit around the Sun by 5° (Fig. 17.8). As in transits, the two points where the Moon's orbit intersects the ecliptic are called nodes. Only when the Sun is at one of the lunar nodes and the Moon is full can an eclipse occur.

The Moon does not disappear during a total eclipse because Earth's atmosphere scatters some sunlight onto the Moon's surface. Total lunar eclipses have a wide range of colors. They can appear dim yellow-white to orange, copper, reddish brown, and nearly black. The color depends on factors such as the amount of dust and clouds in our atmosphere at the time.

Three types of lunar eclipses are possible: total, partial, and penumbral (Fig. 17.9). A **total lunar eclipse** occurs when Earth's umbra covers the entire visible surface of the Moon. The contrast between the brilliant lunar face and the dark portion of the Earth's

> **GLOSSARY TERMS**
>
> **Penumbra** light, outer part of the shadow of Earth or the Moon
>
> **Umbra** dark, inner part of the shadow of Earth or the Moon
>
> **Total lunar eclipse** eclipse of the Moon during which it is totally immersed within Earth's umbra

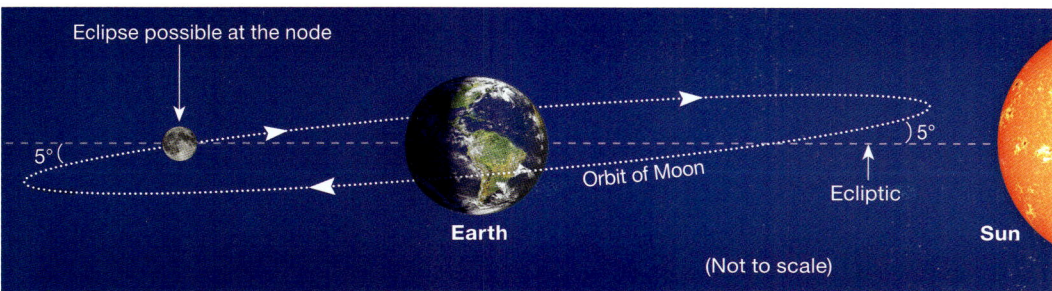

17.8 Moon's orbit at a tilt of 5° to the ecliptic (apparent path of the Sun in our sky). When the Moon is at the node (the intersection of the Moon's orbit with the ecliptic), an eclipse is possible.

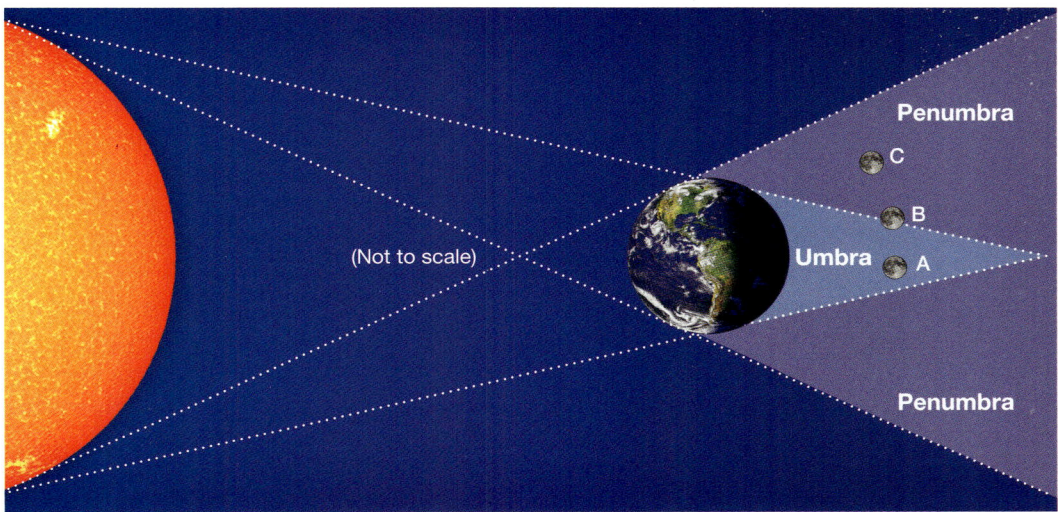

17.9 Three types of lunar eclipses: (**A**) total, where the entire Moon passes through the umbra; (**B**) partial, where a portion of the Moon passes through the umbra; and (**C**) penumbral, where a portion or all of the Moon passes through the penumbra but does not pass through the umbra.

shadow causes the full Moon's circular surface to gradually change as if it is going through phases. Only during and near **totality** does the covered portion of the Moon become visible, and then in color.

The Sun, Earth, and Moon do not have to exactly align for a total eclipse to occur. The Earth's umbral shadow measures approximately 5,600 miles (9,000 kilometers) in diameter at the Moon's distance, so the Moon can pass quite a bit above or below its center and an eclipse still will happen. The closer to the centerline of the umbra the Moon passes, the longer the eclipse will be. Another variable that determines the length of a lunar eclipse is the Moon-Earth distance at the time of the eclipse. If all conditions are right, the total phase of a lunar eclipse can last up to 1 hour and 40 minutes.

A **partial lunar eclipse** occurs when the umbra covers only a fraction of the Moon's surface. At the Moon's distance, the diameter of Earth's penumbral shadow is approximately 10,000 miles (16,000 kilometers). Few images of partial lunar eclipses captivate amateur astronomers. However, the series in Figure 17.10 taken on August 16, 2008, conveys the sense that the Moon is passing through Earth's shadow.

A **penumbral eclipse** occurs when the Moon enters only the penumbra of Earth's shadow. During such an eclipse, the brightness of the Moon's surface gradually decreases and some extremely subtle color changes can occur. The smaller the percentage of the Moon's surface entering the penumbra, the more difficult the eclipse is to see. Most penumbral eclipses are noticeable (and thus worth observing) if the percentage of the Moon's surface covered by the penumbra is greater than about 40 percent.

> **GLOSSARY TERMS**
>
> **Totality** for a lunar eclipse, the time during which the Moon is completely within Earth's umbra; for a solar eclipse, the time during which the Moon completely covers the Sun's bright disk
>
> **Partial lunar eclipse** eclipse of the Moon during which only part of the Moon's disk is covered by Earth's umbra
>
> **Penumbral eclipse** lunar eclipse during which the Moon never enters the darker, inner part of Earth's shadow (the umbra), but remains in the outer, lighter penumbra

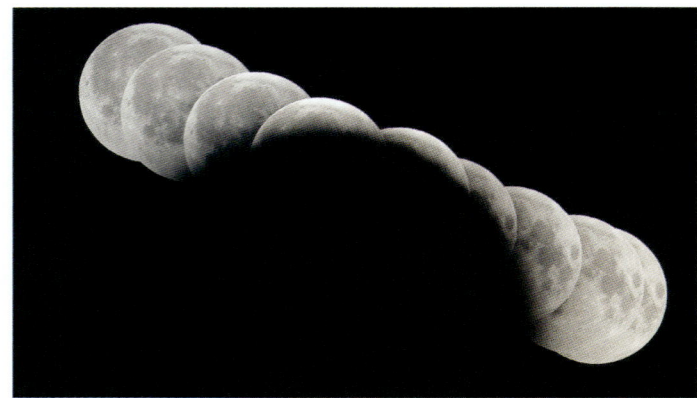

17.10 Passage of the Moon through Earth's umbral shadow during a partial lunar eclipse.

Courtesy of Anthony Ayiomamitis

Observing a Lunar Eclipse

Lunar eclipses are easy and safe to observe. No equipment is necessary, although some eclipse watchers prefer to use binoculars or a telescope with a wide field of view, at least wide enough to encompass the whole Moon.

In the partial phases before and after, Earth's shadow appears dark and has little color. During totality, however, sunlight passes through our atmosphere, which reddens it, and then strikes the Moon. During total lunar eclipses the brightness of the disk at mid-totality appears different. French astronomer Andre Danjon (1890–1967) developed a rating system for the color and brightness of lunar eclipses. The **Danjon scale** (Table 17.3) assigns a value for the luminosity (L) of the fully eclipsed Moon. This is a qualitative observation based on each observer's own judgment. The central moon exposure in Figure 17.11 is showing a Danjon number of 3.

GLOSSARY TERM

Danjon scale number from 0 (darkest) to 4 (lightest) that describes the brightness of a lunar eclipse at the middle of totality

TABLE 17.3 The Danjon Scale

Danjon Number	Overall Coloration	Details
L = 0	Very dark	Moon almost invisible, especially at mid-totality
L = 1	Dark, gray or brownish	Details distinguishable only with difficulty
L = 2	Deep red or rust	Dark central shadow; outer edge of umbra is relatively bright
L = 3	Brick red	Umbral shadow has a bright or yellow rim
L = 4	Very bright copper red or orange	Umbral shadow has a bluish, very bright rim

17.11 Progression of the total lunar eclipse on November 8, 2003.

Courtesy of John Chumack

Unit 4 The Solar System

Imaging a Lunar Eclipse

Lunar eclipse photography is among the easiest anyone can do. You have plenty of time to shoot and can image through a telescope or with a telephoto lens on a camera connected to a tripod or guided with your telescope's drive. Use at least a 300mm telephoto lens. Such a lens will provide a reasonable scale for your images. A 300mm lens has a field of view of 8°, so the full Moon will be approximately $\frac{1}{16}$, or a bit more than 8 percent of the width of the field of view. Table 17.4 shows the projected next five total lunar eclipses, so get out there and try to photograph it!

TABLE 17.4 Future Total Lunar Eclipses

Date	Time of Mid-Eclipse	Duration of Totality	Region Best Seen
September 28, 2015	2h48m UT	1 hour 12 minutes	The Americas, Europe, Africa
January 31, 2018	13h31m UT	1 hour 16 minutes	Asia, Australia, Western North America
July 27, 2018	20h23m UT	1 hour 42 minutes	Asia, Africa, Europe
January 21, 2019	5h13m UT	1 hour 2 minutes	Africa, Europe, the Americas
May 26, 2021	11h20m UT	15 minutes	Asia, Australia, the Americas

Check Your Understanding

2.1 Describe the two parts of the shadows of Earth and the Moon.

2.2 For a lunar eclipse to happen, what must the Moon's phase be?

2.3 How would a total lunar eclipse look if it had a Danjon number of L = 2?

Solar Eclipses

If the authors had one astronomy-related wish for each of you reading this lab manual, it would be that, at some time in your life, you could stand underneath a totally eclipsed Sun (Fig. 17.12). Each eclipse is an amazing testimony to sublime celestial geometry.

Solar eclipses occur at **new Moon**, but not at every new Moon, for the same reason we don't see lunar eclipses at every full Moon. The Moon's orbit tilts to Earth's orbit around the Sun by 5°. As in transits, the two points where the Moon's orbit intersects the ecliptic are called nodes. Only when the Sun is at one of the lunar nodes and the Moon is new can a solar eclipse occur.

Solar eclipses may be total, partial, or annular. Where the Moon's umbra touches Earth the eclipse will be total. Within the penumbra, the eclipse is partial. If our satellite's umbra does not quite reach Earth, the eclipse beneath it will be annular.

Total solar eclipses occur when the Moon's umbra falls on Earth (Fig. 17.13). At best, the diameter of the Moon's shadow is 170 miles (273 km), so it is a small swath of Earth's surface that experiences each total solar eclipse, even though the path of the eclipse is typically 9,300 miles (15,000 km) long. The track of the Moon's shadow across Earth's surface is called the **path of totality.**

> **GLOSSARY TERMS**
>
> **New Moon** phase when our natural satellite cannot be seen because it lies between Earth and the Sun
>
> **Total solar eclipse** eclipse of the Sun during which the Moon completely covers the Sun's disk
>
> **Path of totality** line that represents the track of the Moon's umbra across Earth's surface during a total solar eclipse

17.12 Combined 15-exposure shot of the Sun's corona during a total eclipse imaged through a 4-inch refractor at *f*/5.

Courtesy of Anthony Ayiomamitis

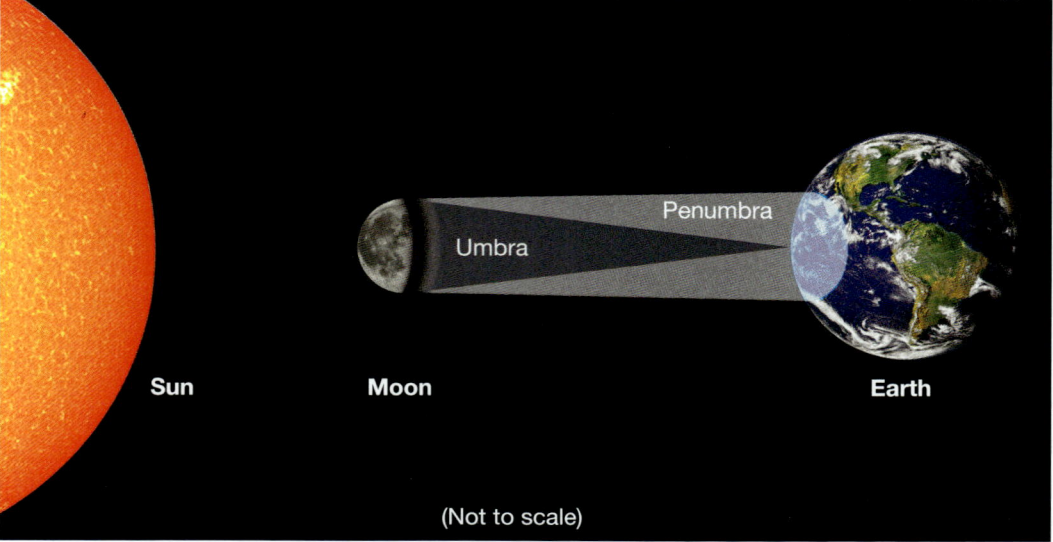

(Not to scale)

17.13 Circumstances of a solar eclipse.

Courtesy of Holley Y. Bakich

332   **Unit 4** *The Solar System*

Partial solar eclipses occur over much larger areas of Earth's surface. A partial eclipse can be partial everywhere if the Moon's umbra misses Earth or it can appear partial due to an observer not being on the centerline of either a total or an annular eclipse.

Annular eclipses (Figs. 17.14A and B) result from the fact that Earth is not always the same distance from the Sun and the fact that the Moon is not always the same distance from Earth. The Earth-Sun distance varies by 3 percent and the Moon-Earth distance by 12 percent. The result of this is that the apparent diameter of the Moon can range from 7 percent larger to 10 percent smaller than that of the Sun.

When the Moon's apparent diameter is smaller than the Sun's, a ring of the bright solar disk appears around the Moon at mid-eclipse. The Latin word for ring is *annulus*, so astronomers call such an event an annular eclipse. In this case, the umbral shadow of the Moon falls short of Earth's surface. Because the sunlight reaching us is still brilliant, we can't see any trace of the corona during annular eclipses. The maximum diameter of the light cone of an annular eclipse is 195 miles (313 km). The maximum length of the central part of an annular eclipse is approximately 12 minutes.

There also is a type of eclipse that astronomers call a hybrid. It occurs when the Moon's shadow contacts Earth's surface only during the mid-point of the eclipse. Earth's spherical shape causes this type of eclipse.

Researchers who study solar eclipses categorize them in terms of their magnitude and obscuration. The magnitude of a solar eclipse is the percentage of the Sun's diameter that the Moon covers during the maximum part of the eclipse. The obscuration is the percent of the Sun's total surface area covered at maximum.

Observing a Total Solar Eclipse

The same precautions you would use to observe the Sun at any time are in effect during eclipses of the Sun. The only exception is that during totality—and only during totality—no safety measures are necessary. Table 17.5 shows the projected next five total solar eclipses. If you are able to witness one of these events, try looking for some of the following features: partial phases, planets and bright stars, shadow bands, the approaching shadow, Baily's beads, the diamond ring, the chromosphere, and the corona.

> **GLOSSARY TERMS**
>
> **Partial solar eclipse** eclipse of the Sun during which the Moon never totally covers the Sun's disk
>
> **Annular eclipse** type of solar eclipse in which the Moon's shadow does not reach Earth's surface; this results in a ring (Latin = *annulus*) of the Sun's disk showing around the Moon

 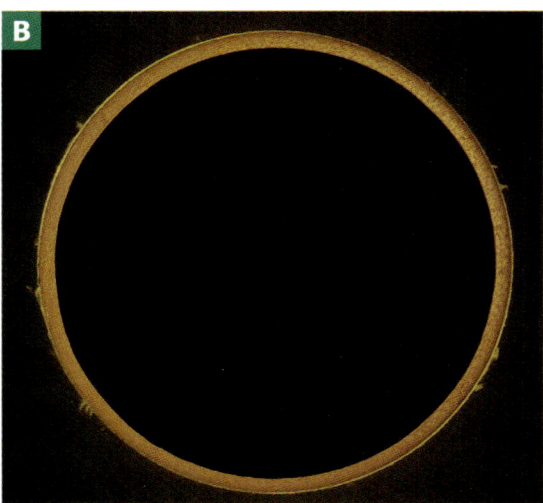

17.14 Annular eclipse on May 20, 2012: **(A)** 24-exposure sequence, and **(B)** single-exposure close-up taken through a hydrogen-alpha filter.

Courtesy of Chris Schur

TABLE **17.5** Future Total Solar Eclipses

Date	Time	Duration of Totality	Best Viewing Location
March 9, 2016	2h UT	4m10s	Sumatra, Borneo, the Pacific
August 21, 2017	18h UT	2m40s	United States
July 2, 2019	19h UT	4m32s	Chile, Argentina, South Pacific
December 14, 2020	16h UT	2m10s	South Atlantic, Chile, Argentina
December 4, 2021	7h UT	1m54s	Southern Ocean, Antarctica

Partial Phases

First contact occurs at the moment the eclipse begins—the moment you first see the black outline of the Moon's disk against the bright circle of the Sun. You can photograph the event during this time at a leisurely pace. Many observers image at regular intervals (most use their digital camera's automatic settings), combining them later as an animation.

You can also try pinhole projection, which is a totally safe way to view the partially eclipsed Sun (or the Sun at any time, really). Take two thin pieces of white cardboard. Poke a pinhole in one and hold them so they line up with the Sun with the pinhole closest to our daytime star. You will see a tiny image of the Sun projected onto the second piece of cardboard. Change the distance between the two cards to see what effect that has. You also can see a natural form of pinhole projection when shafts of the partially eclipsed Sun's light shine through the leaves of a tree. Hundreds of images can appear simultaneously on the ground.

Be sure to notice that, as totality approaches, shadows sharpen and darken. The sharpening is because the Sun, normally a large disk in our sky, is shrinking and pretty much becoming a point source of light, which creates sharp, dark shadows. Shadows blacken during this time because less scattered light reflects into the dark areas (Fig. 17.15).

Planets and Bright Stars

During the early partial stages of the eclipse, the sky is still bright. But as totality approaches, the light fades quickly and you will see the brightest celestial objects. Always find out the positions of the planets and brightest stars in the sky before an eclipse begins. You can see 0 magnitude and brighter objects a minute or two before totality (Venus sometimes is visible as much as 20 minutes before and after totality) and 3rd-magnitude stars with your naked eyes once totality begins.

> **GLOSSARY TERM**
>
> **Shadow bands** faint ripples of light sometimes seen on flat, light-colored surfaces just before and just after totality

Shadow Bands

Just before the total phase begins, while a sliver of the bright Sun remains in the sky, many observers have reported long, straight bands of shadows moving across the ground. These wavy lines of alternating light and dark are called **shadow bands**. They are the result of refraction—sunlight bent by irregularities in Earth's atmosphere. Shadow bands have low contrast. Many observers have spread a large, white sheet in front of them to help them view this phenomenon. A few observers have photographed shadow bands, but those instances are rare. If you are viewing your first eclipse, don't look for shadow bands because to do so means you are looking down, away from the spectacle happening in the sky.

The Approaching Shadow

Occasionally, just moments before totality begins, observers have reported seeing the shadow approach. This is most easily seen from a flat, arid landscape or when the observer is next to a large body of water that lies in the

17.15 Shadows appear more distinct during a solar eclipse.
Courtesy of John Chumack

direction of the shadow's approach. Occasionally, thin, high cirrus clouds in the sky just before totality can help observers see the shadow approaching from overhead.

Baily's Beads

In 1836, the English astronomer Francis Baily (1774–1844) noticed a string of irregularly spaced, brightly lit points at the edge of the Moon just before totality. Now known as Baily's beads (Fig. 17.16), you can see this phenomenon several seconds before and after totality. The beads are actually the last few rays of sunlight shining through valleys on the edge of the Moon.

The Diamond Ring

Just before and just after totality, the last tiny sliver of sunlight is visible with the brightening corona. Early observers gave this combination of one brilliant point of light and the round corona the appropriate name "diamond ring" (Fig. 17.17).

At this point during the eclipse, quickly remove any solar filters from your telescope, binoculars, camera, or other optical equipment. The Sun is temporarily safe to look at until the next diamond ring.

The Chromosphere

Immediately after the diamond ring vanishes, you may be able to glimpse the reddish chromosphere. This is a difficult observation because it is only visible for about one second. With the disappearance of the chromosphere behind the Moon's disk, second contact (the beginning of totality) begins.

At this time, and throughout totality, you may be able to see any number of large solar prominences. Smaller ones will be visible through binoculars or a telescope. The number and size of visible prominences depends upon the solar activity at that time. Observers who have seen a number of total solar eclipses tend to remember each by the size of the largest prominences and by the overall shape of the corona. Take advantage of this view. The only other way to see such features is through a hydrogen-alpha filter, and it is not nearly so dramatic.

The Corona

The wispy outer atmosphere called the corona of the Sun is now in full view (Fig. 17.18). Even experienced eclipse researchers have only a general idea of how the corona will appear prior to totality. Many times, it is a great surprise. Note the shape and extent of the corona. Is it longer in one dimension than the other?

At this point during the eclipse, take a moment and make yourself aware of the bigger picture. The spectacle is not only in the sky, it is also on Earth. You will see that there is a resemblance to the onset of night, although not exactly. Around the horizon you will notice areas much lighter than those in the part of the sky near the Sun. Listen. Usually, any breeze will dissipate and birds (many of which will come in to roost) will stop chirping. A 9–18°F drop in temperature is not unusual. For the annular eclipse of May 30, 1984, from a cemetery in Picayune, MS, observers recorded a temperature drop of almost 13°C (23°F).

The end of totality signals third contact. The above described pattern now works itself in reverse. Finally, the last bit of the lunar disk passes away from the Sun at fourth contact.

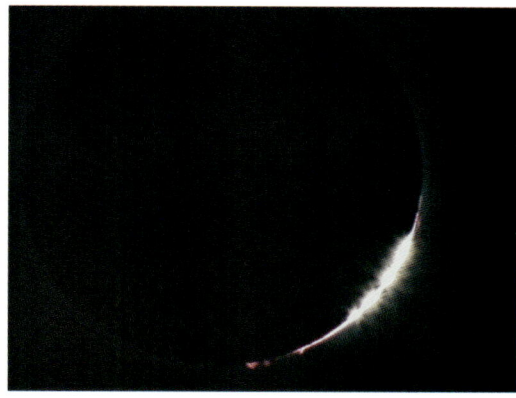

17.16 Baily's beads.

17.17 Diamond ring from a solar eclipse over the waters of the South Pacific on November 14, 2012.

17.18 Corona of the Sun (visible only during the total phase of a solar eclipse).

Imaging a Solar Eclipse

The urge of many first-time eclipse viewers is to record such a fantastic event. But others, even those who have seen many totalities, never image during a total solar eclipse. The spectacle is simply too overwhelming. They want to absorb it all without worrying about shutter speeds and focal ratios and switches and . . . you get the idea. If you are determined to image a total solar eclipse, however, at least wait until your second one. If you want images, more than enough will be available, such as the sequence shown in Figure 17.19. Enjoy your first solar eclipse for what it is: nature's grandest spectacle.

17.19 Total solar eclipse viewed from Novosibirsk, Russia, on August 1, 2008, through a 4-inch refractor.

Courtesy of Anthony Ayiomamitis

Check Your Understanding

3.1 Explain why many more people have observed total lunar eclipses than total solar eclipses.

3.2 Why don't we have a solar (or lunar) eclipse approximately every month?

3.3 What is the maximum length of time the central part of an annular eclipse can last?

3.4 Why do shadows sharpen as totality approaches during a solar eclipse?

3.5 Describe how Baily's beads form.

3.6 Why might you not want to take pictures during your first total solar eclipse?

Exercise 17.1 Calculating Distances Using Transits

Astronomers call the average distance between Earth and the Sun an astronomical unit (AU). Numerically, this equals 92,955,807 miles (149,597,871 kilometers).

Early astronomers wanted to determine the length of the AU, which would allow them to calculate the scale of the solar system; however, determining this number presented many challenges. Edmund Halley, a brilliant mathematician of his era, eventually devised a method astronomers could use during a transit of Venus. Halley found that by obtaining Venus transit observations made at the same time, knowing the distance between the two Earth-based stations, and determining the angle p, anyone could calculate the distance from Earth to Venus, and thus the distance for 1 AU.

In this exercise, you will imagine that two astronomers have sent you observations they made of a transit of Venus from two widely separated sites. They also calculated the parallax (p) in arcseconds. That angle shows the difference between the two observations.

In addition to the measurement of p, you will need to know three quantities in order to calculate the length of the astronomical unit.

1. The baseline (B) between the two sites, in kilometers.

2. The relative distance (in astronomical units) between Earth and Venus (D_{ev}) at the time of the eclipse.

3. The relative distance (in astronomical units) between Earth and the Sun (D_{es}) at the time of the eclipse. We already know this is 1, because astronomers define an astronomical unit as the average distance between Earth and the Sun.

You don't yet know the value of the astronomical unit in kilometers; all you know is relative distances (percentages) in AU. The formula used to calculate the length of the astronomical unit (A) in km is:

$$A = \frac{B(D_{es} - D_{ev})(206265)}{(p)(D_{ev})(D_{es})}$$

Be sure your units are correct before you plug into this formula. B is in kilometers, D_{ev} and D_{es} are both in AU, and **p** is in arcseconds (Fig. 17.20).

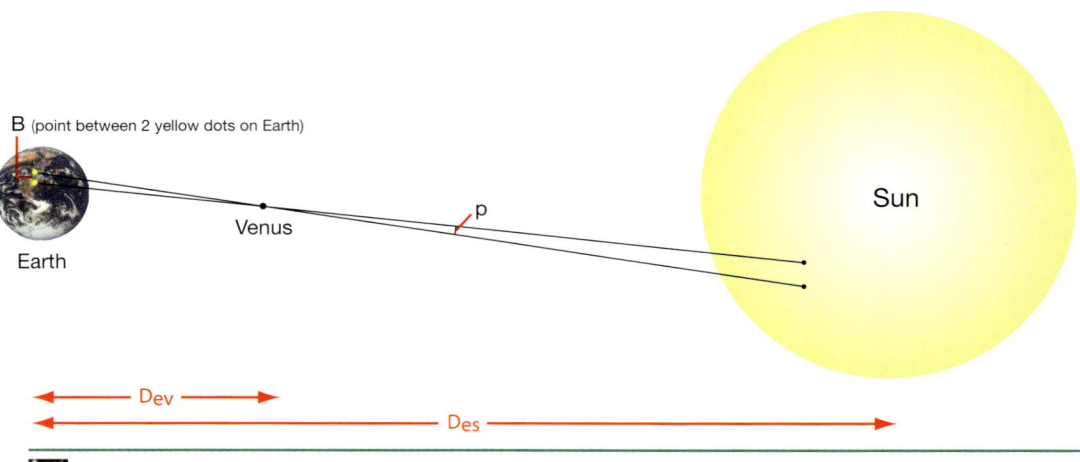

17.20 Edmund Halley's method of determining the astronomical unit.

Courtesy of Holley Y. Bakich

Procedure	Materials
	❏ Scientific calculator

1. Imagine that you work at an astronomical data collection center. For the recent transit of Venus on June 5, 2012, you asked a couple of your friends, both astronomers, to send you the results of their observations. One of the astronomers made his observations from Tucson, Arizona. The other set up his telescope in Aurora, Colorado. When they studied their observations, they reported to you a parallax of 3.86".

2. Assume that during the transit, Venus was 0.737 AU from the Sun. Why can we make that assumption? Remember Kepler's third law from Chapter 15:

$$a^3 = kP^2$$

Where:

  a = the planet's semimajor axis
  P = the orbiting object's orbital period
  k = a constant, referred to as Kepler's constant

Using this law, determine **a** for Venus, which will be its distance from the Sun. For this calculation, the constant (k) equals 1.

$$a^3 = P^2$$

  a The orbital period for Venus is 224 days and 16.8 hours.

   Express this as a decimal: _____ days

  b One Earth year equals approximately 365.25 days. What percentage of an Earth year is a Venus year?

   _____

  c Multiply this number by itself to get P^2.

   _____

  d This number equals a^3. Using a scientific calculator, find the cube root of this number (the number which, multiplied by itself three times, equals P^2).

   _____

   This number is the distance between Venus and the Sun in astronomical units.

   Now you are ready to calculate a value for the astronomical unit.

3. Find the distance between Tucson and Aurora in kilometers.

   _____

4. Using that distance, the above data, and the formula on page 338, calculate the value of the astronomical unit. A = _____ kilometers

5 Now imagine the second astronomer had observed from New Orleans, LA, instead of Aurora, CO.

What is the distance between Tucson and New Orleans? _____ kilometers

What would the parallax have been? _____ "

6 Earth's equatorial diameter is 12,756 km. If two theoretical observers could view the transit from opposite ends of Earth, what would be the measured parallax?

_____ "

7 For observations like this one, astronomers want the observers separated by as much distance as possible. Why do you think that's important?

Exercise 17.2 — Observing a Lunar Eclipse

Procedure

1 Which type of lunar eclipse will this be?

☐ Penumbral ☐ Partial ☐ Umbral

2 How will you be observing the Moon?

☐ Naked eye ☐ Binoculars _____ × _____

☐ Telescope _____

3 How many hours must you add to the time in your time zone to equal Universal Time?

(*Examples:* UT = EST + 5 hours; UT = EDT + 4 hours; UT = CST + 6 hours.)

4 Complete the tables and questions regarding your observations of the lunar eclipse in the data sheet on pages 341–342.

Materials

- ☐ Telescope and accessories, such as eyepieces
- ☐ Eclipse prediction data
- ☐ Timing device, such as a stopwatch
- ☐ GPS or WWV time signals
- ☐ Clipboard to hold your papers
- ☐ Red flashlight for reading the moon chart and instructions at a dark site
- ☐ Camera (optional)

Name _____ Section _____ Date _____

Lunar Eclipse Observations

Exercise 17.2 Data Sheet

1 Time the following events as accurately as you can and record your data in Table 17.6.

TABLE 17.6 Occurrence of Lunar Eclipse Events

Event	Local Time	Universal Time
Penumbral eclipse begins		
Partial (umbral) eclipse begins		
Total (umbral) eclipse begins		
Greatest eclipse		
Total (umbral) eclipse ends		
Partial (umbral) eclipse ends		
Penumbral eclipse ends		

2 Briefly describe your observing method.

_____

_____

3 Estimate the sky brightness during totality and record the data in Table 17.7.

TABLE 17.7 Estimations of Sky Brightness

Event	Time	Estimate of Sky Brightness
Start of totality		
Greatest eclipse		
End of totality		

4 What method did you use to determine sky brightness?

_____

_____

Exercise 17.2 Data Sheet

5 Estimate the brightness of the disk using the Danjon scale and record your data in Table 17.8.

TABLE **17.8** Estimations of Disk Brightness

Event	Time	Method (Naked Eye, Binoculars, Telescope)	Estimate of Disk Brightness
Start of totality			
Greatest eclipse			
End of totality			

6 If you are imaging the eclipse, note the following.

Camera: ☐ Digital   ☐ CCD   ☐ Video

Manufacturer and model: _____

Telescope/lens used: _____

F/ratio: _____

ISO speed: _____

Exposure times: _____

Other details: _____

Attach printouts of images you recorded during the event.

7 Include the details of any sketches you made. Attach originals.

Time of sketch _____   Media _____

Method: ☐ Naked eye   ☐ Binoculars ____ × ____   ☐ Telescope _____

What were your overall impressions during your sketching time?

_____

8 List any other observations (of planets, stars, deep-sky objects, etc.) you made during the eclipse.

Method: ☐ Naked eye   ☐ Binoculars ____ × ____   ☐ Telescope _____

Unit 4 The Solar System

The Deep Sky

Stars

18

LEARNING OBJECTIVES
Upon completion of this chapter, you should be able to:
1. Explain the differences between double stars and variable stars.
2. Identify the two main classes of variable stars.
3. Describe the historical events that led to the discovery of variable stars.
4. Estimate the brightness of various stars.
5. Explain how a light curve represents how stars change in brightness over time.
6. Define all glossary terms.

Open Cluster NGC 3603
Courtesy of NASA/ESA/R. O'Connell (University of Virginia)/F. Paresce (National Institute for Astrophysics, Bologna, Italy)/E. Young (Universities Space Research Association/Ames Research Center)/the WFC3 Science Oversight Committee, and the Hubble Heritage Team STScI/AURA

> **GLOSSARY TERMS**
>
> **Double star** any pair of stars that appear close to one another on the celestial sphere; double stars may be simply chance alignments or they may be physically linked by gravity to one another, in which case they are known as binary stars
>
> **Primary** usually the brightest star in a binary system
>
> **Companion** name sometimes given to the fainter of the two stars in a binary system
>
> **Secondary** in a system of two or more orbiting celestial bodies, any body not closest to the center of mass and any that revolves around the primary
>
> **Separation** distance, expressed in angular measure, between two celestial bodies
>
> **Position angle** in a double-star system, the angle, measured from north through east, of the line joining the primary with the companion star

In this chapter, you will observe double stars and several different types of variable stars, including eclipsing binaries. You also will sketch selected double and variable stars and star fields, estimate star brightnesses, and create light curves representing how the brightness of selected stars changes.

Double Stars

Observing **double stars** is one of the most neglected aspects of stargazing. Even small telescopes can resolve thousands of these tinted jewels, and no two look exactly alike. Also, you can observe double stars on nights that are all but useless for viewing other deep-sky objects because of haze, bright moonlight, or light pollution.

Double-star nomenclature (Fig. 18.1) is simple. The **primary** (A) is the pair's brighter star, and astronomers consider it the system's center. The fainter star is the **companion**, or **secondary** (B). **Separation** measures the distance between the stars in arcseconds. Astronomers measure **position angle** in degrees from north (0°) through east and continuing back to north (360°).

While well-known blue-white star pairs like Mizar (Zeta [ζ] Ursa Majoris) and Castor (Alpha [α] Geminorum) or golden ones like Algeiba (Gamma [γ] Leonis) look beautiful through the telescope, it is those doubles displaying a marked color contrast that most observers find exciting.

You will discover targets among the naked-eye stars that split easily. As author J. Dorman Steele pointed out more than a century ago, "Every tint that blooms in the flowers of summer flames out in the stars at night." While many of these colors are illusory, caused by contrast effects between stars of different brightnesses, in other cases the colors are definitely real.

Following are descriptions of 21 colorful double stars. (They are also listed briefly in Table 18.1). Arranged in order of right ascension, you will be able to see several during any season. For maximum effect, use the lowest magnification that will split each pair. Sometimes ever-so-slightly defocusing the image will help you perceive subtle hues by spreading the light over a greater area of the eye's retina.

A number of writers over the years have looked poetically upon the stars as "flowers in the meadows of heaven." Observing the colorful starry jewels on this list should convince you they are just that.

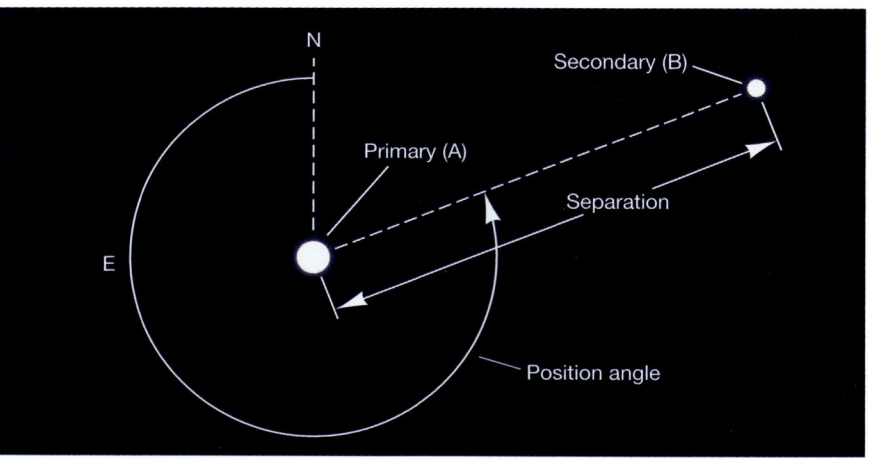

 18.1 Double-star nomenclature.   Courtesy of Holley Y. Bakich

Table **18.1** Colorful Double Stars

Double Star	Constellation	Right Ascension (2000.0)	Declination (2000.0)	Magnitudes	Separation
Struve 3053	Cassiopeia	0h03m	66°06'	5.9, 7.3	15"
Eta Cas	Cassiopeia	0h49m	57°49'	3.4, 7.4	13"
Almach	Andromeda	2h04m	42°20'	2.3, 5.0	10"
Eta Per	Perseus	2h51m	55°54'	3.8, 8.5	28"
32 Eri	Eridanus	3h54m	−2°57'	4.8, 5.9	7"
Gamma Lep	Lepus	5h44m	−22°27'	3.6, 6.3	97"
h3945	Canis Major	7h17m	−23°19'	5.0, 5.8	27"
Iota Can	Cancer	8h47m	28°46'	4.1, 6.0	31"
24 Com	Coma Berenices	12h35m	18°32'	5.1, 6.3	21"
Izar	Boötes	14h45m	27°04'	2.6, 4.8	3"
Xi Boo	Boötes	14h51m	19°06'	4.8, 7.0	6"
Kappa Her	Hercules	16h08m	17°03'	5.1, 6.2	27"
Antares	Scorpius	16h29m	−26°26'	0.9–1.2, 5.4	3"
Ras Algethi	Hercules	17h15m	14°23'	2.7–3.9, 5.4	5"
Omicron Oph	Ophiuchus	17h18m	−24°17'	5.2, 6.6	10"
95 Her	Hercules	18h02m	21°36'	4.8, 5.2	6"
70 Oph	Ophiuchus	18h06m	−2°30'	4.2, 6.2	5"
Albireo	Cygnus	19h31m	27°58'	3.4, 4.7	35"
Omicron¹ Cyg	Cygnus	20h14m	46°44'	3.8, 7.0	106"
Gamma Del	Delphinus	20h47m	16°07'	4.4, 5.0	9"
Delta Cep	Cepheus	22h29m	58°25'	3.4–4.4, 6.1	41"

> **GLOSSARY TERM**
>
> **Magnitude** measure of the amount of light flux (or other radiation) received from a luminous celestial object

Twin Treats of Autumn

Four double stars shine brightly in autumn: Struve 3053; Eta (η) Cassiopeiae; Gamma (γ) Andromedae (Almach); and Eta Persei.

Struve 3053

Cassiopeia holds a little-known gem—Struve 3053 (Fig. 18.2)—whose gold and blue hues remind some observers of the more famous Albireo (in the summer section). Although noticeably fainter than Albireo, Struve 3053 is an easy catch in a 3-inch telescope at 30×. It looks best through an 8-inch at 60×.

Eta (η) Cassiopeiae

Eta (η) Cassiopeiae (Fig. 18.3) is a beautiful double with a yellow primary four **magnitudes** brighter than its ruddy-purple companion. The latter has been described as garnet, lilac, lavender, and even orange by various observers. A 2-inch telescope at 25× shows the fainter star's distinctive hue, but Eta becomes ever more striking as aperture increases. Through a 10- or 12-inch at 60× or more, prepare to be wowed.

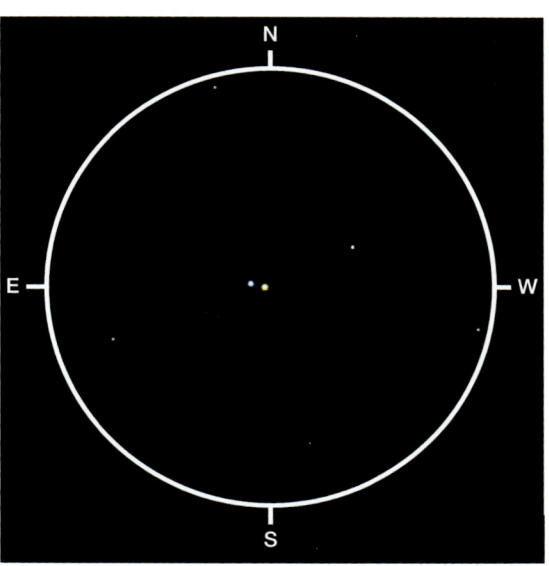

18.2 Sketch of Struve 3053.

Courtesy of Jeremy Perez

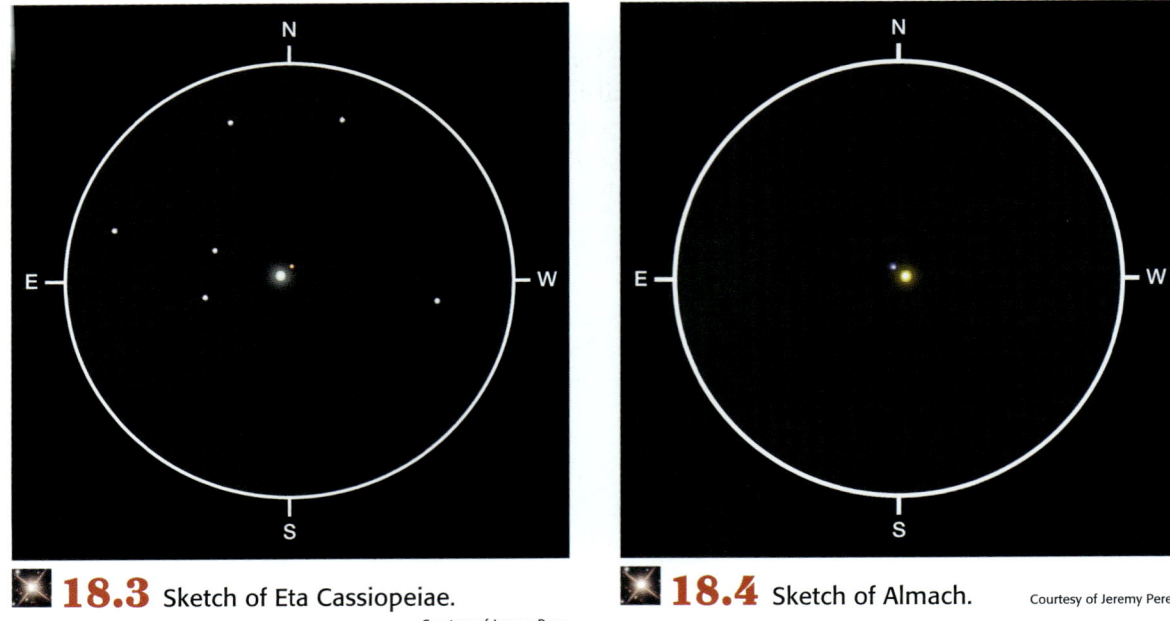

18.3 Sketch of Eta Cassiopeiae.
Courtesy of Jeremy Perez

18.4 Sketch of Almach.
Courtesy of Jeremy Perez

Gamma (γ) Andromedae (Almach)

One of the sky's most colorful double stars is Almach, or Gamma (γ) Andromedae (Fig 18.4). Its primary shines with a topaz hue, and the companion appears aquamarine. These colors are superb through a 4-inch telescope at $50\times$ to $100\times$. Also, this is one double that doesn't lose its impact through large telescopes. Observers tend to use higher magnifications with larger telescopes, which often unduly spread out wider pairs.

Eta Persei

Perseus contains an often-overlooked orange-and-blue double that offers observers a pleasing contrast in both color and brightness. Although Eta Persei has a companion nearly five magnitudes fainter than the primary, the tint difference is strikingly obvious through a 5-inch telescope at $50\times$.

Winter Beauties

The three double stars to watch for in winter are 32 Eridani, Gamma Leporis, and h3945.

32 Eridani

32 Eridani comprises a vividly tinted yellow star and a blue-green star that appear lovely through anything from a 2-inch glass at $25\times$ to a 14-inch at $150\times$. Italian astronomer Angelo Secchi (1818–1878) pronounced the colors of this double as "magnifici, superbi," and so they are.

Gamma Leporis

Below Orion, in Lepus, is Gamma Leporis (Fig. 18.5), an attractive wide pair for low-power telescopes. This pair displays a yellowish primary and a garnet secondary. *Astronomy* columnist Phil Harrington describes this double as "awash in vivid color" and considers it one of the finest binocular duos in the sky. While a pretty sight through a 2-inch refractor at $25\times$, it is too spread out for optimum effect at higher magnifications.

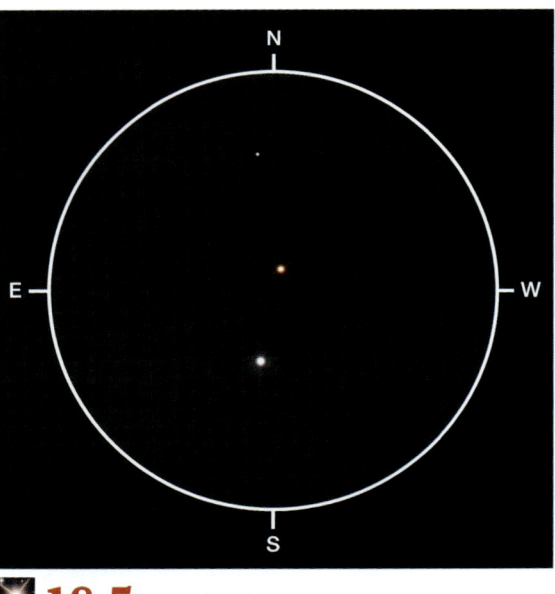

18.5 Sketch of Gamma Leporis.
Courtesy of Jeremy Perez

h3945

Observers have called h3945 in Canis Major the "winter Albireo," and the name seems to have caught on. This magnificently tinted combo shines fiery red and greenish blue. Both stars appear striking through a 3-inch telescope at 30× and superb in a 6-inch at 50×. The "h" prefix indicates this is one of English astronomer John Herschel's (1792–1871) many double-star discoveries.

Spring's Special Pairs

Spring is a great time to go outside and look for these four double stars: Iota (ι) Cancri; 24 Comae Berenices; Epsilon (ε) Boötis (Izar); and Xi (ξ) Boötis.

Iota (ι) Cancri

Iota (ι) Cancri (Fig. 18.6) is a striking orange and blue pair you can just resolve through binoculars. It is a lovely sight through even the smallest telescope. Well-known Pennsylvania double-star observer Sissy Haas thinks this pair has the most striking color-contrast in the sky, as seen through her 2.4-inch refractor. She even claims it is better than Albireo. People's color perceptions vary, so be sure to check out Iota yourself.

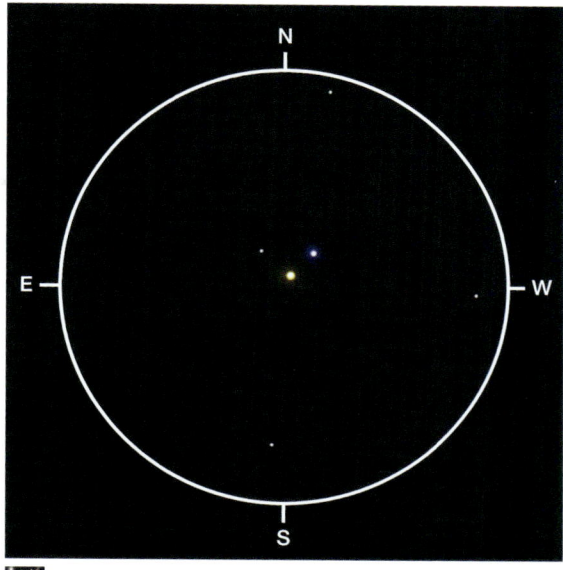

18.6 Sketch of Iota Cancri. *Courtesy of Jeremy Perez*

24 Comae Berenices

Vivid orange and emerald seem to best describe the tints of 24 Comae Berenices. It is an exquisite sight through a 4-inch telescope at 45×, and it is waiting among the many faint lights of this galaxy-rich constellation.

Epsilon (ε) Boötis (Izar)

Called Pulcherrima, the "most beautiful one," by Russian double-star observer Wilhelm Struve (1793–1864), Izar, or Epsilon (ε) Boötis (Fig. 18.7), combines the light of magnitude 2.6 and 4.8 stars. If you have trouble finding it, just look 10° northeast of Arcturus (Alpha Boötis). The

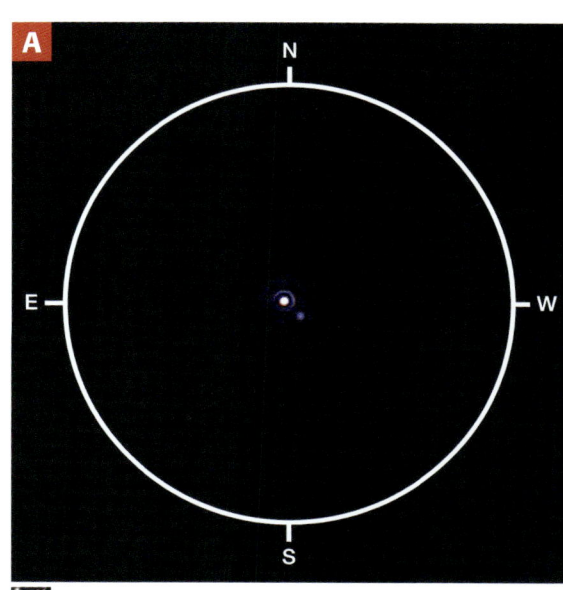

18.7 Pulcherrima, the "most beautiful one": **(A)** Izar (Epsilon [ε] Boötis), and **(B)** star chart to locate Izar (Epsilon [ε] Boötis). *(A) Courtesy of Jeremy Perez; (B) courtesy of Richard Talcott and Roen Kelly, Astronomy magazine*

components of this double star lie only 3" apart, so use high magnification to split them. A 5-inch telescope at 100× does it. A 3-inch at 150× shows its golden-yellow and blue-green disks nearly in contact, a truly awesome sight!

Xi (ξ) Boötis

Xi (ξ) Boötis (Fig. 18.8) is a deep-yellow and reddish-purple pair. Exquisite through a 6-inch telescope at 50×, these tints hold up well in big telescopes, which often saturate the colors of brighter doubles because larger instruments collect too much light. In such cases, stopping down the aperture helps.

Hot Colors of Summer

Summer offers the best double-star viewing. The list includes: Kappa (κ) Herculis, Alpha (α) Scorpii (Antares), Alpha Herculis (Ras Algethi), Omicron (o) Ophiuchi, 95 Herculis, 70 Ophiuchi, Beta (β) Cygni (Albireo), Omicron[1] (o[1]) Cygni, Gamma (γ) Delphini, and Delta (δ) Cephei.

Kappa (κ) Herculis

Kappa (κ) Herculis appears as a wider clone of Xi (ξ) Boötis. An easy split through a 2-inch telescope at 25×, the colors are yellow and garnet, with the latter perhaps tinged with a bit of orange.

Alpha (α) Scorpii (Antares)

Antares (Alpha [α] Scorpii) is a difficult pair to observe, but through a large telescope it is the most spectacular pair in our lineup (Fig. 18.9). The fiery-red primary is closely attended by a companion that can be described only as emerald green. But is the green color real?

English observer and author William H. Smyth (1788–1865), saw the companion emerge from an occultation (a term used by astronomers to indicate that one celestial body moves in front on another) by the Moon before the primary. The green hue Smyth saw did not result from a contrast effect with the primary. Similar recent observations attribute a blue color to the companion.

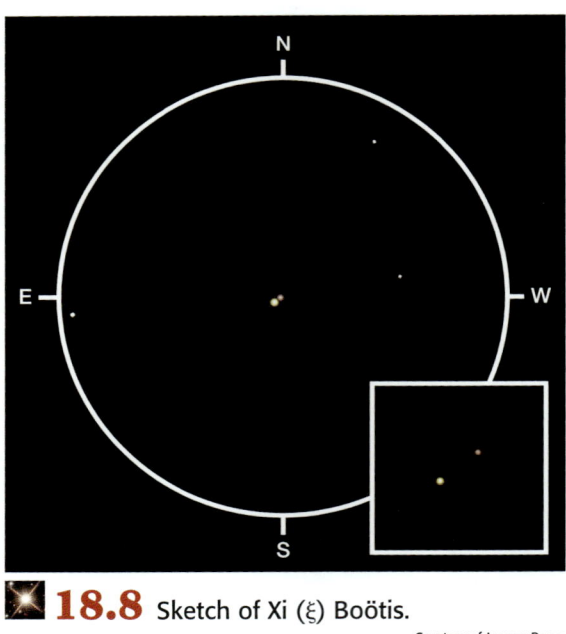

18.8 Sketch of Xi (ξ) Boötis.
Courtesy of Jeremy Perez

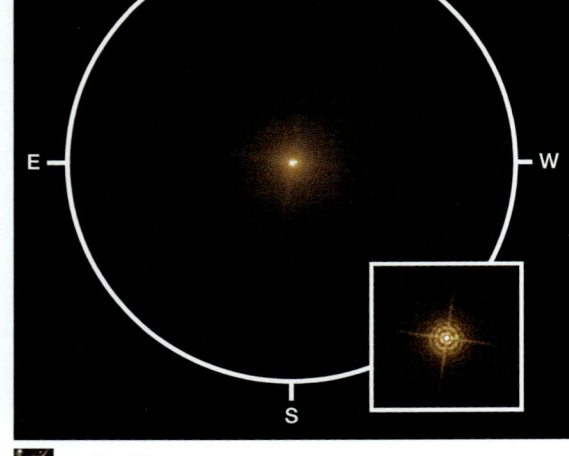

18.9 Sketch of Antares.
Courtesy of Jeremy Perez

Unit 5 The Deep Sky

You can split Antares through a 5-inch telescope at 100× on steady nights. Seen in a 13-inch refractor at 190×, it is perhaps the most spectacular colored double star in the entire sky!

Alpha Herculis (Ras Algethi)

The orange and blue-green pair comprising Ras Algethi, or Alpha Herculis (Fig. 18.10), reminds some observers of an easier-to-split Antares. A 2.4-inch telescope at 75× will divide these two stars nicely, and they make a lovely sight through an 8-inch at 100×. The magnification needed to resolve Ras Algethi depends on the changing brightness of the variable primary. Repeated observations reveal the primary varies in magnitude from 2.7 to 3.9, while the companion remains a steady magnitude 5.4.

Omicron (o) Ophiuchi

Omicron (o) Ophiuchi tends to be overshadowed by the many globular clusters found in this region, but be sure to seek it out. These pale-orange and clear-blue jewels are superb through a 3-inch telescope, even at 30×.

18.10 (A) Ras Algethi (Alpha [α] Herculis), and (B) star chart to locate Ras Algethi (Alpha [α] Herculis).

(A) Courtesy of Jeremy Perez; (B) courtesy of Richard Talcott and Roen Kelly, *Astronomy* magazine

95 Herculis

You have to see 95 Herculis (Fig. 18.11) to believe it! The colors are just as Smyth described them: apple green and cherry red. These amazing hues glow subtly but persistently in all apertures, but they are best seen through a 6-inch telescope at $50\times$.

70 Ophiuchi

70 Ophiuchi (Fig. 18.12) is a close double with an entirely different kind of color contrast than most of those on this list: yellow and red. It combines a magnitude 4.2 yellow primary with a red companion that shines at magnitude 6.2. Snug in a 3-inch telescope at $100\times$, a 10-inch at double that magnification separates the pair beautifully. Some observers report seeing a persistent touch of violet in the 6th-magnitude companion, but the hue you detect depends on your eyes' color perception.

Beta (β) Cygni (Albireo)

Albireo (Beta [β] Cygni) marks the head of Cygnus the Swan. Albireo may be the best-known double star in the sky (Fig. 18.13). It bears the moniker "everyone's favorite double star," and for good reason. Its glorious topaz-orange and sapphire-blue colors glow at magnitudes 3.4 and 4.7, respectively. They are unmistakable even in a 2-inch glass at $20\times$, and can actually be seen through 10×50 binoculars. And the colors are definitely real, as you can readily prove to yourself by attaching a narrow strip of aluminum foil across your eyepiece's field stop. Hiding either star behind the strip has no effect on the color of the other!

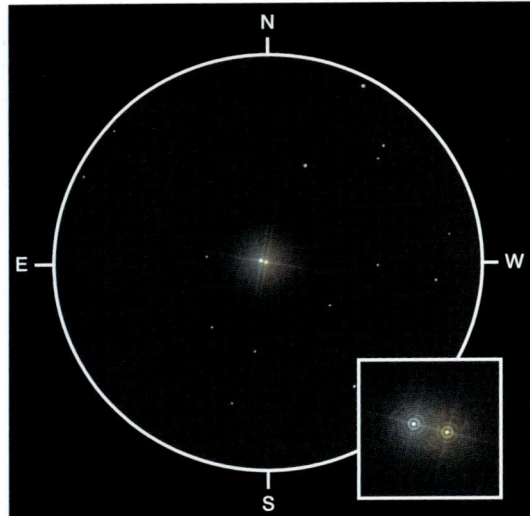

18.11 Sketch of 95 Herculis.
Courtesy of Jeremy Perez

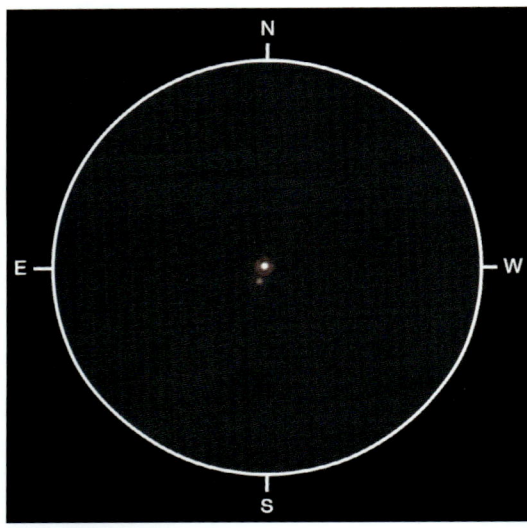

18.12 70 Ophiuchi. Courtesy of Jeremy Perez

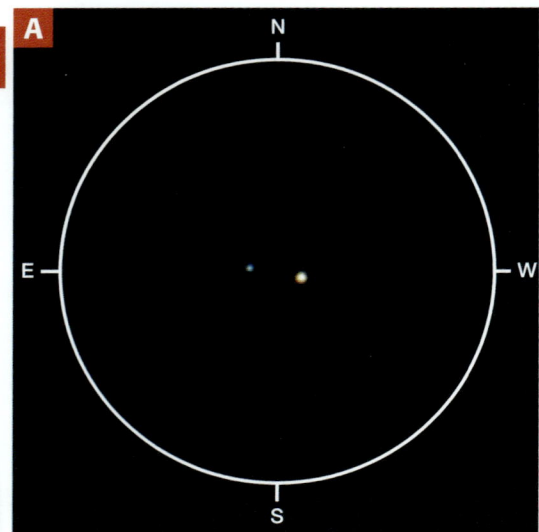

18.13 "Everyone's favorite double star": (**A**) Albireo (Beta [β] Cygni), and (**B**) star chart to locate Albireo (Beta [β] Cygni).

(A) Courtesy of Jeremy Perez; (B) courtesy of Richard Talcott and Roen Kelly, *Astronomy* magazine

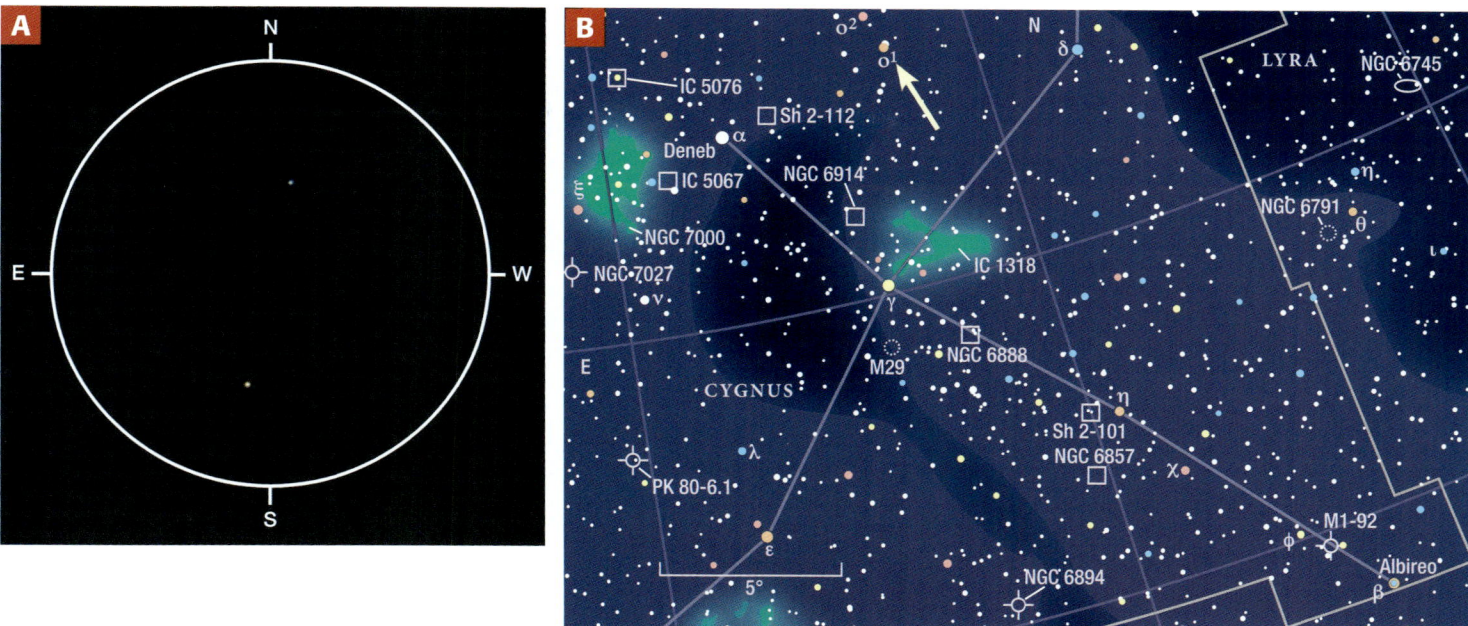

18.14 (**A**) Omicron¹ (o¹) Cygni, and (**B**) star chart to locate Omicron¹ (o¹) Cygni.

(A) Courtesy of Jeremy Perez; (B) courtesy of Richard Talcott and Roen Kelly, *Astronomy* magazine

Omicron¹ (o¹) Cygni

As if Albireo isn't enough, Cygnus offers another highly tinted double that is also striking through small telescopes: Omicron¹ (o¹) Cygni (Fig. 18.14). At low magnification, the hues here mimic the yellow and blue of its famous neighbor, Albireo, but the two suns are much farther apart so the contrast isn't quite as dramatic. They are a widely separated pair (106") located 5° west-northwest of Deneb (Alpha Cygni). There is also a 4.8-magnitude bluish-white star 331" distant, making this an apparent triple system.

Gamma (γ) Delphini

The yellow and pale-green double Gamma (γ) Delphini (Fig. 18.15) sits 15° east-northeast of the brilliant star Altair (Alpha Aquilae) at the tip of the parallelogram that forms the small constellation Delphinus. It is neatly resolved in a 3-inch telescope at 45×. Through

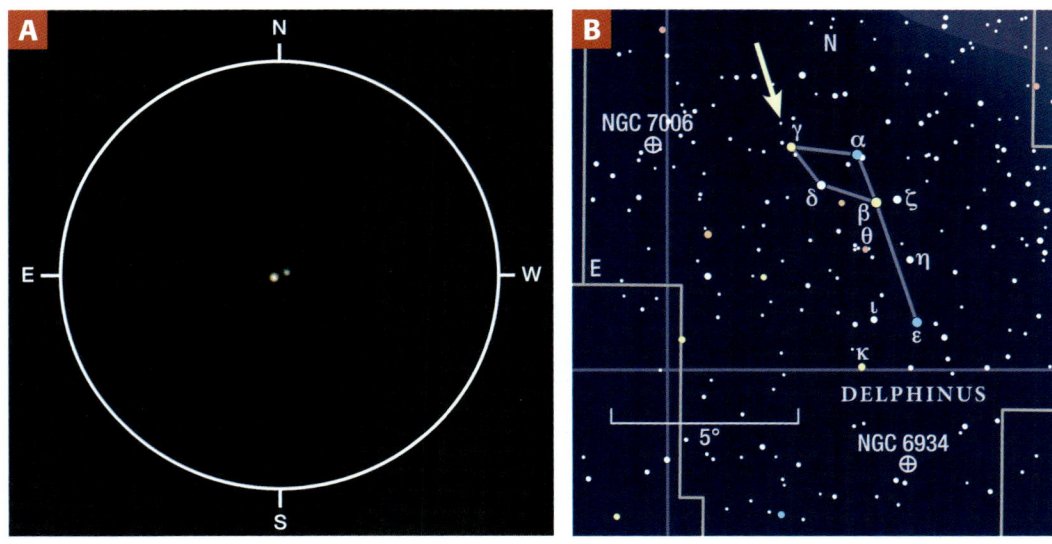

18.15 (**A**) Gamma (γ) Delphini, and (**B**) star chart to locate Gamma (γ) Delphini.

(A) Courtesy of Jeremy Perez; (B) courtesy of Richard Talcott and Roen Kelly, *Astronomy* magazine

6-inch and larger telescopes, it is one of the finest doubles in the sky. Yellow and lime green are the colors typically used to describe this pair. There is also a ghostly looking double lurking close by in the same field of view. Designated Struve 2725, its magnitude 7.6 and 8.4 components lie just 6" apart and contribute to the fascination of the scene here.

Delta (δ) Cephei

Delta (δ) Cephei happens to be the prototype of the famed Cepheid variables that astronomers use to gauge distances in space. Unfortunately, as such, it is often overlooked as an attractive double star. Its pale yellowish-orange and blue components are a lovely sight through a 4-inch telescope at magnifications as low as $16\times$.

Check Your Understanding

1.1 In a double star system, what is the difference between the primary and the secondary?

1.2 Why do you think different observers describe the colors of double stars differently?

1.3 Observers who describe double stars often compare them to Albireo. Considering this star's description on page 350, why do you think it is so famous?

1.4 How did the astronomer William H. Smyth decide that the companion of Antares was green, and that the color didn't just result from a contrast effect with the red primary?

Variable Stars

"Change in all things is sweet," said Aristotle. In astronomy, objects called variable stars provide sweet views to observers. Variable stars are those whose brightnesses change over time. One night, a star field might contain several dozen similarly bright stars. A week later, one star might shine brighter than the rest. Has it exploded? Probably not. More often, the star will fade and, after a period of time, brighten again. By knowing where to look, you can follow these changes through even a small telescope.

Discovery and Naming Conventions

The first variable star discovered was Mira (Omicron [o] Ceti), in 1596 by German astronomer David Fabricius (1564–1617). Mira's variability caused astronomers to think about stars in different ways. Before then, the universe seemed unchanging. Other variable-star discoveries followed. Italian astronomer Geminiano Montanari (1633–1687) noted the variability of Algol (Beta [β] Persei) in 1669. Dutch-born British astronomer John Goodricke (1764–1786) discovered Delta (δ) Cephei, the prototype for all Cepheid variables, in 1784. He also explained why Algol varies in brightness: It is a double-star system in which one star revolves around another, an **eclipsing binary**, the first to be found.

By the mid-nineteenth century, astronomers had discovered so many variable stars that they didn't have the resources to follow them all. German astronomer Friedrich Wilhelm August Argelander (1799–1875) and American astronomer Benjamin Apthorp Gould (1824–1896) called upon amateurs worldwide to submit observations of variable stars.

Argelander also began a naming convention for variable stars. He assigned the uppercase Roman letters R through Z to variable stars within a constellation. After these nine letters, Argelander used double uppercase letters (RR–RZ, SS–SZ . . . AA–AZ, BB–BZ . . . QQ–QZ).

Because no J was used to begin these pairs, Argelander could denote a total of 334 stars in this fashion. So, R CrB is the first variable star in Corona Borealis and QZ Ori is the 334th variable star in Orion. If a constellation requires even more designations (and many do), astronomers use a single uppercase V plus numbers after 334. Therefore, Plaskett's Star carries the designation of V640 Mon, the 640th variable star in the constellation Monoceros.

Novae are designated initially by constellation and year, for example, Nova Cygni 1975. When the brightness of the event diminishes, such objects receive a variable-star designation. So, for example, astronomers now refer to Nova Cygni 1975 as V1500 Cyg.

The American Association of Variable Star Observers (AAVSO) also employs a numerical system for variable stars. By this numbering, R Leporis has the additional AAVSO designation of 0455–14. These numbers represent the star's approximate right ascension (4h55m) and declination (–14°) for the year 1900.

> **GLOSSARY TERM**
>
> **Eclipsing binary** binary star system in which chance alignment of the orbits means that from Earth the stars periodically pass in front of one another causing eclipses, resulting in a variability of the light from the system

Observing Variable Stars

Seasoned observers telescopically compare target variable stars with nearby, similarly bright stars. Good estimates have an error of about 0.1 magnitude. If you are just starting out, however, you won't come close to this level of precision. Sometimes, you can check in on a few favorite variables just to say, "Wow! That's a lot brighter than it was three days ago." Or, if the star is near minimum brightness, it may have vanished.

One solution in the quest for precision measurements is to take images of the target's star field with either a digital single-lens reflex (DSLR) or charge-coupled device (CCD) camera. With this technology, you can measure variable-star brightnesses to a level of accuracy astronomers could only dream about a century ago. Medium-size telescopes equipped with CCD cameras typically detect 16th-magnitude stars. Add to such a system a computerized drive and software running a control script containing a night's list of variable stars, and the amount of data you can collect will amaze you, all while you are either looking through a different telescope or asleep.

Stars are point sources, and even relatively faint ones record quickly. Use a DSLR or CCD camera, experiment with exposure times, and discard any images you don't like. Once you settle on an exposure time, however, it is important that you use the same interval for each image you take of the variable throughout its entire period. Label each image with the date and time.

If you don't have a digital SLR or CCD camera and you want to record your observations, use the tried-and-true method of sketching what you see through the eyepiece. Take care to portray different star brightnesses accurately, either by larger dots or by pressing lighter or harder with your pencil. As with images, label each sketch with the date and time.

Classifying Variable Stars

Two main types of variables exist: intrinsic and extrinsic. Each type has two classes. Pulsating variables and eruptive variables comprise the intrinsic type. Rotating variables and eclipsing binaries are the two classes within the extrinsic type. Rotating variable stars may expose different areas to us as they spin. A large number of sunspots on one side of the star can greatly affect its brightness. Eclipsing binaries also drop in brightness when one star in the system eclipses the other.

Eclipsing binaries are good candidates for graphing a light curve. A **light curve** (Fig. 18.16) is a simple graph astronomers use to plot a variable star's brightness over time. Observers record many measurements, and software determines the curve that best fits the data. Each point on the curve represents the star's magnitude at a specific time. The more points used to make a light curve—and the smaller the intervals between them—the more accurate the curve will be. Once they have accumulated many points, astronomers use a mathematical formula to smooth out the points and generate a curve. You will do this in Exercise 18.3.

Many subclasses exist within the two major variable-star classifications. Following is an alphabetized list of some familiar types you can observe. These objects possess the names they do because astronomers labeled most variable star types after the first star discovered in its class, known as the prototype.

> **GLOSSARY TERM**
>
> **Light curve** graph showing the change in brightness of a celestial object plotted against time

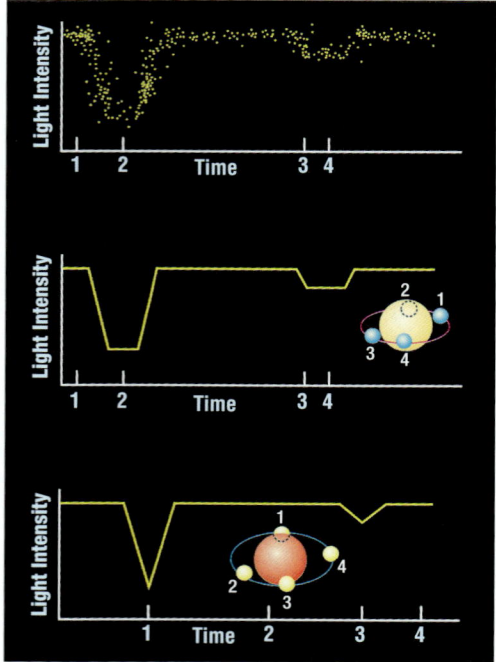

18.16 Example of light curves.
Courtesy of Roen Kelly, *Astronomy* magazine

Algol-Type Variable Stars

The classic eclipsing binary Algol (Fig 18.17) makes an easy celestial target because you can see its brightness vary with your naked eyes. The two stars in the system are far enough apart that, most of the time, their combined light is constant. For observers, the best part happens when the light drops more than a magnitude in about five hours, waits 20 minutes, then begins to climb again.

You can compare Algol to other nearby stars, such as Gamma (γ) Andromedae. At maximum, Algol is just (0.15 magnitude) brighter than Gamma. At minimum, it glows more than a magnitude fainter. Table 18.2 lists several examples of Algol-type variable stars.

Most Algol variables have periods of a few days, but exceptions exist. One is Epsilon (ε) Aurigae, which has a period of 27 years.

TABLE **18.2** Algol-Type Variable Stars

Star	Right Ascension	Declination	Period	Magnitude Range
Beta (β) Persei (Prototype)	3h08m	40°57'	2.87 days	2.1 to 3.4
U Cephei	1h02m	81°52'	2.5 days	6.8 to 9.2
U Sagittae	19h19m	19°37'	3.4 days	6.5 to 9.3
VW Hydri	4h09m	−71°18'	2.7 days	10.5 to 14.1

18.17 Algol (Beta (β) Persei). *Courtesy of Richard Talcott and Roen Kelly, Astronomy magazine*

Beta (β) Lyrae-Type Variable Stars

The Beta Lyrae system's stars (Fig. 18.18) are so close they are either touching or nearly so. Their shapes are so distorted by gravity that they look like eggs. At no point on their light curve is their light output steady. When observing the prototype, β Lyr, compare it to nearby magnitude 3.2 Gamma (γ) Lyrae, which lies 2° to the east-southeast. At its brightest, β is equal to γ in brightness, and it glows a magnitude fainter when dimmest. Table 18.3 lists several examples of Beta Lyrae-type variable stars.

TABLE **18.3** Beta Lyrae-Type Variable Stars

Star	Right Ascension	Declination	Period	Magnitude Range
Beta Lyrae (Prototype)	18h50m	33°22'	12.91 days	3.3 to 4.4
Zeta (ζ) Andromedae	0h47m	24°16'	17.8 days	3.9 to 4.1
UW Canis Majoris	7h19m	−24°33'	4.4 days	4.8 to 5.3
V Puppis	7h58m	−49°15'	1.45 days	4.4 to 4.9

18.18 Beta (β) Lyrae. *Courtesy of Richard Talcott and Roen Kelly, Astronomy magazine*

GLOSSARY TERMS
Absolute magnitude apparent magnitude of a celestial object if the object were at a distance of 10 parsecs; the true brightness of a celestial object; luminosity
Apparent magnitude brightness of a celestial object as seen from Earth irrespective of its true brightness; usually just called "magnitude"

Delta (δ) Cepheid-Type Variable Stars

Delta Cepheids (Fig. 18.19), the "measuring sticks" of astronomy, are usually shortened to Cepheids. Cepheids have an amazing characteristic: Their periods relate directly to their luminosities. Once a Cepheid's period is measured, astronomers can calculate its **absolute magnitude** (how bright it would look from a standard distance of 32.6 light-years). Comparing that quantity with the star's **apparent magnitude** (how bright it appears) allows astronomers to determine its distance. Physically, Cepheids are young stars of several solar masses and about 10,000 times the Sun's luminosity. Cepheids pulsate regularly because helium in the stars' outer layers ionizes, which makes them opaque. As a Cepheid darkens, the gas absorbs more energy, and the star expands, with an increase in luminosity. As it expands, however, the helium cools, causing the star to shrink. Table 18.4 lists several examples of Delta Cepheid-type variable stars.

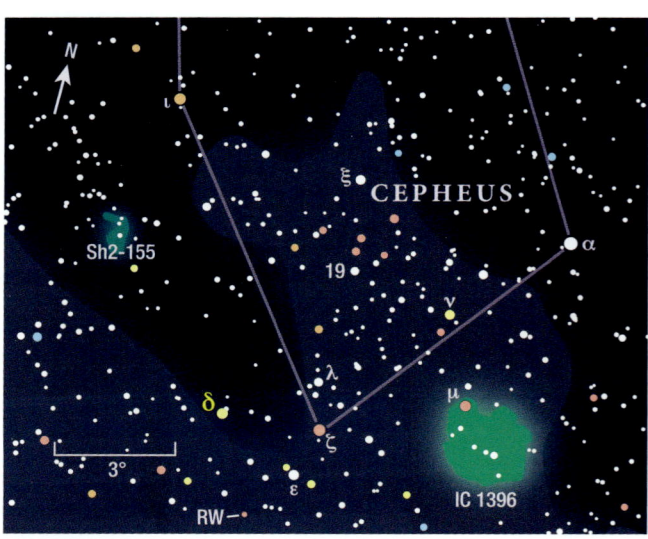

18.19 Delta (δ) Cepheid.
Courtesy of Richard Talcott and Roen Kelly, *Astronomy* magazine

TABLE **18.4** Delta Cepheid-Type Variable Stars

Star	Right Ascension	Declination	Period	Magnitude Range
Delta Cephei (Prototype)	22h29m	58°24'	5.37 days	3.5 to 4.4
Eta (ε) Aquilae	19h52m	1°00'	7.18 days	3.5 to 4.4
RT Aurigae	6h29m	30°30'	3.73 days	5.0 to 5.8
Zeta (ζ) Geminorum	7h04m	20°34'	10.15 days	3.6 to 4.2

Mira-Type Variable Stars

Mira-type variables (Fig. 18.20) are red giants with periods of up to three years. Keeping track of this type of star can be rewarding, however, because the best examples have a wide brightness range. Astronomers discovered many Mira-type variables after Fabricius found their prototype in 1596, and their designations (noted by the letter R) show that, in most northern constellations, a Mira-type variable was the first variable seen. Check out the other examples in Table 18.5 and see a star become invisible (and then reappear).

18.20 Mira (Omicron [o] Ceti).
Courtesy of Richard Talcott and Roen Kelly, *Astronomy* magazine

TABLE 18.5 Mira-Type Variable Stars

Star	Right Ascension	Declination	Period	Magnitude Range
Omicron (o) Ceti (Prototype)	2h19m	−2°59'	333.8 days	2 to 10 (both vary)
R Andromedae	0h24m	38°34'	409 days	5.6 to 14.9
R Aquilae	19h06m	8°14'	284 days	5.5 to 12.0
R Cassiopeiae	23h58m	51°23'	430 days	4.7 to 13.5
R Hydrae	13h30m	−23°17'	389 days	4.0 to 10.9
R Leonis	9h47m	11°26'	310 days	4.4 to 11.3
R Leporis	5h00m	−14°49'	427 days	5.5 to 11.7
R Serpentis	15h51m	15°08'	356 days	5.7 to 14.4
R Trianguli	2h37m	34°16'	267 days	5.4 to 12.6

RR Lyrae-Type Variable Stars

Stars in the RR Lyrae class (Fig. 18.21) all have roughly the same intrinsic brightness, around 60 times the Sun's luminosity. Because of this, these stars make good distance indicators, similar to Cepheids, although astronomers can detect Cepheids at much greater distances. Astronomers find many RR Lyrae stars in globular clusters. Try to observe the prototype's cycle in one night. Depending on your location (and the time of year), you will see either the full range of variability or nearly all of it. Once you have observed the prototype, try looking for some of the other RR Lyrae-type stars listed in Table 18.6.

18.21 RR Lyrae.
Courtesy of Richard Talcott and Roen Kelly, *Astronomy* magazine

TABLE 18.6 RR Lyrae-Type Variable Stars

Star	Right Ascension	Declination	Period	Magnitude Range
RR Lyrae (Prototype)	19h25m	42°47'	13h 36m	7.1 to 8.1
AC Andromedae	23h18m	48°47'	2.1 days	10.6 to 11.2
AR Herculis	16h00m	46°55'	11.3 hours	10.3 to 10.8
RS Boötis	14h34m	31°45'	4 hours	9.2 to 10.6

RV Tauri-Type Variable Stars

Although the class is RV Tauri, the prototype is R Scuti (Fig. 18.22), a star discovered to be variable in 1795. Stars of RV Tau type are luminous yellow giants. The brightness minimum, slightly below 8th magnitude, has two separate levels, alternating between a shallow and a deep minimum. Astronomers define the period of an RV Tau star as the length of time between two of its deep minima. The brightness of the prototype—even at minimum—places it well within the reach of any telescope. RV Tauri-type variable stars are listed in Table 18.7.

TABLE 18.7 RV Tauri-Type Variable Stars

Star	Right Ascension	Declination	Period	Magnitude Range
R Scuti (Prototype)	18h47m	−5°42'	140 days	4.5 to 8.2
U Monocerotis	7h31m	−9°47'	92.3 days	5.1 to 7.1
AC Herculis	18h30m	21°52'	75.5 days	6.4 to 8.7
RV Tauri	4h47m	26°11'	76.7 days	8.8 to 12.3

18.22 R Scuti, the prototype for RV Tauri-type variable stars.
Courtesy of Richard Talcott and Roen Kelly, *Astronomy* magazine

W Ursae Majoris-Type Variable Stars

The W Ursae Majoris (W UMa) type of variable star (Fig. 18.23) resembles the Beta Lyrae type. In both cases, contact binaries make up the systems. W UMa variables are cooler and less massive than β Lyr variables. Table 18.8 lists several examples of W Ursae Majoris-type variable stars.

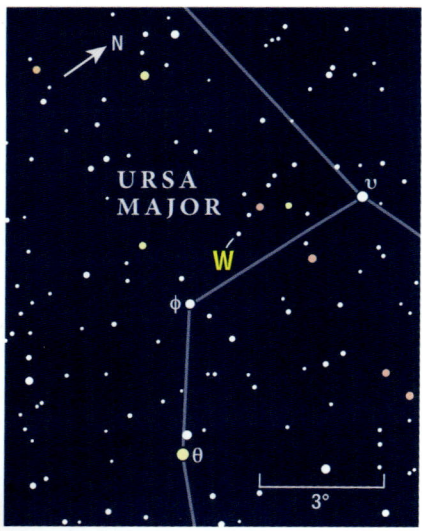

TABLE 18.8 W Ursae-Type Variable Stars

Star	Right Ascension	Declination	Period	Magnitude Range
W Ursae Majoris (Prototype)	9h44m	55°58'	4 hours	7.6 to 8.4
44 Boötis B	15h04m	47°39'	6.43 hours	5.8 to 6.4
VW Cephei	20h37m	75°36'	6.7 hours	8.1 to 8.4
S Antliae	9h32m	−28°38'	15.6 hours	6.3 to 6.8

18.23 W Ursae Majoris.
Courtesy of Richard Talcott and Roen Kelly, *Astronomy* magazine

Check Your Understanding

2.1 How would you define the term *variable star*?

2.2 Why does Algol (Beta [β] Persei) change its brightness?

2.3 Which was the first group founded to observe variable stars?

2.4 How do astronomers designate the 10th variable star in a constellation?

2.5 What does the variable star designation "V500" signify?

2.6 What is a light curve?

2.7 How would you make a light curve more accurate?

2.8 What are the two main types of variable stars?

Exercise 18.1 Observing Double Stars

 **Procedure**

Materials
- ❏ Telescope
- ❏ Clipboard (to hold your papers and for sketching)
- ❏ Red flashlight (for reading the procedural instructions, data sheets, and sketching outside at night)

1. Upon arrival at your destination, your first task is to assess the quality of your observing site. Answer the questions on the data sheet, page 361, at the start of your observing session.

2. Your instructor will ask you to observe up to 6 double stars. Using the data sheets on pages 362–364, sketch the eyepiece's field of view of each double star. Place the binary in the center of the circle. Draw any other noticeable stars that you observed.

Exercise 18.2 Observing Variable Stars

 Procedure

Materials
- ❏ Telescope
- ❏ Clipboard (to hold your papers and for sketching)
- ❏ Red flashlight (for reading the procedural instructions, data sheets, and sketching outside at night)

1. Upon arrival at your destination, your first task is to assess the quality of your observing site. Answer the questions on the data sheet, page 365, at the start of your observing session.

2. Your instructor will ask you to observe up to 6 variable stars. Using the data sheets on pages 366–368, sketch the eyepiece's field of view of each variable star. Place what you think is your target in the center of the circle, and mark it with an arrow. Draw any other noticeable stars that you observed. Indicate if they were much brighter, brighter, fainter, or much fainter than your target star by labeling them with the abbreviations mb, b, f, and mf.

Name _____ Section _____ Date _____

Double Star Observations

Quality of Observing Site

1 How dark does the site seem to you? Explain how you came to your conclusion.

2 How far away from the nearest large city are you? How far from the nearest small city?

3 Are there any "light domes" visible? These bright areas of the sky (brightest near the horizon) originate with artificial lights.

4 Are there any clouds? Approximately how much of the sky do they cover?

5 Is the Moon visible? It is best to choose a moonless night to observe, but maybe the nights have been cloudy. What is the Moon's phase?

6 What is your estimate of your site's seeing? Explain your answer.

7 What is your estimate of your site's transparency? Explain your answer.

Exercise 18.1 Data Sheet

Exercise 18.1 Data Sheet

1 Star _____   Date _____   Time _____

Telescope _____

Eyepiece _____

Magnification _____

Label the brighter star A and the fainter one B.

How easy was it to split the pair?

_____

Record the colors you saw:

A _____

B _____

---

2 Star _____   Date _____   Time _____

Telescope _____

Eyepiece _____

Magnification _____

Label the brighter star A and the fainter one B.

How easy was it to split the pair?

_____

Record the colors you saw:

A _____

B _____

Unit 5 *The Deep Sky*

Name _____   Section _____   Date _____

Double Star Observations

3 Star _____   Date _____   Time _____

   Telescope _____

   Eyepiece _____

   Magnification _____

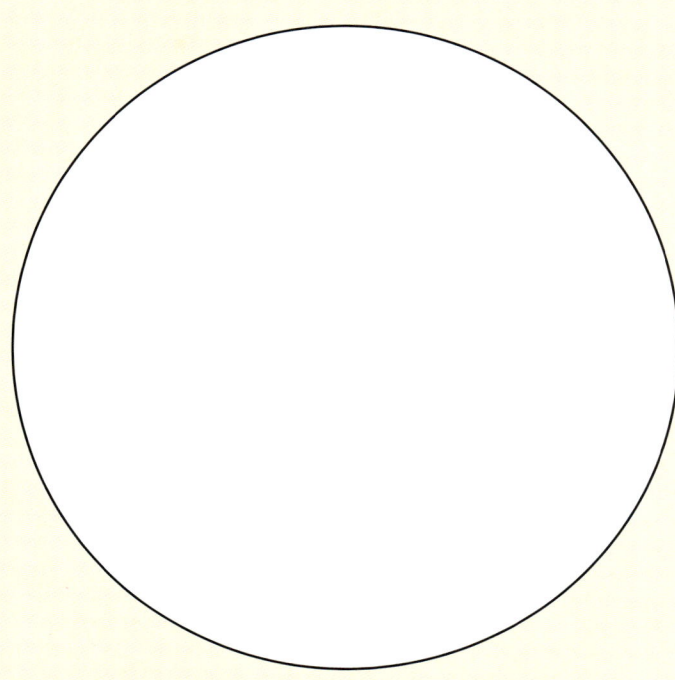

Label the brighter star A and the fainter one B.

How easy was it to split the pair?

_____

Record the colors you saw:

A _____

B _____

---

4 Star _____   Date _____   Time _____

   Telescope _____

   Eyepiece _____

   Magnification _____

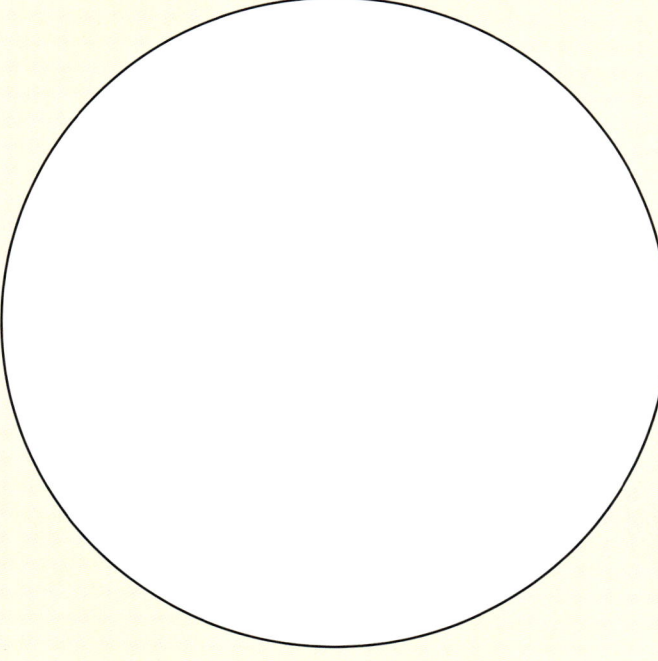

Label the brighter star A and the fainter one B.

How easy was it to split the pair?

_____

Record the colors you saw:

A _____

B _____

Exercise 18.1 Data Sheet

Exercise 18.1 Data Sheet

5 Star _____ Date _____ Time _____

Telescope _____

Eyepiece _____

Magnification _____

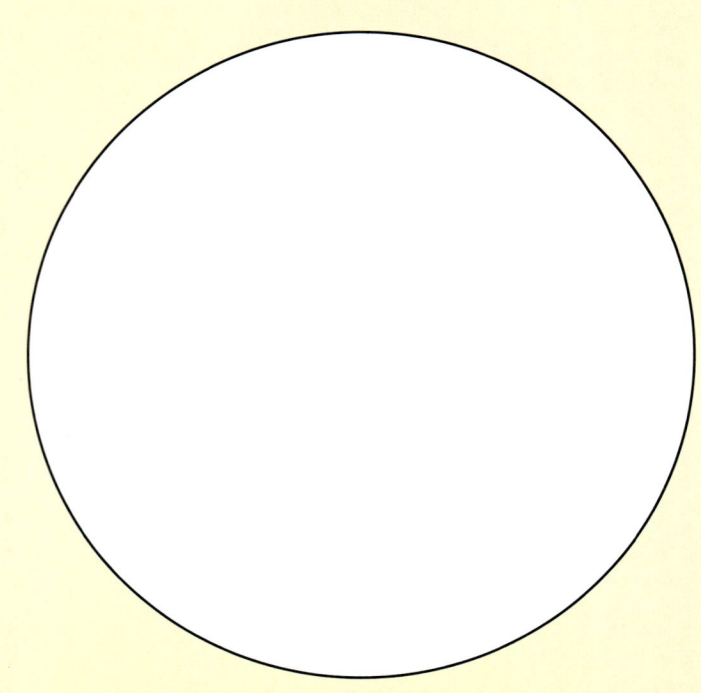

Label the brighter star A and the fainter one B.

How easy was it to split the pair?

_____

Record the colors you saw:

A _____

B _____

---

6 Star _____ Date _____ Time _____

Telescope _____

Eyepiece _____

Magnification _____

Label the brighter star A and the fainter one B.

How easy was it to split the pair?

_____

Record the colors you saw:

A _____

B _____

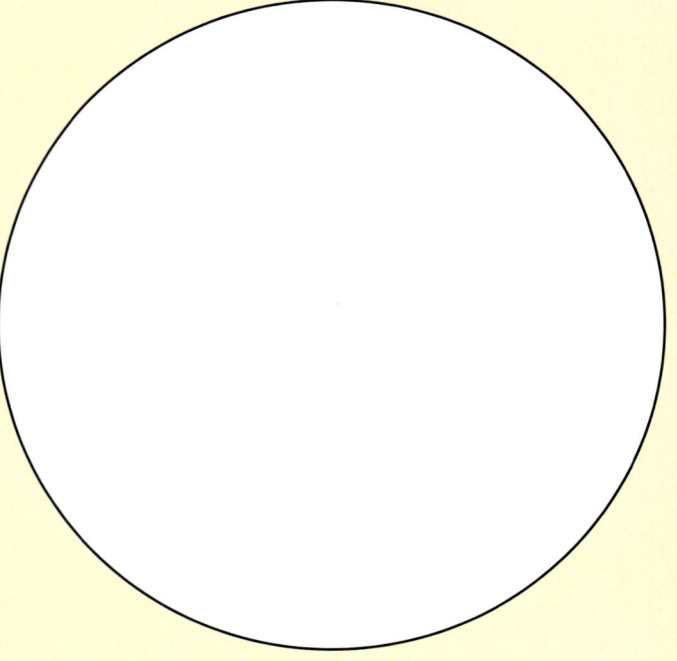

364   Unit 5 *The Deep Sky*

Name _____   Section _____   Date _____

Variable Star Observations

Quality of Observing Site

1 How dark does the site seem to you? Explain how you came to your conclusion.

2 How far away from the nearest large city are you? How far from the nearest small city?

3 Are there any "light domes" visible? These bright areas of the sky (brightest near the horizon) originate with artificial lights.

4 Are there any clouds? Approximately how much of the sky do they cover?

5 Is the Moon visible? It is best to choose a moonless night to observe, but maybe the nights have been cloudy. What is the Moon's phase?

6 What is your estimate of your site's seeing? Explain your answer.

7 What is your estimate of your site's transparency? Explain your answer.

Exercise 18.2 Data Sheet

1 Star _____ Period _____

Date 1 _____ Date 2 _____ Time _____

Telescope _____

Eyepiece _____

Magnification _____

Record the brightness change in the star:

_____

_____

---

2 Star _____ Period _____

Date 1 _____ Date 2 _____ Time _____

Telescope _____

Eyepiece _____

Magnification _____

Record the brightness change in the star:

_____

_____

Name _____ Section _____ Date _____

Variable Star Observations (continued)

3 Star _____ Period _____

Date 1 _____ Date 2 _____ Time _____

Telescope _____

Eyepiece _____

Magnification _____

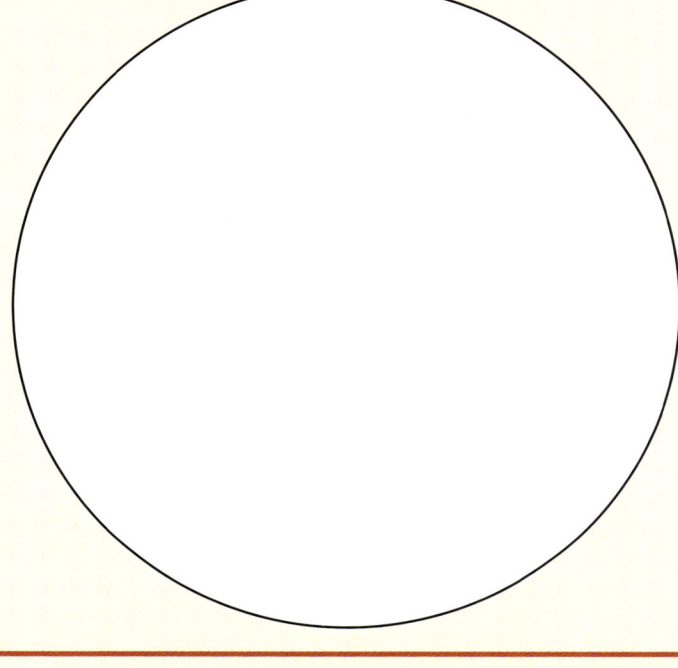

Record the brightness change in the star:

_____
_____

---

4 Star _____ Period _____

Date 1 _____ Date 2 _____ Time _____

Telescope _____

Eyepiece _____

Magnification _____

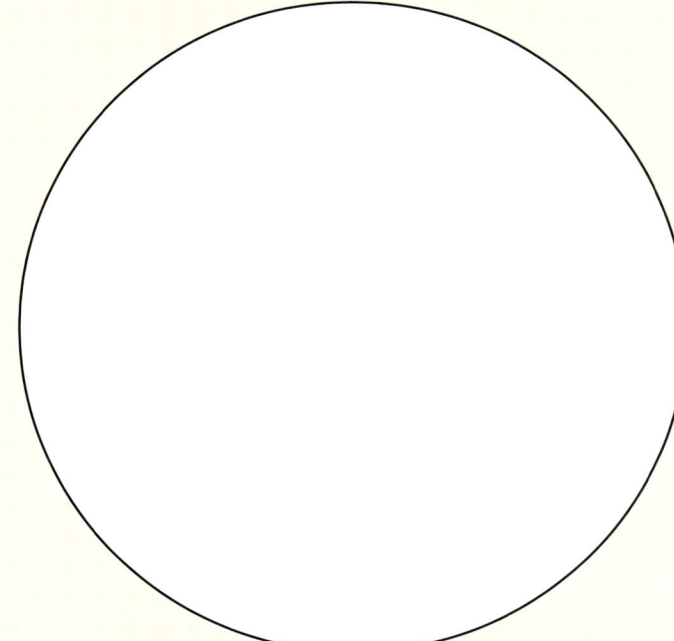

Record the brightness change in the star:

_____
_____

Exercise 18.2 Data Sheet

5 Star _____ Period _____

Date 1 _____ Date 2 _____ Time _____

Telescope _____

Eyepiece _____

Magnification _____

Record the brightness change in the star:

_____

_____

---

6 Star _____ Period _____

Date 1 _____ Date 2 _____ Time _____

Telescope _____

Eyepiece _____

Magnification _____

Record the brightness change in the star:

_____

_____

Creating a Light Curve for Algol

Exercise 18.3

Although an observer discovered this star's variation in 1667, people knew of it long before that. It may be the reason for the star's given Arabic name: *ra's al-ghūl*. Scholars usually translate this as "demon star."

The cause of the sudden drop in brightness is a stellar eclipse. Algol is a close double star whose components orbit each other every two days 20 hours and 49 minutes (2.86736 days). The companion to the blue-white visually observed star is a yellow-orange giant star with a luminosity just 2.5 percent that of the brighter star. The brighter star, with 2.9 times the Sun's diameter, is smaller than the cooler star, which has a diameter of 3.5 times that of the Sun.

Normally, the system shines at magnitude 2.1. During each orbit, however, when the dimmer, larger star passes in front of the brighter star, we see a deep eclipse that drops the brightness to magnitude 3.4. Between the deep eclipses is a smaller dip in brightness that occurs when the bright star passes partially in front of the dim one, but you can't detect that one with your naked eyes. The eclipses of Algol were first accurately recorded photo-electrically around 1910 with the historic 12-inch refractor at the University of Illinois Observatory.

Procedure

Materials
❏ Graph paper

1. To record a light curve, first you must create a graph that shows how Algol's brightness varies over time.

2. For the graph, make "time" the X-axis (the horizontal one), and label its divisions in hours. Start with 0 and let each division represent five hours. Because the data covers three days of observations (72 hours), you can end with the 75-hour mark.

3. Make "brightness" the Y-axis (the vertical one), and label its divisions in magnitude. This star system varies from magnitude 2.1 to magnitude 3.4, so use labels that contain those limits. Making the brightest mark 2.0 and the faintest 3.5 will work fine. Use 0.1 magnitude as your divisions. Also, remember that with magnitudes, smaller numbers are brighter, so put 2.0 at the top of your Y-axis.

4. Plot the data points from Table 18.9 on your graph. For this lab, assume that your imaginary observer started recording Algol's brightness at midnight on a given date (the actual date doesn't matter). Other observers around the world joined in at times, so the star never went out of view because of daylight. Place a dot as far right as you need to go to get to the correct time (in hours) and as high as you need to go to record the star's brightness at that time. And remember, 1:00 a.m. on Day 2 is hour 25, and 1:00 a.m. on Day 3 is hour 49.

5. Connect the dots by drawing a line from the first dot to the second, and so on to the last one.

6. On your light curve, label the beginning and ending of the passage of the fainter, large star in front of the brighter, small star.

7. Likewise, label the beginning and ending of the passage of the brighter, small star in front of the fainter, large star.

8. How do you think the light curve would look if the two stars were the same size and brightness?

TABLE 18.9 Brightness Data for Algol

Day 1		Day 2		Day 3	
Time	Brightness	Time	Brightness	Time	Brightness
12:00 a.m.	2.1	12:00 a.m.	2.3	12:00 a.m.	2.1
1:00 a.m.	2.1	1:00 a.m.	2.2	1:00 a.m.	2.1
2:00 a.m.	2.1	2:00 a.m.	2.1	2:00 a.m.	2.1
3:00 a.m.	2.1	3:00 a.m.	2.1	3:00 a.m.	2.1
4:00 a.m.	2.1	4:00 a.m.	2.1	4:00 a.m.	2.1
5:00 a.m.	2.1	5:00 a.m.	2.1	5:00 a.m.	2.1
6:00 a.m.	2.1	6:00 a.m.	2.1	6:00 a.m.	2.15
7:00 a.m.	2.1	7:00 a.m.	2.1	7:00 a.m.	2.2
8:00 a.m.	2.1	8:00 a.m.	2.1	8:00 a.m.	2.15
9:00 a.m.	2.1	9:00 a.m.	2.1	9:00 a.m.	2.1
10:00 a.m.	2.1	10:00 a.m.	2.1	10:00 a.m.	2.1
11:00 a.m.	2.1	11:00 a.m.	2.1	11:00 a.m.	2.1
12:00 p.m.	2.1	12:00 p.m.	2.1	12:00 p.m.	2.1
1:00 p.m.	2.1	1:00 p.m.	2.1	1:00 p.m.	2.1
2:00 p.m.	2.1	2:00 p.m.	2.1	2:00 p.m.	2.1
3:00 p.m.	2.1	3:00 p.m.	2.1	3:00 p.m.	2.1
4:00 p.m.	2.2	4:00 p.m.	2.1	4:00 p.m.	2.1
5:00 p.m.	2.45	5:00 p.m.	2.1	5:00 p.m.	2.1
6:00 p.m.	2.8	6:00 p.m.	2.1	6:00 p.m.	2.1
7:00 p.m.	3.0	7:00 p.m.	2.1	7:00 p.m.	2.1
8:00 p.m.	3.4	8:00 p.m.	2.1	8:00 p.m.	2.1
9:00 p.m.	3.0	9:00 p.m.	2.1	9:00 p.m.	2.1
10:00 p.m.	2.9	10:00 p.m.	2.1	10:00 p.m.	2.1
11:00 p.m.	2.6	11:00 p.m.	2.1	11:00 p.m.	2.1

The Deep Sky
Non-Stellar Objects

LEARNING OBJECTIVES
Upon completion of this chapter, you should be able to:

1. Compare and contrast the three main types of galaxies.
2. Classify different types of galaxies using both the Hubble tuning fork and deVacouleurs systems.
3. Explain how planetary nebulae form and why they have different shapes.
4. Classify planetary nebulae by shape.
5. Describe the history of the classification of galaxies and planetary nebulae.
6. Define all glossary terms.

19

Bode's Galaxy (M81) Courtesy of Adam Block/
Mount Lemmon SkyCenter/University of Arizona

The cosmos is filled with many wonders, including **galaxies**; nebulae of all types such as planetary, diffuse, and dark nebulae; star clusters; and supernovae. In this chapter you will learn about two of these wonders: galaxies, which are "star cities" that contain up to a trillion individual stars, and planetary nebulae, which are clouds of ionized gas glowing by fluorescence and powered by a dying star. You will observe and learn the major characteristics of different types of each. You also will classify an assortment of galaxies and planetary nebulae.

Galaxies

Hubble Classification

American astronomer Edwin Hubble (1889–1953) is credited with developing a simple classification scheme for galaxies. Hubble first mentioned this classification scheme in a paper he wrote in 1922. He expanded it and added some illustrations four years later. Finally, in 1936, Hubble provided a slightly expanded explanation of the classification scheme in his book *The Realm of the Nebulae*. It was in this book that the famous "tuning fork" diagram first appeared (Fig. 19.1). Although the Hubble tuning fork diagram is now out of date, it remains the most widely recognized classification scheme for galaxies.

Hubble described several different main types of galaxies. His scheme has three types: **elliptical**, **spiral**, and barred spiral. Figure 19.2 shows Arp 116, which comprises an elliptical galaxy (M60) and a spiral galaxy (NGC 4647). Figure 19.3 shows two examples of barred spiral galaxies: the Southern Whirlpool Galaxy (M83) and NGC 5584.

> **GLOSSARY TERM**
>
> **Galaxy** collection of up to thousands of billions of stars, dust, and gas held together by gravity
>
> **Elliptical galaxy** galaxy with an ellipsoidal shape and no spiral arms
>
> **Spiral galaxy** galaxy in which a central bulge of older stars is surrounded by a flattened galactic disk containing a spiral pattern of young, hot stars

19.1 Hubble tuning fork diagram.
Courtesy of Holley Y. Bakich

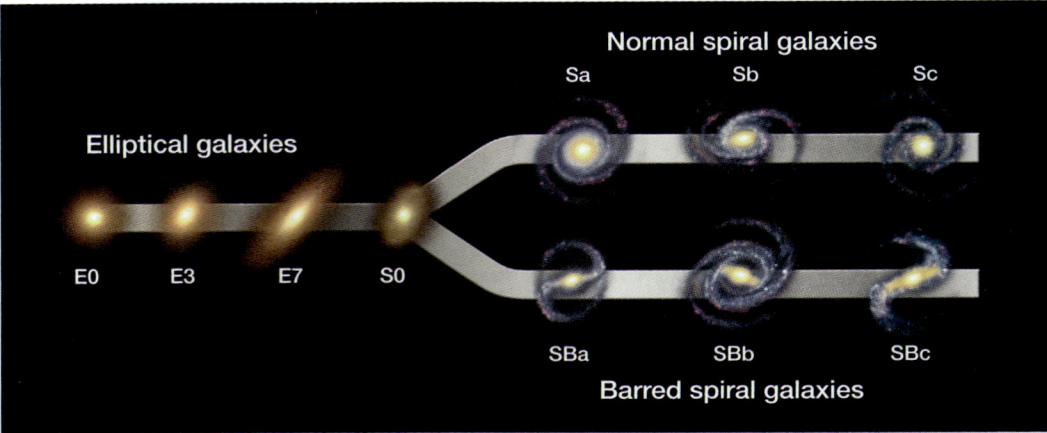

19.2 Arp 116 comprises an elliptical galaxy (M60) and a spiral galaxy (NGC 4647).
Courtesy of NASA/ESA/the Hubble Heritage (STScI/AURA)–ESA/Hubble Collaboration

19.3 Examples of barred spiral galaxies: **(A)** Southern Whirlpool Galaxy (M83), and **(B)** NGC 5584.

Courtesy of Adam Block/Mount Lemmon SkyCenter/University of Arizona

An argument could be made for a fourth Hubble type (off the tuning fork) that would contain **irregular galaxies**, which are every galaxy not on the fork. Figure 19.4 shows the Cigar Galaxy (M82), an example of an irregular galaxy.

Spiral Galaxies

Spirals like our own galaxy fall into several classes, depending on their shape and the relative size of the bulge. Ordinary spirals are labeled Sa, Sb, and Sc while those that have developed a bar in the interior region of the spiral arms are SBa through SBc. The "a" describes a spiral with tightly wound arms (Fig. 19.5). The "b" indicates a spiral with arms not as tightly wrapped (Fig. 19.6). The "c" is a description of a galaxy with loosely wound spiral arms (Fig. 19.7). Initially, as with many descriptive classifications, whether a spiral galaxy was rated as a, b, or c depended on the astronomer classifying it.

19.4 Example of an irregular galaxy, the Cigar Galaxy (M82).

Courtesy of Adam Block/Mount Lemmon SkyCenter/University of Arizona

GLOSSARY TERM

Irregular galaxy one of the main galaxy classifications; a galaxy whose shape is neither elliptical or spiral, but random and unordered; irregular galaxies are generally young and smaller than either ellipticals or spirals

19.5 Example of an Sa galaxy, Bode's Galaxy (M81).

Courtesy of Adam Block/Mount Lemmon SkyCenter/University of Arizona

19.6 Example of an Sb galaxy, Whirlpool Galaxy (M51).

Courtesy of Adam Block/Mount Lemmon SkyCenter/University of Arizona

Looking Up

In 1845, William Parsons, 3rd Earl of Rosse, used his 72-inch reflector (the world's largest telescope until the early twentieth century) to view the spiral structure in M51 for the first time.

19.7 Example of an Sc galaxy, M101.
Courtesy of Adam Block/Mount Lemmon SkyCenter/University of Arizona

Spiral galaxies are characterized by the presence of gas in their disks, which indicates that star formation remains active, hence the younger population of stars. Spirals are usually found in an area of low galaxy density where their delicate shape can avoid disruption by tidal forces from neighboring galaxies.

There are other features to look for when observing a galaxy. The **core**, also known as the nucleus, forms the center of the galaxy. Figure 19.8 shows the prominent bright cores of Galaxy M106 and the Andromeda Galaxy, M31. **Dust lanes** are easy to see in edge-on galaxies such as NGC 4565, the Needle Galaxy (Fig. 19.9A), but are also visible in face-on galaxies. M63, the Sunflower Galaxy (Fig. 19.9B), which is tipped 60 percent to our line of sight, also has visible dust lanes.

Something else to notice when looking at galaxies is **surface brightness**. For example, NGC 6946, the Fireworks Galaxy (Fig. 19.10), has a magnitude of 8.9, but it spreads this brightness over a circle 11' across. So, although it ranks as one of the brightest galaxies, its surface brightness is low. By comparison NGC 2403 in the constellation Camelopardalis the Giraffe (Fig. 19.11) ranks as a high-surface-brightness galaxy.

GLOSSARY TERM

Core central part of a galaxy; also called the nucleus

Dust lane dark band (composed of gas and dust) that blocks out the light from stars in a galaxy

Surface brightness luminosity of any celestial object (or of the sky itself) per unit area; often expressed as magnitude per square arcsecond

19.8 Prominent bright cores of: (**A**) M106 and (**B**) Andromeda Galaxy (M31).
Courtesy of Adam Block/Mount Lemmon SkyCenter/University of Arizona

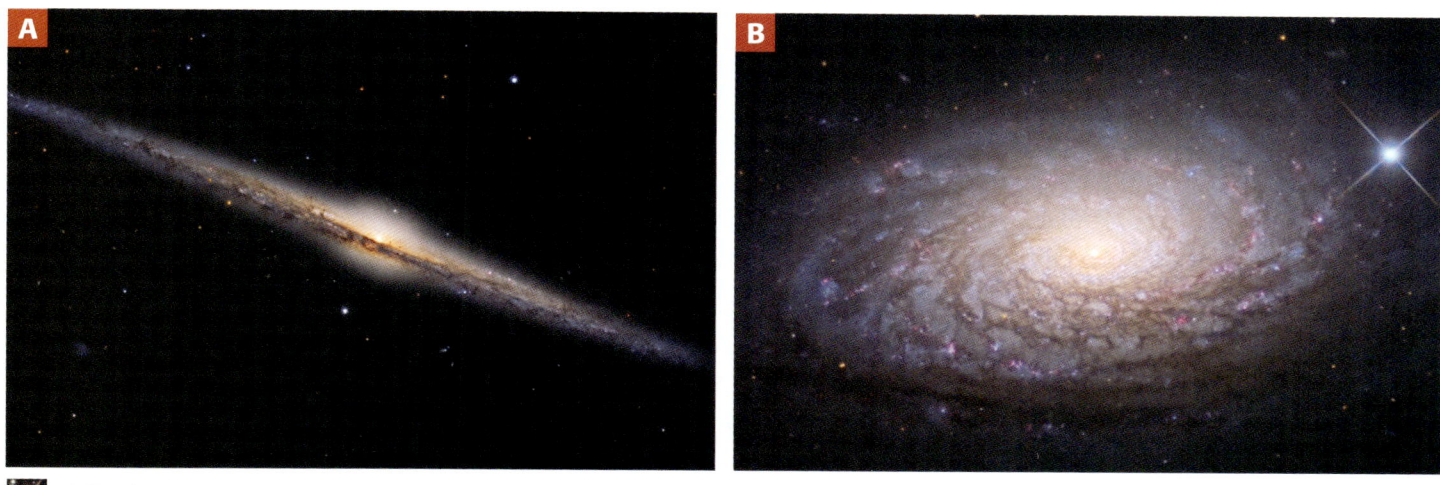

19.9 Dust lanes of (**A**) the Needle Galaxy (NGC 4565), and (**B**) the Sunflower Galaxy (M63).

Courtesy of Adam Block/Mount Lemmon SkyCenter/University of Arizona

19.10 Fireworks Galaxy (NGC 6946), which shows low surface brightness.

Courtesy of Adam Block/Mount Lemmon SkyCenter/University of Arizona

19.11 Spiral galaxy NGC 2403, which shows high surface brightness.

Courtesy of Adam Block/Mount Lemmon SkyCenter/University of Arizona

Elliptical Galaxies

Ellipticals are placed in the categories E0, E1, and through E7, depending on their degree of ellipticity. They have a uniform brightness and are similar to the bulge in a spiral galaxy, but with no disk. The stars are old and there is no gas present. Ellipticals are usually found in areas of high galaxy density, or at the center of clusters.

de Vaucouleurs System

Today, if you ask for an example of one of the types of galaxies represented on Hubble's tuning fork diagram, astronomers who study galaxies will tell you why that system is no longer used. Unfortunately, Hubble based his system mostly on luminous, symmetrical galaxies, which are in the minority in nearby space. The reason the "tuning fork" isn't good enough is that Hubble hadn't looked at enough galaxies!

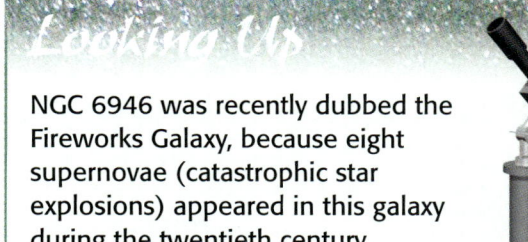

NGC 6946 was recently dubbed the Fireworks Galaxy, because eight supernovae (catastrophic star explosions) appeared in this galaxy during the twentieth century.

Non-Stellar Objects **CHAPTER 19**

In 1959 French astronomer Gerard de Vaucouleurs (1918–1995) published an update of Hubble's classification scheme. His was a much a more detailed analysis of types of galaxies and includes four classes. Figure 19.12 shows the diagram de Vaucouleurs developed, which is sometimes called the "lemon." At the top are the ordinary spirals (labeled A). Toward the bottom are the barred spiral families (labeled B). On the near side are the s-shaped varieties (labeled s). On the far side are the ringed varieties (labeled r). On the far left, E represents ellipticals. He made the Magellanic class (of which the Large Magellanic Cloud is the prototype) its own group, separate from the irregulars, which can be seen on the far right, where I stands for the Magellanic-type irregulars. The percentage of types of galaxies was also revised (Table 19.1). The latest data indicates that about one-third of all galaxies have bars, about one-third have none, and about one-third are mixed cases.

De Vaucouleurs' system essentially takes the "fork" and spins it along the handle, so instead of two tines there is a circle of transitional classes. Sweep one way around from the ordinary to the barred galaxies, and you go through the transitional classes with inner rings. In the other direction are the transitional classes without the inner rings (or lenses).

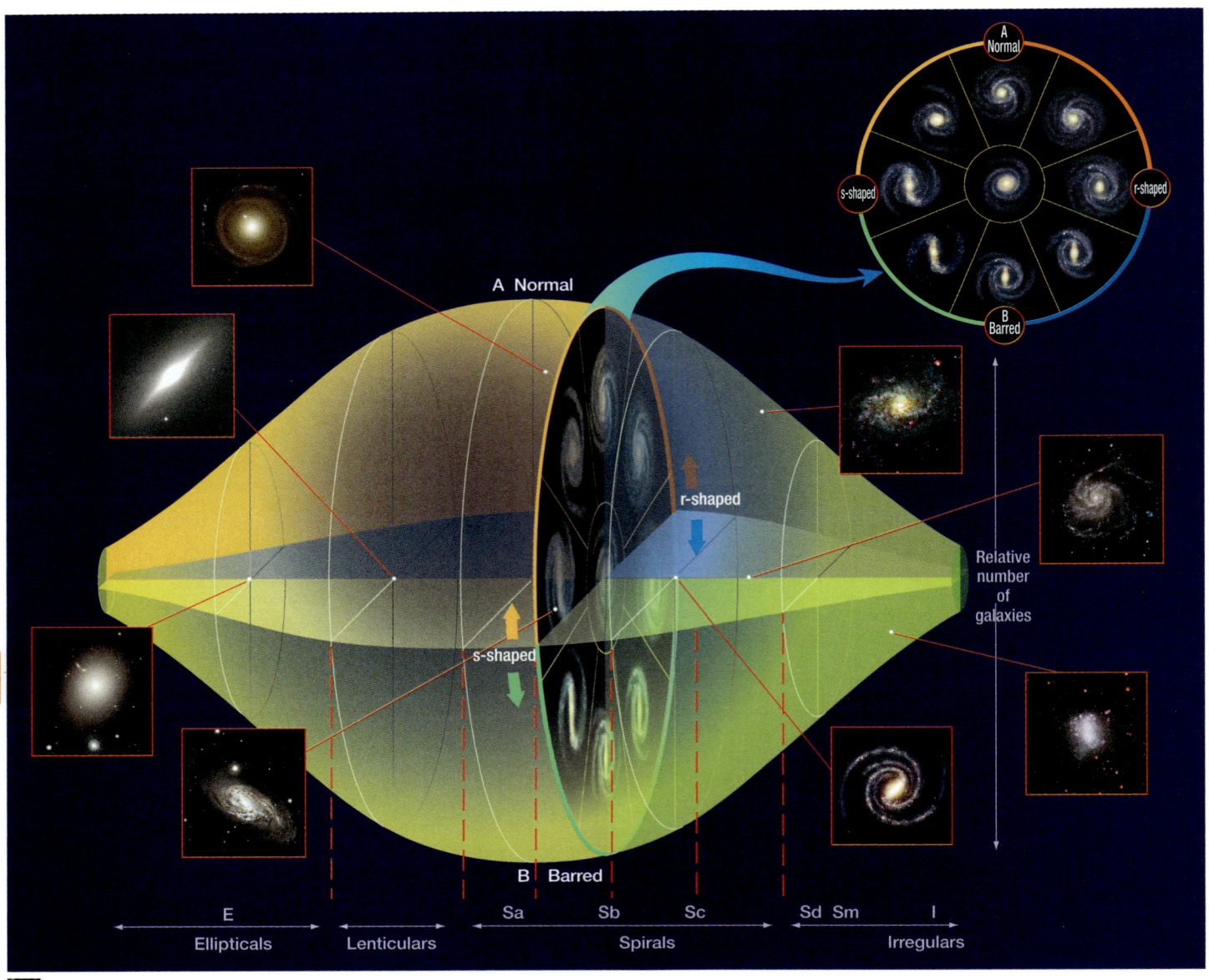

19.12 Galaxy classification scheme of Gerard de Vaucouleurs.

TABLE **19.1** Original Hubble versus de Vaucouleurs Classification Schemes

ORIGINAL HUBBLE CLASSIFICATION						
Type	E	S0	SA SB	SB	Irregular	Peculiar
Frequency (%)	23.4	21.0	24.4	26.3	3.4	1.5
DE VAUCOULEURS CLASSIFICATION						
Type	E	Sa, SBa	Sb, SBb	Sc, SBc	I	
Frequency (%)	17	19	25	36	3	

Best Galaxies to View in the Northern Sky

Galaxies are best seen in the fall and spring because the Milky Way's stars, gas, and dust no longer block them from view. Nineteenth-century astronomers termed the lack of galaxies within the Milky Way's luminous band the "zone of avoidance."

The galaxies listed in Table 19.2 are not limited to one area of the sky or during one season, making them relatively easy to observe. All that is needed is a high-quality telescope and a dark sky. Change eyepieces frequently; this will allow you to view at a range of magnifications, and each should show different details.

If you are a budding amateur astronomer, this list is a great starter course. More experienced observers can look at this list as a refresher.

Look at the image of NGC 4631 below. Can you guess what its common name is?

Courtesy of Adam Block/Mount Lemmon SkyCenter/University of Arizona

If you guessed Whale Galaxy, you are correct!

TABLE **19.2** Best Galaxies to View in the Northern Sky

Name	Right Ascension	Declination	Magnitude	Size
NGC 2403	7h37m	65°36'	8.4	18' by 11'
NGC 6946	20h35m	60°09'	8.9	11'
M81	9h56m	69°04'	6.9	21' by 10'
M31	0h43m	41°16'	3.4	180' by 65'
M63	13h16m	42°02'	8.6	10' by 6'
NGC 4565	12h36m	25°59'	9.6	16' by 3'
M51	13h30m	47°12'	8.4	11' by 7'
M101	14h03m	54°21'	7.9	22'
NGC 4631	12h42m	32°32'	9.8	17' by 4'
M82	9h56m	69°41'	8.4	9' by 4'

Check Your Understanding

1.1 Why do you think astronomers don't see many galaxies when they point their telescopes toward the Milky Way?

1.2 How can a galaxy (or any other celestial object) emit a lot of light but still appear faint?

1.3 Some galaxies, including several on the list in Table 19.2, have dark bands stretching in front of their bright stars. What do you think causes this?

1.4 A spiral galaxy has arms that wind outward from its center. As they do, they appear fainter the farther from the galaxy's core you see them. List a couple possible reasons for this.

Planetary Nebulae

> **GLOSSARY TERM**
>
> **Planetary nebula** outer, gaseous layers of a red giant star that have been gently blown off into space, and which glow because the gas is excited by radiation from the central, collapsing star
>
> **Red giant** any star that is burning helium in its core and has a spectral classification of K or M
>
> **Fusion** nuclear process in the cores of stars that generates energy

Planetary nebulae (Fig. 19.13) are the last gasps of dying stars. These clouds of ionized gas glowing by fluorescence, powered by the hot ember of the stellar corpse within, fascinate backyard observers because many are bright and relatively easy to see through small telescopes, even from moderately light-polluted areas. Their short life spans, only about 50,000 years out of the billions of an average star's life, offer an intimate glimpse into stellar evolution. They hold the keys to the story of how our Sun will end its days. Nearly all stars become planetary nebulae. And yet astronomers have only recently unraveled the details of how such objects form and evolve.

In one key area of research, astronomers have resolved the mystery of planetary nebula shapes, which is a first step in understanding other more complex mysteries. Throughout the history of exploring the world and the cosmos, scientists first catalog objects, classify them, and then move on to deeper understanding.

The cataloging of planetary nebulae is mostly completed, at least within our Milky Way. Astronomers now know of more than 3,000 planetaries in the galaxy, 500 of which are bright enough to view or image with backyard telescopes.

In short, a planetary nebula develops when a star of similar mass to our Sun exhausts its nuclear fuel and begins an old-age phase. After the normal sequence of events that takes a star like ours through a detour as a **red giant**, the star's core becomes an Earth-sized lump of carbon. It begins **fusion**, which occurs in fits and starts, ejecting shells, or bullets, of gas into space surrounding the star.

The shells of gas, expanding at high velocities into the surrounding space, form a beautiful show as they glow and crash into the surrounding interstellar medium, creating a large number of intricate shapes and patterns. The question is how we can make sense of these objects in some coherent, organized way.

History

For 225 years the question of what planetary nebulae are and how they form, remained murky. To underscore the confusion over these objects, they began with an unusual name; planetary nebulae have nothing to do with planets.

The name arose during the eighteenth century when many observers who stumbled upon these objects believed their disklike structures, some of which appeared strongly blue green, resembled planets as seen through their telescopes. NGC 3242 is an example of a nebula that resembles a blue-green planet when seen through small telescopes (Fig. 19.14). Its common name, the Ghost of Jupiter, reflects its similarity to a planet. This planetary nebula glows slightly brighter than 8th magnitude.

The famous French comet hunter Charles Messier observed and cataloged the first planetary nebula in 1764 when he spotted the Dumbbell Nebula in Vulpecula and recorded it as Object 27 in the list of deep-sky objects that he compiled to warn other astronomers, many of whom were comet hunters, that those objects were not comets. Other discoveries followed, although they were slow relative to other deep-sky objects mostly because planetary nebulae are small compared to star clusters, emission nebulae, and galaxies. William Herschel (1738–1822) made the biggest gains in finding planetaries, even if he didn't know what they were, and applied the name to his class of small, round nebulae that reminded him of the planet he discovered: Uranus.

The British amateur astronomer William Huggins (1824–1910) pushed the study of nebulae forward in the mid-nineteenth century by turning a spectroscope toward many nebulae. In 1864 Huggins systematically studied stars and nebulae and found that they resemble the Sun in gaseous structure. He carefully examined NGC 6543 in Draco and found "a single bright line only." He remarked in his notes, "The riddle of the nebulae was solved. The answer, which had come to us in the light itself, read: Not an aggregation of stars, but a luminous gas."

By the twentieth century, with the dawn of modern astrophysics, the study of planetary nebulae catapulted forward. Even as recently as 1918, however, fewer than 100 planetaries were known, and astronomers didn't yet understand the nature of galaxies. Few professional astronomers specialized in studying planetary nebulae, so a generalized list of the objects did not exist until recently. Planetaries lay scattered throughout various catalogs.

In 1967 this deficiency evaporated when astronomers Lubos Perek and Lubos Kohoutek published their *Catalogue of Galactic Planetary Nebulae*, which contains 1,036 objects. This act marked the first step in modern planetary nebula research and remains the standard reference catalog on planetary nebulae.

Shapes of Planetary Nebulae

Research went on with the normal types of discovery, analysis, publication of papers, estimates of ages, sizes, compositions, and so on. But no one until the 1990s, with the aid of the Hubble Space Telescope, began to crack the mystery of how planetary nebulae are shaped.

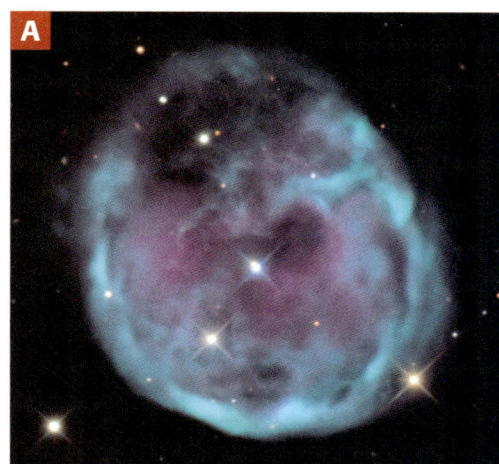

19.13 Examples of planetary nebulae: (**A**) NGC 246 in the constellation Cetus the Whale, and (**B**) NGC 4361 in the constellation Corvus the Crow.

Courtesy of Adam Block/Mount Lemmon SkyCenter/University of Arizona

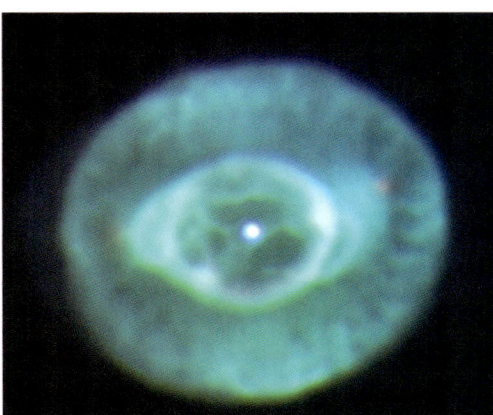

19.14 Ghost of Jupiter (NGC 3242) in the constellation Hydra the Water Snake.

Courtesy of Adam Block/Mount Lemmon SkyCenter/University of Arizona

Enter Bruce Balick and the group of astronomers focusing on nebulae research. The conundrum was that planetary nebulae as they are seen appear to have a huge array of shapes, sizes, brightnesses, and so on, such that making a coherent story out of their physical nature was daunting.

The solution came down to classifying the basic structural types and then understanding the complexities thrust upon them because we are viewing the nebulae from many different angles as we see them in the sky.

Balick and his collaborators proposed that, unlike the myriad varieties seen, just three main classes of planetaries exist: round, elliptical, and butterfly shapes. Figure 19.15 shows examples of round planetaries, Figure 19.16 shows examples of elliptical planetaries, and Figure 19.17 shows an example of a butterfly-shaped planetary. Within each of these classes, degrees exist based on the nebula's age so that early, middle, and late types look different from one another. Aside from this matrix of nine possible classes (Fig. 19.18), all the other variations in planetaries exist because the nebulae tilt at different angles to our line of sight.

Once this solution evolved, the picture of planetaries and what they mean became clear. Astronomers could see that some sun-like stars die and expel round spheres of material that form halos of glowing gas, with brighter rings inside them. Others blow off a shell of gas and then the helium flash, a final dying gasp from the star that creates an elongated shot of gas and dust. This shoots out at high speed through the older, slower halo. Still others are constricted by a disk that lies within, tidal forces that focus high-speed gas like a cosmic sprinkler, or magnetic forces that shape a planetary into a highly elongated butterfly shape.

Scientists now know that these three basic shapes and their variations characterize all the planetaries seen in the sky. On your next night out under the stars, spend time looking carefully at planetaries, and see what they have to offer. If you are out during the summer months, search for two of the greatest such nebulae: the Ring Nebula (M57) in the constellation Lyra the Harp, and the Dumbbell Nebula (M27) in Vulpecula the Fox. Find these objects at low power, and then increase the magnification to see additional details.

Dozens of other planetary nebulae are waiting to be viewed. Many observers claim that planetaries are among the most fascinating objects to observe from your backyard because of their brightness, visibility from less than perfect skies, and strange array of shapes and sizes. And you can be among the first generation of amateur astronomers that gazes on a planetary nebula and also understands how it formed and why it is shaped the way it is.

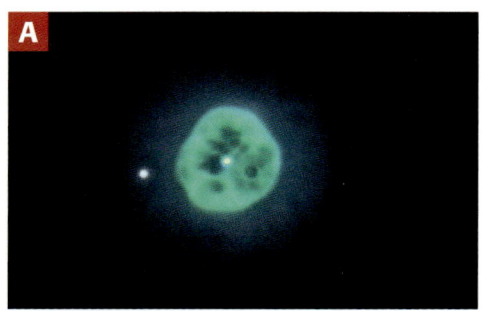

19.15 Round planetary nebulae: **(A)** NGC 1535 in the constellation Eridanus the River, and **(B)** NGC 6781 in the constellation Aquila the Eagle.
Courtesy of Adam Block/Mount Lemmon SkyCenter/University of Arizona

19.16 Elliptical planetary nebulae: **(A)** NGC 7008 in the constellation Cygnus the Swan, and **(B)** NGC 6563 in the constellation Sagittarius the Archer.
Courtesy of Adam Block/Mount Lemmon SkyCenter/University of Arizona

19.17 Butterfly-shaped planetary nebula: Little Dumbbell Nebula (M76) in the constellation Perseus the Rescuer of Andromeda.

Courtesy of Adam Block/Mount Lemmon SkyCenter/University of Arizona

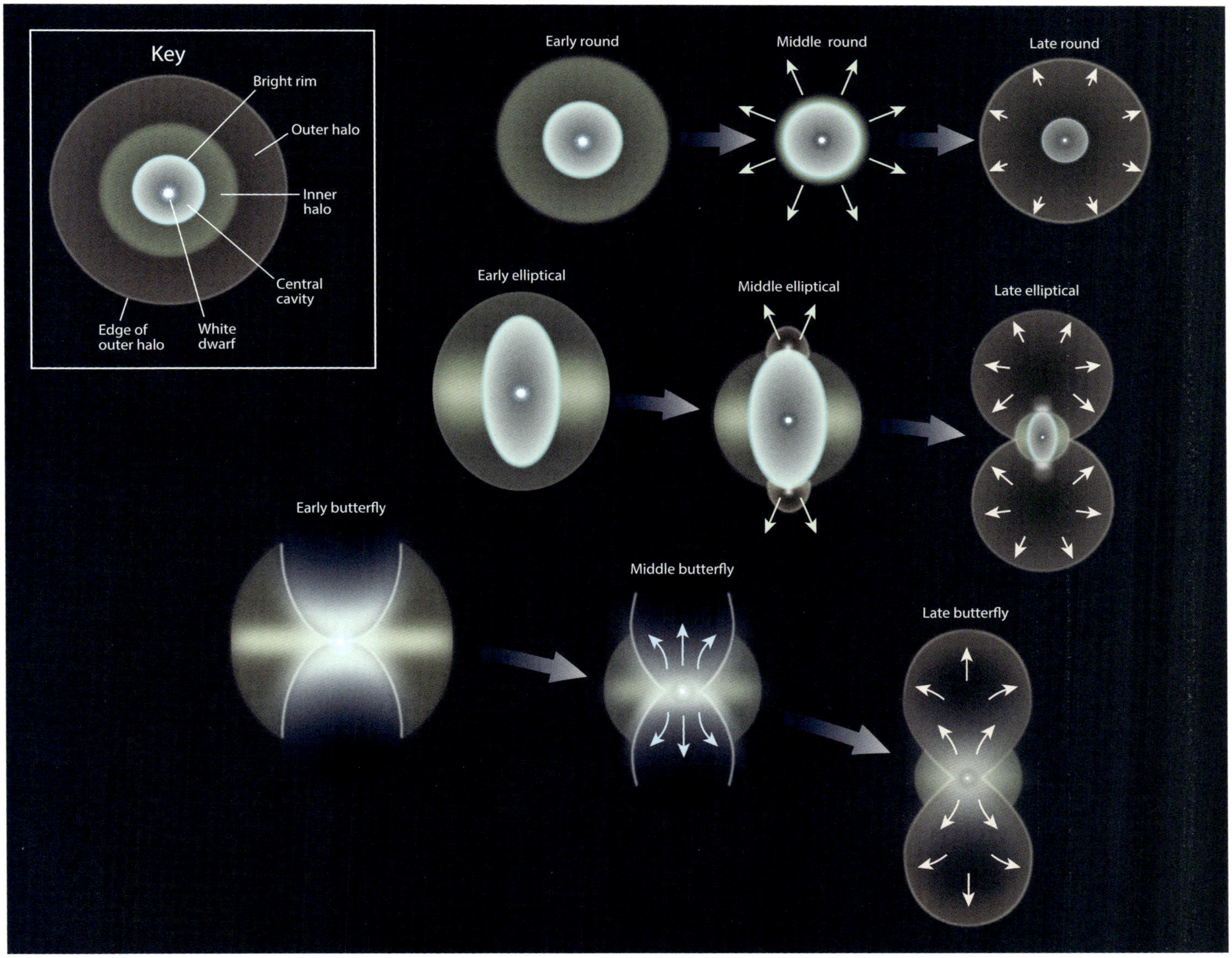

19.18 Matrix of nine possible classes of planetary nebulae.

Courtesy of Roen Kelly, *Astronomy* magazine

Non-Stellar Objects CHAPTER 19   381

Check Your Understanding

2.1 Approximately how long do planetary nebulae live?

2.2 How many planetary nebulae does the Milky Way contain?

2.3 What kind of stars form planetary nebulae?

2.4 Why are these objects called "planetary" nebulae?

2.5 Why do planetary nebulae seem to have such complex structures?

2.6 How many basic classes of planetary nebulae do astronomers now recognize?

Exercise 19.1 — Classifying Galaxies

Procedure

1. Study the images of the 15 galaxies shown in Figures 19.19–19.33 on the data sheet, pages 383–384.

2. In the space provided below each image on the data sheet assign a Hubble classification to each, referencing Figure 19.1, page 372.

3. Also assign each a de Vaucouleurs classification, referencing Figure 19.12, page 376.

Name _____ Section _____ Date _____

Galaxy Classification

19.19 M58.

Hubble classification: _____

de Vaucouleurs classification: _____

19.20 M100.

Hubble classification: _____

de Vaucouleurs classification: _____

19.21 M104.

Hubble classification: _____

de Vaucouleurs classification: _____

19.22 M64.

Hubble classification: _____

de Vaucouleurs classification: _____

19.23 M66.

Hubble classification: _____

de Vaucouleurs classification: _____

19.24 M77.

Hubble classification: _____

de Vaucouleurs classification: _____

Figures 19.19 through 19.33 courtesy of Adam Block/Mount Lemmon SkyCenter/University of Arizona

Exercise 19.1 Data Sheet

19.25 M87.

Hubble classification: _____

de Vaucouleurs classification: _____

19.26 NGC 1357.

Hubble classification: _____

de Vaucouleurs classification: _____

19.27 NGC 157.

Hubble classification: _____

de Vaucouleurs classification: _____

19.28 NGC 2841.

Hubble classification: _____

de Vaucouleurs classification: _____

19.29 NGC 4216.

Hubble classification: _____

de Vaucouleurs classification: _____

19.30 NGC 5248.

Hubble classification: _____

de Vaucouleurs classification: _____

19.31 NGC 7606.

Hubble classification: _____

de Vaucouleurs classification: _____

19.32 NGC 7741.

Hubble classification: _____

de Vaucouleurs classification: _____

19.33 NGC 7814.

Hubble classification: _____

de Vaucouleurs classification: _____

Unit 5 The Deep Sky

Classifying Planetary Nebulae

Exercise 19.2

Procedure

1 Study the images of the 15 planetary nebulae shown in Figures 19.34–19.48 on the data sheet, pages 387–388.

2 In the space provided below each image on the data sheet, classify each using one of the nine categories shown in Figure 19.18, page 381.

3 List descriptive comments in the space provided below each image on the data sheet, referencing the key in Figure 19.18. (Example: "White dwarf easily visible" or, "Outer halo barely seen.")

Name _____ Section _____ Date _____

Planetary Nebulae Classification

 **19.34** Abell 21.

 **19.35** Abell 31.

 **19.36** Abell 33.

 **19.37** Abell 39.

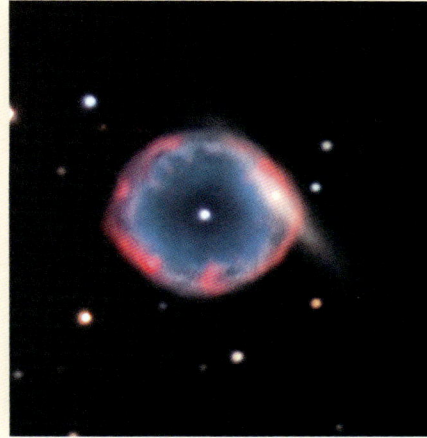 **19.38** Abell 70.

 **19.39** Abell 78.

Figures 19.34 through 19.48 Courtesy of Adam Block/Mount Lemmon SkyCenter/University of Arizona

Exercise 19.2 Data Sheet

Exercise 19.2 Data Sheet

19.40 M76.

19.41 NGC 246.

19.42 NGC 2346.

19.43 NGC 2438.

19.44 NGC 6302.

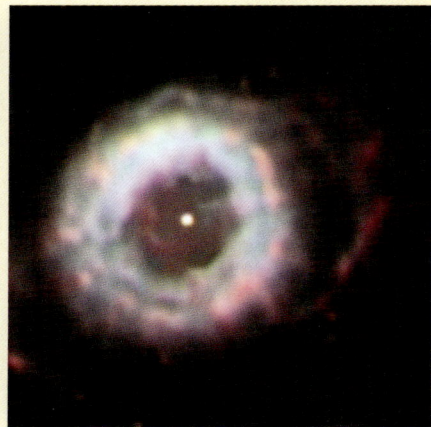
19.45 NGC 6369.

19.46 NGC 6543.

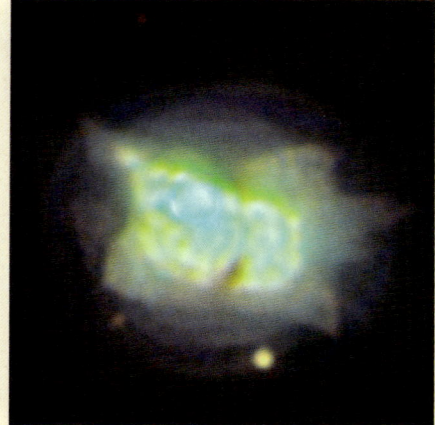
19.47 NGC 7027.

19.48 PK 164+31.1.

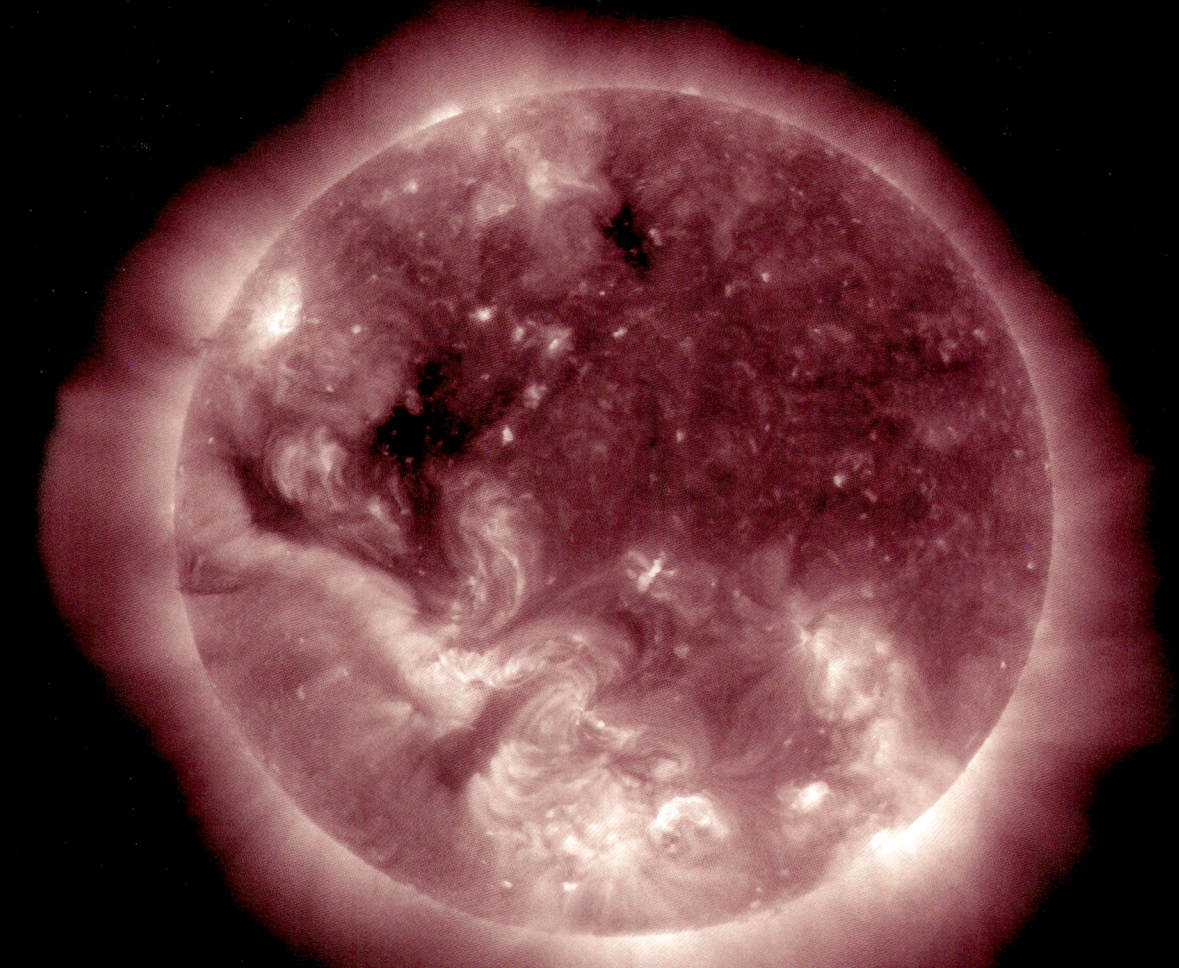

Classifying Celestial Objects

Spectroscopy

LEARNING OBJECTIVES

Upon completion of this chapter, you should be able to:

1 Identify the three main types of spectra.
2 Explain how spectra allow astronomers to determine the composition of celestial objects.
3 Observe various spectra from the Sun and a variety of artificial light sources.
4 Compare the spectra of elements viewed in the lab to other light sources.
5 Define all glossary terms.

20

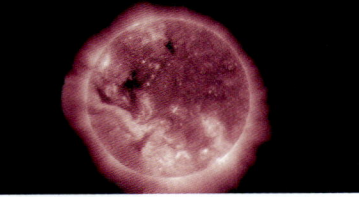

Sun imaged through ultraviolet filter.
Courtesy of NASA/Solar Dynamics Observatory

GLOSSARY TERM

Spectroscopy study of spectra

Electromagnetic spectrum referring to radiation in the universe: the full range of wavelengths; referring to an object: the distribution of radiation emitted or absorbed

Wavelength distance over which the shape of a wave repeats

Spectrum (pl. spectra) plot of the range of colors (wavelengths) observed when a source's light disperses through a prism or diffraction grating

Prism transparent material that breaks light into its component colors

When most people hear "astronomy," a picture comes to mind. It may be of a total solar eclipse, Halley's comet, or perhaps just the Moon. Magazine readers and those with access to the Internet might envision a Hubble Space Telescope image of a far-flung galaxy. But as jaw-dropping as Hubble's photographs have been, it is **spectroscopy**—not photography—that has unraveled most of the universe's mysteries.

When astronomers learned the secrets sunlight contains, they took a major step closer to understanding the cosmos. Spectroscopy reveals how matter absorbs and emits radiation. It tells astronomers what objects are made of, how hot they are, and even how fast the universe is expanding.

In this chapter, you will encounter the **electromagnetic spectrum** and **wavelengths**. You will learn how astronomers create spectra as well as the basic stellar spectral sequence (O, B, A, F, G, K, and M) and what it represents. Finally, by examining the **spectrum** of a distant object, you will learn to calculate its relative motion in the cosmos.

Historical Discoveries

English physicist Sir Isaac Newton (Fig. 20.1) used a **prism** to break sunlight into its component colors in 1666. While he wasn't the first to do this, he did perform scientific analysis on sunlight and by using a second (reversed) prism to recombine the colors proved that white light is made of different colors, which we now know have particular wavelengths. He also coined the term *spectrum* to describe the rainbow-like pattern.

Not much was done with Newton's findings until the nineteenth century. In 1802 English chemist William Hyde Wollaston (1766–1828) first noted, and pretty much ignored, a few dark lines crossing the Sun's spectrum. But it was German physicist Joseph von Fraunhofer (1781–1825) who, by spreading out the solar spectrum in 1814, discovered a vast number of fine lines (Fig. 20.2). Scientists now refer to these features as Fraunhofer lines and to the tool he used to study them as the spectroscope.

Most important to astronomy, Fraunhofer used his spectroscope with telescopes. He studied light from stars and planets, breaking it into spectra he could study. In doing this, he laid the foundation for astrophysics.

Despite all his firsts, Fraunhofer never understood what produced the dark lines he saw. In 1835 English scientist Charles Wheatstone (1802–1875) began the process of

20.1 Sir Isaac Newton (1643–1727).
Courtesy of Wikimedia Commons

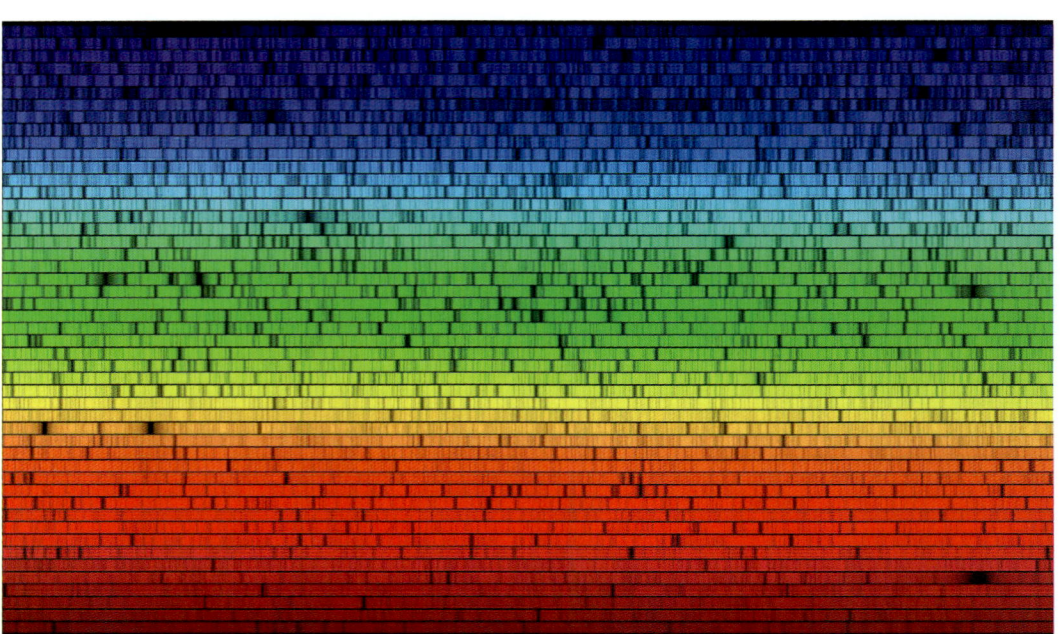

20.2 Spectrum of the Sun divided into rows from 400 nm (upper left) to 700 nm (lower right).
Courtesy of N. A. Sharp/NOAO/NSO/Kitt Peak FTS/AURA/NSF

390 Unit 6 Classifying Celestial Objects

spectral analysis by observing **emission** spectra. He identified twelve metals by the bright lines in their spectra, which he created by vaporizing a small quantity of each using an electric arc. In 1848 French physicist Léon Foucault (1819–1868) observed **absorption** spectra. He detected that a flame created by burning sodium absorbed the yellow light emitted by an electric arc behind it.

But it was the work of two German chemists, Gustav Kirchhoff (1824–1887) and Robert Bunsen (1811–1899), starting in 1859 that refined and systematized spectral analysis. They used a spectroscope of their own design to classify the unique spectra of sodium, lithium, and potassium. Also, because their device showed them lines they could not identify, they discovered two new elements (cesium and rubidium).

In 1868 Swedish physicist Anders Jonas Ångström (1814–1874) combined a spectroscope and a camera, which demonstrated the new technology of photography, to create the first spectrograph. His resulting map contained more than 1,000 dark lines. To measure the wavelengths of the light that produced them, he developed a unit equal to 10^{-10} of a meter, now called the **ångström**, whose abbreviation is the Swedish letter Å. Since the development of the International System of Units in 1960, however, most researchers use the **nanometer** (10^{-9} of a meter) to define wavelength. Ten ångströms are equal to 1 nanometer.

By the mid-nineteenth century, scientists had observed three types of spectra: those made of bright lines; those with dark lines superimposed; and continuous, which shows neither type of line. Kirchhoff explained how each type formed, and we now refer to his concepts as Kirchhoff's three laws of spectroscopy:

1. A hot solid, liquid, or dense gas produces a **continuous spectrum**.
2. A hot, low-density gas produces a spectrum with bright lines (an emission spectrum).
3. A low-density gas in front of a hotter (continuous spectrum) source produces a spectrum showing dark lines (an absorption spectrum).

Kirchhoff's laws showed scientists how they could reveal what the universe was made of, and not just by studying light. Kirchhoff explained that the dark lines in the Sun's spectrum form when cooler gases made of various chemical elements surrounding the hot core absorb specific radiation. The amount of radiation absorbed by each element shows how much of the element the Sun contains.

In 1873 Scottish physicist James Clerk Maxwell (1831–1879) published the theory of electric and magnetic forces, summarized in his famous four equations. They showed that electricity and magnetism were two aspects of the same force. The equations predicted that electromagnetic radiation could exist with any wavelength. Maxwell's work led to the expansion of studies beyond visible wavelengths and to the coining of the term *electromagnetic spectrum* (Fig. 20.3).

> **GLOSSARY TERM**
>
> **Emission** occurs when a particle gives off a photon
>
> **Absorption** occurs when the energy of a photon is transferred to a particle
>
> **Ångström** unit of length equal to 10^{-10} meter; formerly how astronomers measured wavelengths of light, superceded by the nanometer
>
> **Nanometer** unit of length equal to 10^{-9} meter; the current measure of light wavelengths
>
> **Continuous spectrum** one that contains all visible wavelengths

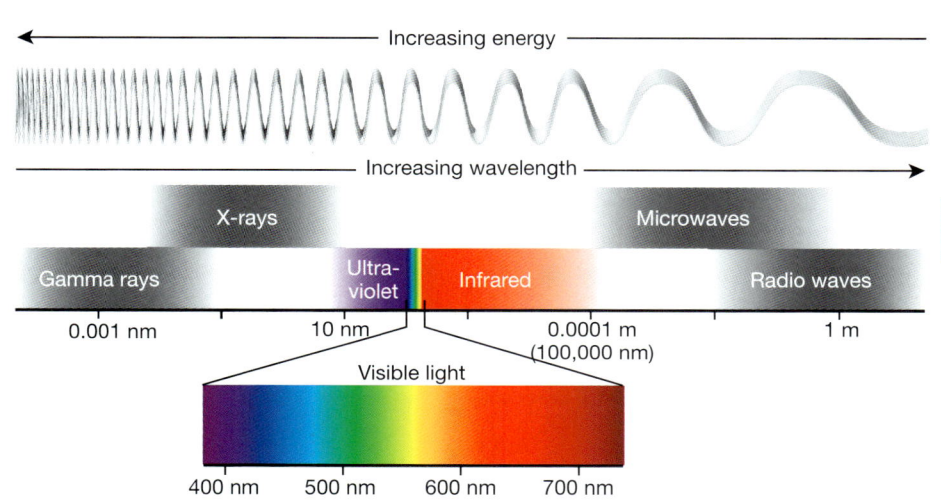

20.3 Electromagnetic spectrum.

Courtesy of Roen Kelly, *Astronomy* magazine

Check Your Understanding

1.1 What is spectroscopy?

1.2 How did Isaac Newton prove that white light is made up of all colors?

1.3 Who first combined a spectroscope and a telescope?

1.4 How does the formation of emission spectra differ from that of absorption spectra?

1.5 How does Kirchhoff's third law help astronomers identify what the Sun is made of?

1.6 How did Kirchhoff prove how absorption lines form?

1.7 What is a spectrograph?

Recent Discoveries

Why Spectra Form

As the twentieth century dawned, scientists found themselves with a tremendous amount of spectral data but still no theory to explain how the lines formed. That changed in 1913 when Danish physicist Niels Bohr (1885–1962) published his model of the structure of atoms. Bohr's theory placed electrons at discrete energy levels around the atom's nucleus. These particles can absorb radiation, but by doing so they must move to higher levels. Likewise, electrons release radiation as they move to lower energy levels.

It is this process of absorption and emission that creates the dark lines in the Sun's spectrum. This occurs because the core of our star emits a continuous spectrum (one that contains all visible colors, or wavelengths), but atoms of gas in the solar atmosphere absorb the light the core produces. Certain atoms absorb only certain wavelengths of light (Fig. 20.4). Their electrons move up one, two, or more energy levels, depending on how much radiation they absorb. When electrons jump up a level (or more), they absorb energy and create an **absorption spectrum** (Fig. 20.5). But the electrons cannot stay at the higher levels for long. They almost immediately fall back to their original levels. Because most of the light in any particular wavelength that was originally headed toward us is

> **GLOSSARY TERM**
> **Absorption spectrum** one that shows a series of dark lines superimposed on a brighter background

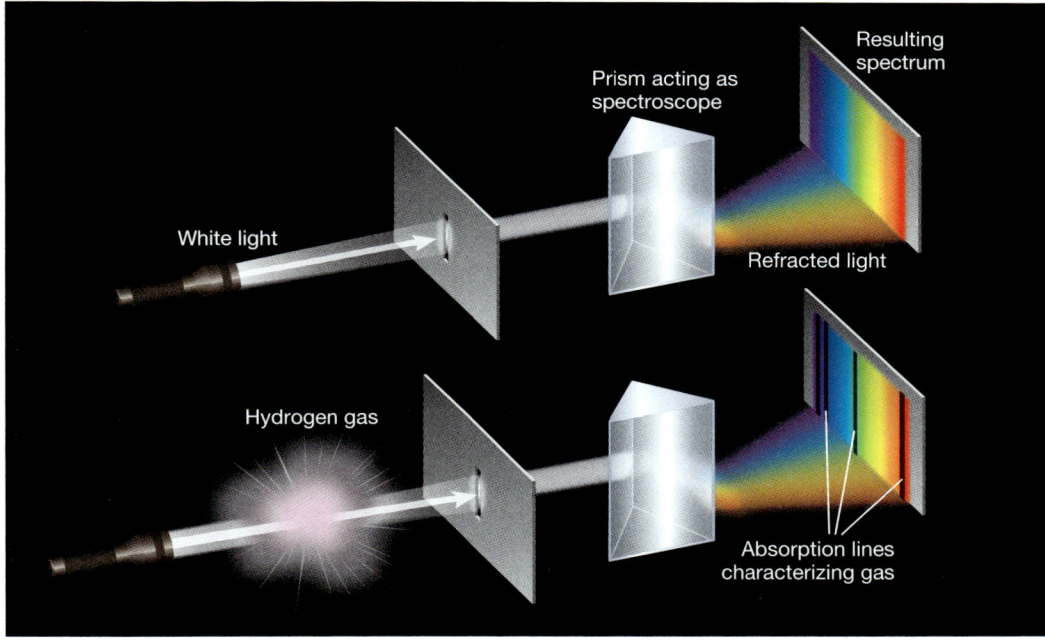

20.4 Electrons in hydrogen absorb certain wavelengths of light and then re-emit it in random directions, resulting in a dip in intensity that we call absorption lines.
Courtesy of Roen Kelly, Astronomy magazine

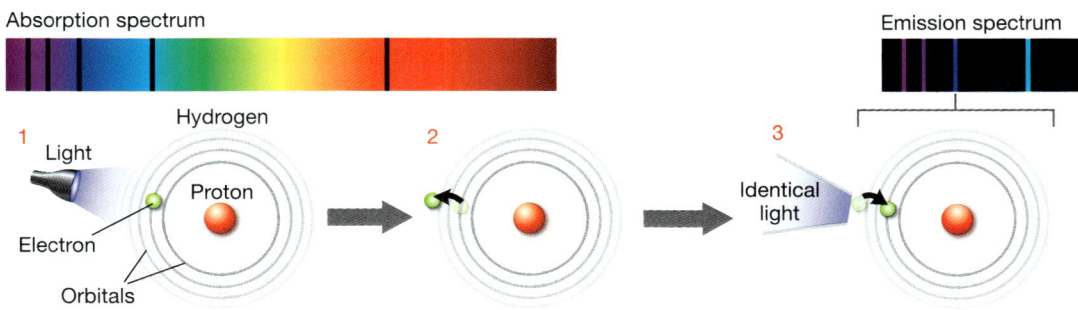

20.5 Creation of absorption and emission spectra. *Courtesy of Roen Kelly and Alison Mackey, Astronomy magazine*

> **GLOSSARY TERM**
>
> **Emission spectrum** one that shows a series of bright lines
>
> **Luminosity** actual light output of a celestial object irrespective of how bright it appears

now moving in random directions, the point on the Sun's spectrum representing that wavelength shows a dip in strength: a dark (absorption) line.

Electrons prefer to stay in their original energy level, however, so after the briefest of intervals, they resume their normal levels and emit the light in a random direction. When electrons release the energy on their way back to their original level, they create an **emission spectrum** (Fig. 20.5). An emission line at the same wavelength as an absorption line would look bright (or colored), while the absorption line at the same point on the spectrum would be a dark area.

How Spectra are Classified

In modern classification, spectra are organized according to their absorption lines. Each element in a star absorbs light at wavelengths specific to that element and its atomic composition. The result is black lines on a continuous spectrum, shown in Figure 20.6. This eventually led to the main letter designations (from hottest stars to coolest) currently in use: O, B, A, F, G, K, and M (Fig. 20.7). Additional numbers from 0 to 9 (0 being hottest and 9 coolest) further subdivide the classes. For example, a star of spectral class A5 lies halfway from class A to F. In this system, our Sun is a G2 star.

The final refinement to spectral classification is the MKK system that introduces five main **luminosity** classes: supergiants (designated I); bright giants (II); normal giants (III); subgiants (IV); and dwarfs (V), also called main sequence stars.

These divisions were added because lines in the spectrum of a giant star have different widths and intensities than those in a dwarf, even though the same element is responsible for them. In this scheme, our Sun is a G2V star.

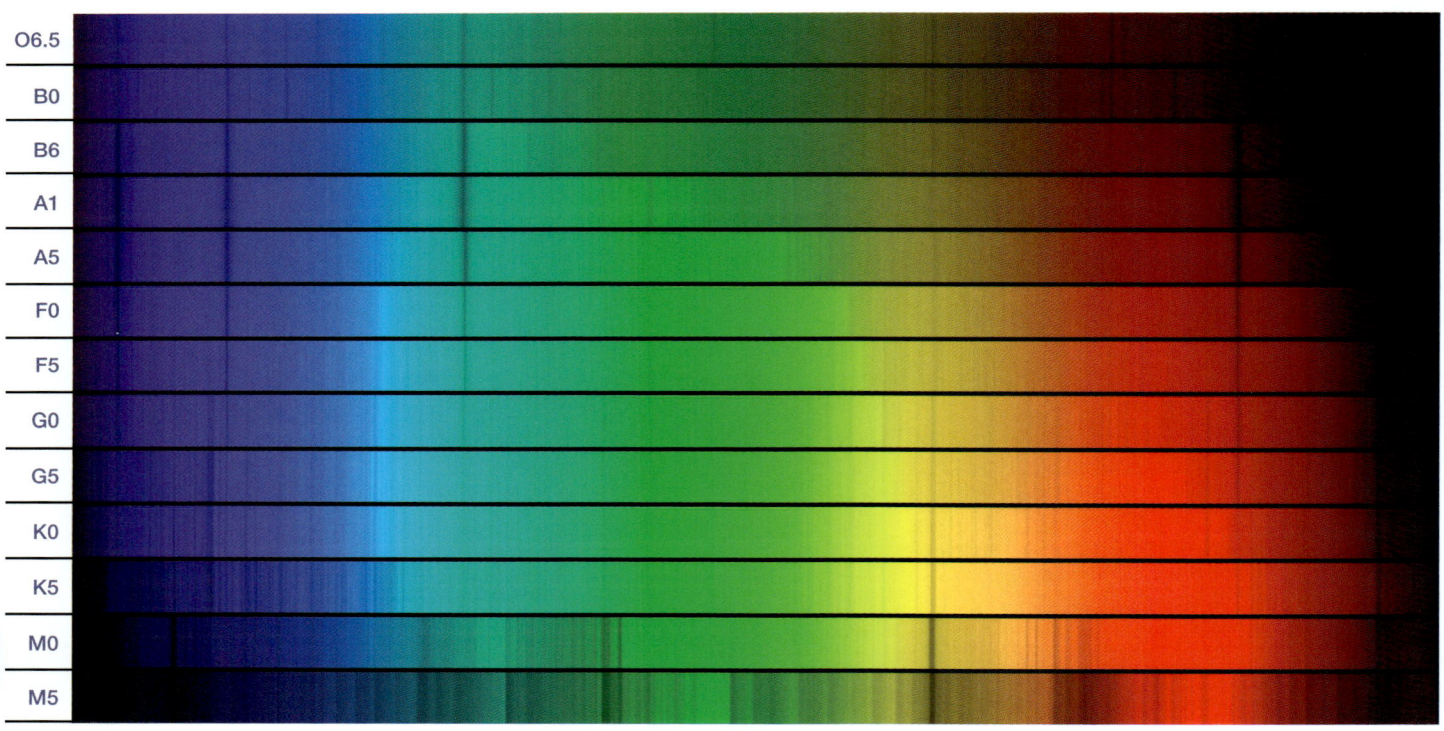

20.6 Examples of spectral types.

Courtesy of NOAO/AURA/NSF

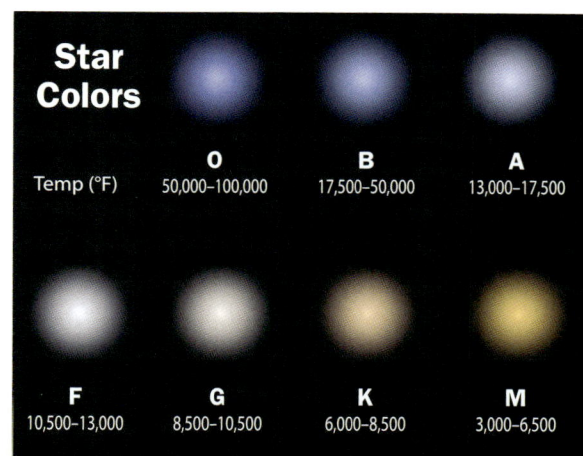

20.7 Correspondence of star colors to spectral types.

Courtesy of Roen Kelly, *Astronomy* magazine

 *Check Your Understanding*

2.1 When do electrons release radiation?

2.2 According to the MKK system of spectral classification, what spectral type is the Sun?

A Closer Look at Spectra

The Visible Spectrum

Astronomers can learn a great deal about a celestial object by studying its visible spectrum. Perhaps the easiest facts they glean are composition and temperature. By comparing the number and positions of the object's absorption lines to a spectrum created by heating a sample in the lab, scientists immediately know what elements are present and at what temperatures.

Today, most spectra do not come in the form of a rainbow diagram but rather as a graph that plots relative brightness (intensity of the line) against wavelength. The spectrum in Figure 20.8 indicates the presence of the molecules methane (CH_4), carbon monoxide (CO), and water (H_2O) in the planet WASP-12b, a super-hot gas giant that orbits near its star.

Lines that appear split or broadened indicate the presence of a strong magnetic field. A strange spectrum that shows lines characteristic of both hot and cool stars usually proves the existence of a double star too close for optical telescopes to split. Observers call such pairs spectrum binaries.

Spectra even reveal if an object is moving toward or away from us, and at what speed. Imagine the spectrum of a star that shows normal but shifted absorption lines when compared with a laboratory spectrum. If the lines shift to the blue end of the spectrum (known as blueshift), the object is moving toward us. If (as in most cases) the

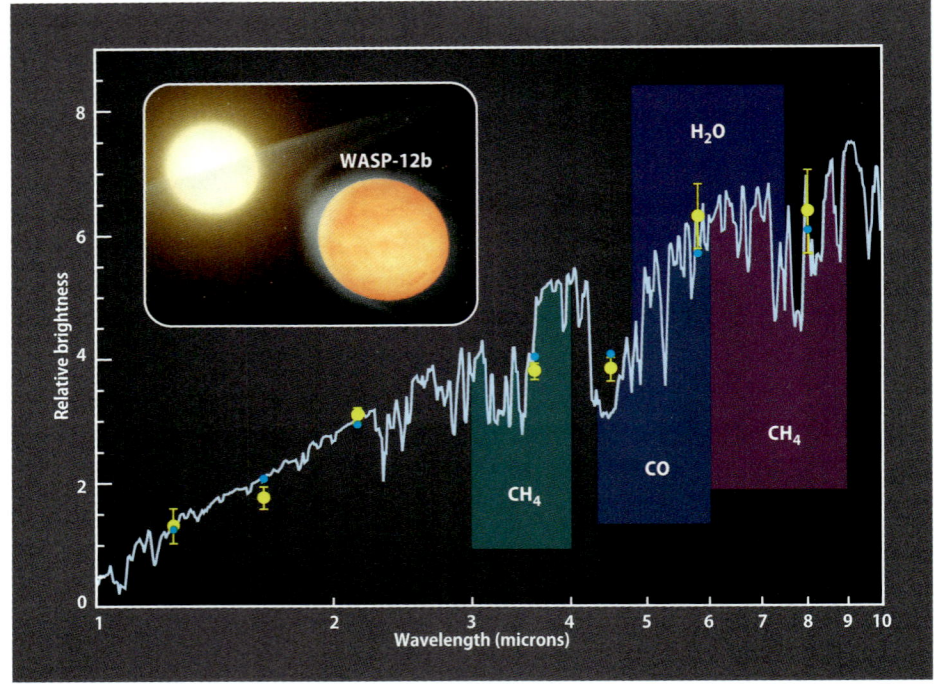

20.8 Spectrum of the planet WASP-12b.

Courtesy of NASA/JPL-Caltech, N. Madhusudhan (Princeton University)

shift is to the red (a redshift), the motion is away from us. And the percentage of the shift gives the star's velocity in terms of the speed of light.

Understanding redshift helped astronomers solve one particularly troubling mystery, that of quasars, which are a type of active galaxy. In the late 1950s scientists collected spectra of a class of objects they dubbed "quasi-stellar radio sources," or quasars. Their spectra contained absorption and emission lines never before seen. In 1962 Dutch astronomer Maarten Schmidt (1929–) used the 200-inch Hale telescope on Palomar Mountain in California to obtain the spectrum of quasar 3C 273. It also contained the unknown lines until he realized that he was looking at normal hydrogen lines redshifted nearly 16 percent. Quasars were flying away from us at high velocities.

The top spectrum in Figure 20.9 is what a typical quasar spectrum might look like if it were stationary.

Astronomers use the shift of spectral lines to measure the velocity of astronomical objects moving toward (blueshift) or away from (redshift) Earth. As redshift increases, the green line is shifted into the yellow region, and then into the red region. The other lines also shift, but the relative spacing remains constant.

20.9 Spectra of four quasi-stellar objects (quasars).

Courtesy of C. Pilachowski, M. Corbin, NOAO/AURA/NSF

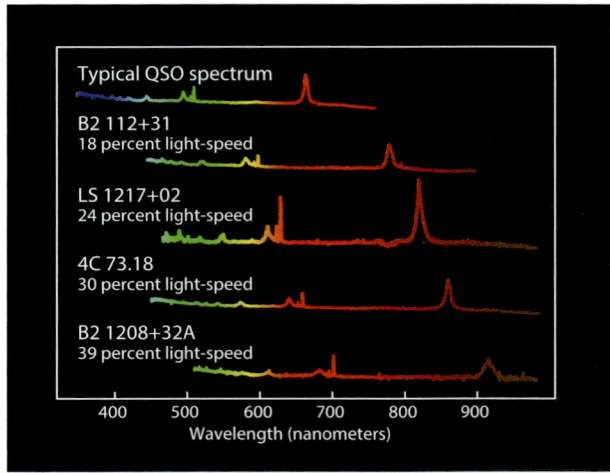

396 Unit 6 Classifying Celestial Objects

Check Your Understanding

3.1 List four things astronomers can learn by studying a star's spectrum.

3.2 What does it mean if a star's spectrum lines are redshifted?

3.3 The first quasar astronomers discovered, 3C 273, had spectral lines redshifted by 16 percent. What did this tell astronomers?

Wave Characteristics of Light

Scientists have long recognized that color provides useful information about many visual objects. Over the years they have developed devices, referred to collectively as **spectrometers** (Fig. 20.10), for analyzing the characteristics of light.

The spectrometers contain **diffraction gratings** that separate light into its component colors. A diffraction grating consists of many parallel slits or grooves, each a specific distance (d) apart, arranged in a plane. There are thousands of slits per centimeter. Gratings, "ruled" on glass, plastic, etc., are mounted so that incident light—the direct light that falls on a surface—is parallel to the grating.

The Dutch scientist Christiaan Huygens (1629–1695) noted that when a wave strikes a barrier with perforations, each perforation produces its own wavelet (smaller wave) with the same wavelength and speed as the parent wave. As each wavelet spreads out from the barrier, it runs into wavelets from other perforations. The **interference** (addition or superposition of two or more waves that results in a new wave pattern) between these wavelets produces a set of diffraction patterns.

20.10 Spectrometer.

> **GLOSSARY TERM**
>
> **Spectrometer** device used to measure the positions of bright or dark spectral lines
>
> **Diffraction grating** transmissive or reflective surface upon which many equally spaced lines have been ruled to break light into its component colors
>
> **Interference** occurs when two light waves superimpose; in constructive interference, the waves are in phase and the result is a bright line; destructive interference means the waves are out of phase and results in a dark line

> **GLOSSARY TERM**
> **Blackbody** an idealized physical body that absorbs all electro-magnetic radiation

Wien's Law and Blackbody Curves

Wien's law relates the color of an incandescent object to its temperature. The law states that the wavelength of the brightest color, λ_{max}, is inversely proportional to the temperature (T) or $\lambda_{max} \cong \frac{1}{T}$.

At the peak of the radiation curve is λ_{max}, a plot of light intensity versus wavelength, for an object at temperature (T). Shorter wavelength (bluer) light corresponds to higher energy waves. More energy equals more relative intensity. An idealized radiation curve for a given temperature is called a **blackbody curve** (Fig. 20.11). The radiation curves of stars closely resemble blackbody curves. Blackbody radiation is electro-magnetic radiation in thermal equilibrium with a blackbody at a given temperature.

When scientists normalize (smooth out) the spectrum of a star, the resulting curve appears similar to that from a black-body of the star's temperature. The three spectral curves shown in Figure 20.12 are for two stars in the constellation Orion—Betelgeuse and Rigel—and our Sun. All three have different surface temperatures. Betelgeuse's color peaks in the red (Fig. 20.12A), Rigel's peaks in the ultraviolet so it appears blue (Fig. 20.12B), and our Sun's color peaks in the green, although because of the way our eyes respond to color, and because our eyes receive a full complement of red, green, and blue light, we see it as a yellow-white star (Fig. 20.12C).

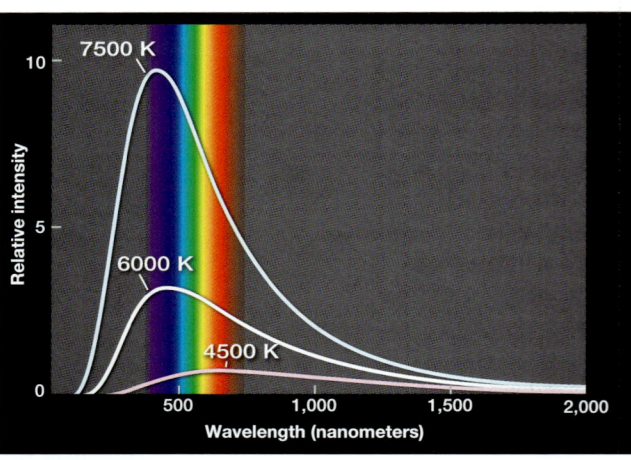

20.11 Blackbody curves for objects at three different temperatures.
Courtesy of Roen Kelly and Alison Mackey, Astronomy magazine

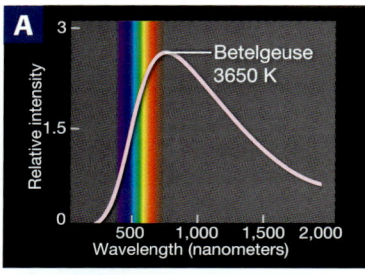

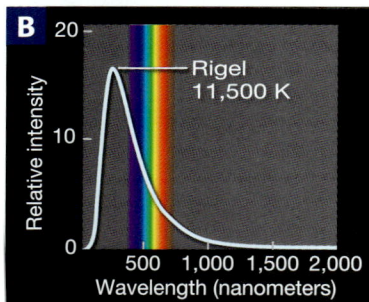

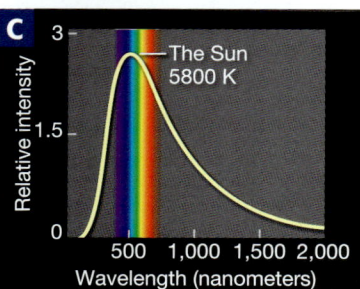

20.12 Blackbody curves of: **(A)** Betelgeuse, **(B)** Rigel, and **(C)** our Sun.
Courtesy of Roen Kelly and Alison Mackey, Astronomy magazine

Stefan-Boltzmann Law

There are a number of ways to examine the characteristic of any wave, including light. The most used are the wave's speed, or velocity, its wavelength, frequency, and the period of the wave (Table 20.1). These characteristics can apply to ocean waves as well as light.

As shown in Table 20.1, there is a relationship between velocity, wavelength, and frequency, as well as between the period and frequency. The speed of light, designated **c**, is 3×10^8 m/s.

The Stefan-Boltzmann law states that the power (**P**) emitted per unit of area of an object is proportional to the fourth power of its temperature (**T**). Mathematically this is written as:

$$P \sim T^4$$

TABLE 20.1 Wave Characteristics

Characteristic/Quantity	Symbol	Formula (units in metric)
Speed (velocity) Rate of movement (in a specific direction)	v	$v = f\lambda$ m/s, *meters per second* Light, $c = 3 \times 10^8$ m/s
Wavelength Distance between similar points in a wave	λ	$\lambda = v/F$ m, *meters*
Frequency Number of waves passing a given point each second	f	$F = v/\lambda$ 1/s = Hz, *Hertz*
Period Time required for a wave to pass a given point	T	$T = 1/f$ s, *seconds*

Luminosity is the total power (energy emitted per unit time) of an object. The luminosity of an object also increases with the object's surface area, so a bigger star will be more luminous than a smaller one if their surfaces are the same brightness. Luminosity is not a property that depends on the object's distance, and to emphasize this it is often called **absolute magnitude**, which is the brightness a star would have at a distance of 32.6 light-years from Earth.

The apparent brightness of an object, on the other hand, depends on both its luminosity and its distance. A nearby low-luminosity star might appear brighter than more distant stars with higher luminosities to our eyes (or other detectors). If, however, two stars were at the same distance, the more luminous one always will appear brighter.

> **GLOSSARY TERM**
> **Absolute magnitude** how bright a star would appear if it were at a distance of 10 parsecs (32.6 light-years) from Earth

Check Your Understanding

4.1 What element in a spectrometer separates light into colors?

4.2 What does Wien's law state?

4.3 At what color does our Sun peak? Why don't we see it as a star with that color?

4.4 What are the two terms in the Stefan-Boltzmann law, and how do they relate to one another?

Exercise 20.1 — Exploring Spectra

In this exercise, you will use a simple but effective spectrometer. After answering the following questions, you will observe and then compare and contrast the spectra of three light sources: incandescent, fluorescent, and sunlight. Make certain you complete your observation of sunlight before sunset. Then you will observe at least four different spectrum tubes of known elements/compounds with the spectrometer, noting the lines visible and at what wavelengths. Use colored pencils to indicate the positions and colors of the lines on the spectrometer scale for each light source and spectrum tube.

Procedure

Materials
- ❏ Colored pencils
- ❏ Spectrometer
- ❏ Spectrum tubes
- ❏ Light sources (incandescent and fluorescent)

1 Answer the following questions:

  a What material is used in this exercise to separate light? How does it work?

  b List four characteristics of light waves and the formula for each.

  c What is luminosity?

2 Collect a spectrometer as directed by your instructor. Work individually; report any problems with your spectrometer.

3 Hold the spectrometer so the printed side is facing upward to the right. The diffraction grating is at one end, the slit (used to make the spectral lines narrow) is at the opposite end (toward the light source). You will also see a scale to the left of the slit when you look through the spectrometer. The bottom part of the spectrometer's scale is the wavelength in nanometers (nm): 1 nm = 10^{-9} m. When observing, bring one eye to the slit, and close your other eye. You should see the scale as well as colors inside the spectrometer.

4 Notice that the plastic disk with the attached diffraction grating can rotate a small amount. Also note that the colors move as you rotate the disk. Set the disk so you can see all the colors in a horizontal line. The colors should appear between the two lines of numbers on the spectrometer's scales.

5 Keep the following tips in mind:

▶ Do not block the slit or the scale adjustment window with your hand.

▶ Align the slit to the light source; failure to do this, especially with the spectrum tubes, will produce no or poorly visible spectral lines.

▶ You might see a secondary, bright-line spectra as well as the primary spectra.

6 You are ready to begin by observing the incandescent bulb through the spectrometer. Remember to carefully aim the slit at the bulb and observe the spectrum on the scale.

7 You should see a continuous spectrum of colors. On the scale below the spectrum, indicate the colors you see, noting any prominent lines.

___

___

8 Read the number on the scale corresponding to the light farthest to the right that you can see and the number corresponding to the light farthest to the left you can see.

a The observed spectrum extends from _____ nm to _____ nm.

b The colors at these places on the scale are _____ and _____.

9 Next, observe a fluorescent light source through the spectrometer. You should see an emission spectrum with discrete bands of color. On the scale below the spectrum, indicate the colors you see, noting any prominent lines.

___

___

10 Read the number on the scale corresponding to the light farthest to the right that you can see and the number corresponding to the light farthest to the left you can see.

a The observed spectrum extends from _____ nm to _____ nm.

b The colors at these places on the scale are _____ and _____.

11 How are the incandescent and fluorescent light source spectra similar? How are they different?

___

___

___

12 Following all safety warnings for observing the Sun; observe sunlight outdoors through the spectrometer. On the scale below the spectrum, indicate the colors you see, noting prominent color lines, if any.

CAUTION!
Do not look directly at the Sun with the spectrometer.

___

13 Read the number on the scale corresponding to the light farthest to the right that you can see and the number corresponding to the light farthest to the left you can see.

  a The observed spectrum extends from _____ nm to _____ nm.

  b The colors at these places on the scale are _____ and
  _____.

14 How did the solar spectra compare to the incandescent and fluorescent spectra observed?

_____

_____

15 Now make spectrum tube observations. In the lab, you will find at least four different spectrum tubes (Fig. 20.13) set up. Each of the tubes contains a specific element, such as hydrogen (H), or a molecule, such as carbon dioxide (CO_2). A spectrum tube's power supply uses electricity to excite the gas within the glass tube, which gives off light, or glows. When glowing, the gas emits an emission spectrum. Good examples of this are neon signs. The glass tubes in such signs contain neon (Ne) gas. When the power is turned on, the tube glows. When observed through a spectrometer, each specific element or molecule produces a spectrum. You should easily notice the bright lines each spectrum tube produces. It is these specific lines and their locations that allow physicists, chemists, and astronomers to determine an object's composition.

16 Record which element/molecule you are observing. On the scale for each element/molecule, indicate the lines/colors you see with the corresponding colored pencils.

  a Element or molecule: _____

  Spectral lines:

  b Element or molecule: _____

  Spectral lines:

  c Element or molecule: _____

  Spectral lines:

  d Element or molecule: _____

  Spectral lines:

17 Compare and contrast the spectra of the elements/molecules you observed in step 16.

_____

_____

20.13 Spectrum tube.

18 How did the spectra of the elements/molecules compare to the spectra of sunlight and the incandescent and fluorescent sources?

19 Return your spectrometer to the assigned location when you are finished with the lab.

Stellar Spectra Classification

Exercise 20.2

With the relatively easy access to online information and data today, a number of online experiences, referred to as *Citizen Science*, have been developed. These projects allow participants to review scientific data and participate in some way in the reduction and refinement of the data. Perhaps the best-known astronomy citizen science projects are Galaxy Zoo (www.galaxyzoo.org) and SETI@Home (http://setiathome.ssl.berkeley.edu/).

The Pisgah Astronomical Research Institute (PARI), www.pari.edu, located in Rosman, North Carolina, has developed a stellar spectroscopy classification citizen science project. SCOPE (Fig. 20.14), or Stellar Classification Online Public Exploration, (www.pari.edu/programs/public/scope) has participants classifying stellar spectra from the vast PARI astronomical data archive, a collection of spectra (on glass plates) taken over a period of about 100 years. Various astronomers at many of the world's major observatories took these plates. About 33,000 of these plates archived in the PARI collection have stellar spectra data and information for 1 million or more stars.

SCOPE first takes you through a simple registration process and a practice session. The SCOPE site contains details about stellar spectroscopy as well as an excellent tutorial on stellar spectra, star structure, and the spectral classification process. The site is a PHP platform (Java-free) and loads relatively quickly, depending on your type of connection and speed.

20.14 SCOPE (Stellar Classification Online Public Exploration).

Once you complete the registration and practice session, you can select plates from which you can classify specific stellar spectra. Many of these spectra have never been previously classified, so you are contributing to science and astronomy.

 **Procedure**

1 Go to the PARI/SCOPE site and register. Your instructor may provide specific username and password directions to follow.

2 Complete the practice session as directed.

3 Choose a plate to classify. Your instructor will tell you the minimum number of stars to classify for this exercise.

4 Print your classified stars from the PARI/SCOPE site. To print, click on one of the two print buttons at the top of the MY STARS page and print as you normally would.

5 Answer the following questions about the spectra you classified.

   a What spectral types of stars did you find on your plate?

   List the star types and number found in Table 20.2.

TABLE **20.2** Star Spectral Type Data

Star Spectral Type	Number Found

   b What does each type of star represent, with respect to its stellar characteristics?

   c What star type did you find the most of? What type did you find the least of?

   d Speculate as to why you found the distribution noted in question c.

Classifying Celestial Objects
The Hertzsprung-Russell Diagram

LEARNING OBJECTIVES

Upon completion of this chapter, you should be able to:

1. Locate the groups of stellar types on the Hertzsprung-Russell diagram.
2. Describe the relationships between a star's luminosity and temperature.
3. Describe how the Hertzsprung-Russell diagram is one of the most powerful analytical tools available to astronomers.
4. Identify stars that are hotter, cooler, brighter, and fainter than our Sun by using the H-R diagram.
5. Plot some of the nearest and brightest stars on the H-R diagram and determine any patterns.
6. Define all glossary terms.

21

Rosette Nebula (NGC 2237—9/46)
Courtesy of Adam Block/Mount Lemmon SkyCenter/University of Arizona

> **GLOSSARY TERM**
>
> **Star** self-luminous sphere of gas that generates energy by means of nuclear fusion reactions in its core
>
> **Hertzsprung-Russell (H-R) diagram** graph that has the absolute magnitudes (or luminosities) of stars plotted against their spectral classification (or effective temperatures)
>
> **Luminosity** total energy radiated into space every second by a celestial object such as a star
>
> **Kelvin temperature scale** temperature scale that has absolute zero as its starting point; to convert Kelvin to Celsius temperature, subtract 273.15. So 300 Kelvins = 26.85°C
>
> **Spectral class** designation a star receives based on its temperature, chemical composition, and its type
>
> **Photosphere** surface of a star, including the Sun, which is the layer emitting visible light, about 500 kilometers deep
>
> **Absolute magnitude** apparent magnitude of a celestial object if the object were at a distance of 10 parsecs; the true brightness of a celestial object; luminosity
>
> **Main sequence** region on the H-R diagram in which stable, middle-aged, hydrogen-burning stars are found
>
> **Giant** type of star much larger than the Sun, and also more massive, but not proportionately so, having a very thin outer atmosphere
>
> **White dwarf** remains of a stellar core following the cessation of nuclear fusion, composed of electron degenerate matter
>
> **Supergiant** star with a higher luminosity and a larger radius than a giant of the same spectral classification; typically has 100 times the luminosity of a giant and almost certainly will become a supernova

Look up at the night sky from a dark site, and you will see thousands of burning orbs of gas. Just one of those twinkling dots we call **stars** could be a behemoth with a mass 80 times that of our own Sun. At its core sits a cauldron of nuclear reactions that power the star, allowing us to see it glowing from hundreds of light-years away.

What could hold such a massive object together, and how does the pent-up energy not blow it apart? Over the past century, astronomers have learned an immense amount about stars. They have pieced together the life cycles of different stars to learn what is happening within them and how we, as observers, view each stage.

Research has shown that a star's mass dictates almost everything about the object, from the core temperature, to how long the star lives, to how it dies. While the Sun and a star 10 times its mass may have similarities during the "adult" stages of their lives, they couldn't be more different as they reach the later stages. Why do some end their lives in spectacular blasts while others puff away their outer shells and fade slowly? This chapter introduces a powerful tool—the **Hertzsprung-Russell diagram**—which allows astronomers, at a glance, to tell where a star is in its life cycle.

Luminosity and Temperature

The Hertzsprung-Russell diagram (Fig. 21.1) is the essential guide when it comes to understanding stars. In the early 1900s two astronomers, Ejnar Hertzsprung (1873–1967) and Henry Norris Russell (1877–1957), independently discovered a relationship between a star's **luminosity** and its temperature. This is the basic physical relationship for the Hertzsprung-Russell (H-R) diagram.

Temperature is the X-axis (horizontal) on the graph and is usually plotted as temperature using the **Kelvin temperature scale,** or as the star's **spectral class**. A star's temperature relates to its color: cooler stars are red while the hottest stars are blue. With this scheme, the coolest temperatures are on the graph's right side but not at what is typically the zero (0) coordinate. A star's spectral class provides astronomers a measurement of the temperature at the star's **photosphere**, so the hottest classes, O and B, are on the left side and the cooler classes, K and M, lie to the right.

The Y-axis (vertical) on the graph is luminosity, where the luminosity of the Sun is denoted L_{sun} = 1. Luminosity is the total amount of light emitted by the star. Both it and **absolute magnitude** are logarithmic scales. When astronomers measure luminosity relative to the Sun, the scale for stars in our Milky Way galaxy range from $10^5 L_{Sun}$ to $10^{-5} L_{Sun}$, where the Sun = 1. You will notice that more luminous stars tend to be larger. This is because luminosity depends partly on surface area.

When you examine the H-R diagram, you will see that nearly 90 percent of observed stars lie along a band that stretches from the upper left to the lower right. This group of stars is the **main sequence**. Main sequence stars range from hot, bright stars at the upper left to cool, dim stars at the lower right. The temperature range is from 3,000 K to around 30,000 K, which is a tenfold increase. Yet the stellar luminosities for the same stars range over ten powers of 10 [10^{-5}LSun to $10^5 L_{Sun}$], a factor of 1 billion!

The range of stars along the main sequence is due to differences in their masses, which is the key characteristic to understanding how a star operates. Another characteristic of stars is the relationship between luminosity and mass. Stellar masses and luminosities increase from the lower right of the main sequence to the upper left. Because this relationship exists, astronomers can estimate the mass of a main sequence star by its position on the H-R diagram. Unfortunately there is no easy relationship between luminosity and mass for the remainder of the stars (about 10 percent) not on the main sequence.

The upper right corner of the H-R diagram contains cool but bright giant stars. Small, hot stars, whose low luminosities contradict their high temperatures, are **white dwarfs**. Finally, **supergiants** occupy the region where a star's luminosity is greater than 10,000 times that of our Sun.

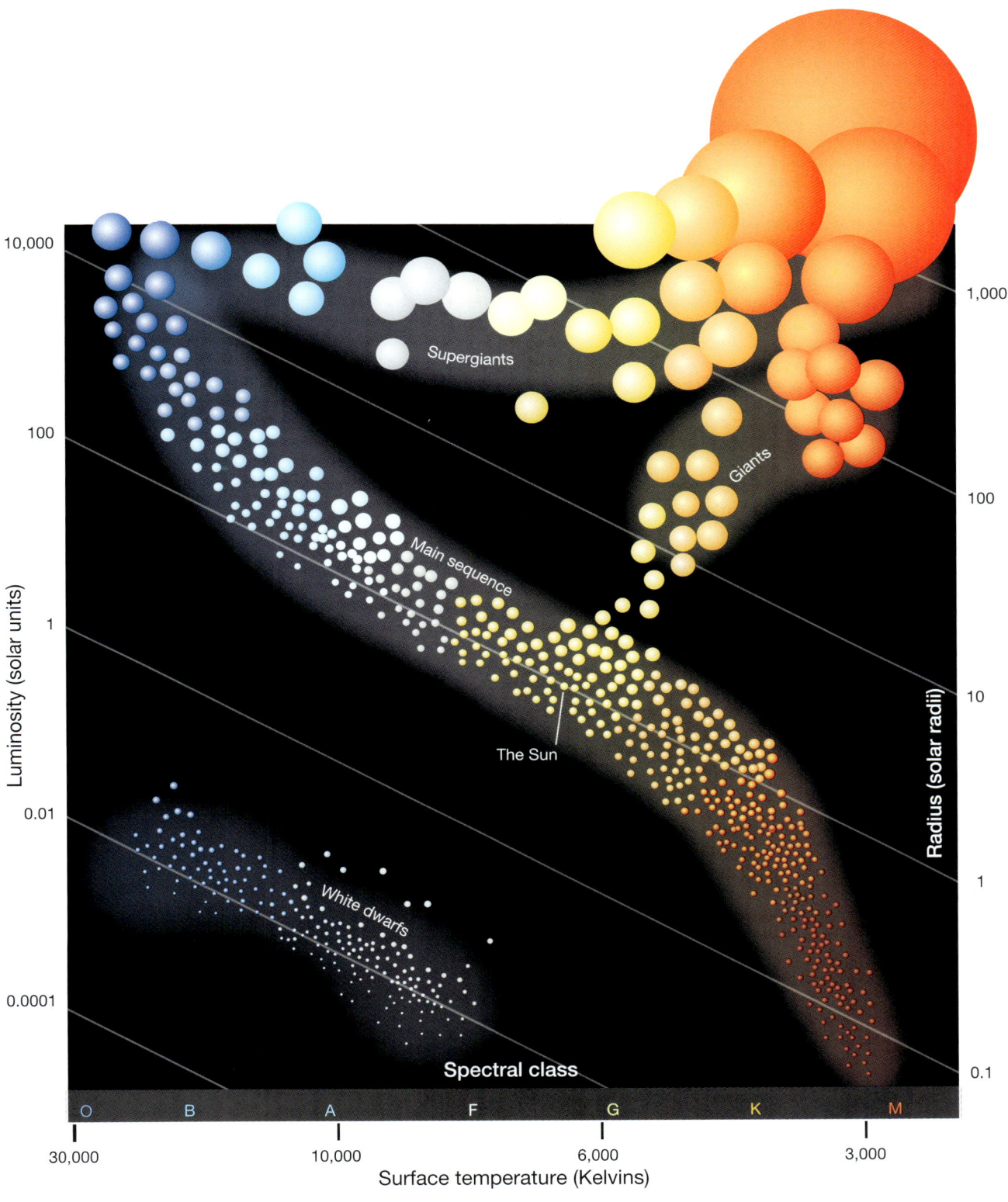

21.1 Hertzsprung-Russell (H-R) diagram.

Courtesy of Roen Kelly, *Astronomy* magazine

Check Your Understanding

1.1 What is the main factor that governs nearly everything about a star?

1.2 The Hertzsprung-Russell diagram plots what two characteristics about stars?

1.3 Which spectral types are the hottest? Which are the coolest?

1.4 Why are the most luminous stars usually larger?

1.5 Where on the H-R diagram do most stars lie?

Reactions in a Main Sequence Star

Astronomers now understand that all the stars on the main sequence generate energy by the same mechanism: they fuse hydrogen into helium. Fusion is the most common mechanism that powers main sequence stars. Fusion of elements heavier than hydrogen becomes more important in main sequence stars with cores much hotter than our Sun's core.

Nuclear Reactions

Gravity pulls a star's gas toward the center. There, the pressure and temperature are so high that nuclear fusion occurs. The outward radiative pressure from fusion balances the inward gravitational force (Fig. 21.2).

This conversion from hydrogen to helium liberates energy from hydrogen nuclei (protons). Nuclear fusion of hydrogen occurs within stars, but not all stars are the same temperature. Fusion in a star like the Sun begins with two hydrogen atoms and results in one normal helium nucleus (left of center in Fig. 21.3). Hotter stars can fuse helium nuclei into beryllium and lithium, also resulting in normal helium (below center in Fig. 21.3). In even hotter stars, beryllium and hydrogen create boron, which decays into beryllium that also decays into normal helium.

Three hydrogen-to-helium conversion chains are important in stars containing less than about 1.5 solar masses (M_{sun}). In addition to producing helium, the reactions spit out high-energy radiation (gamma rays) and neutral tiny mass particles called neutrinos.

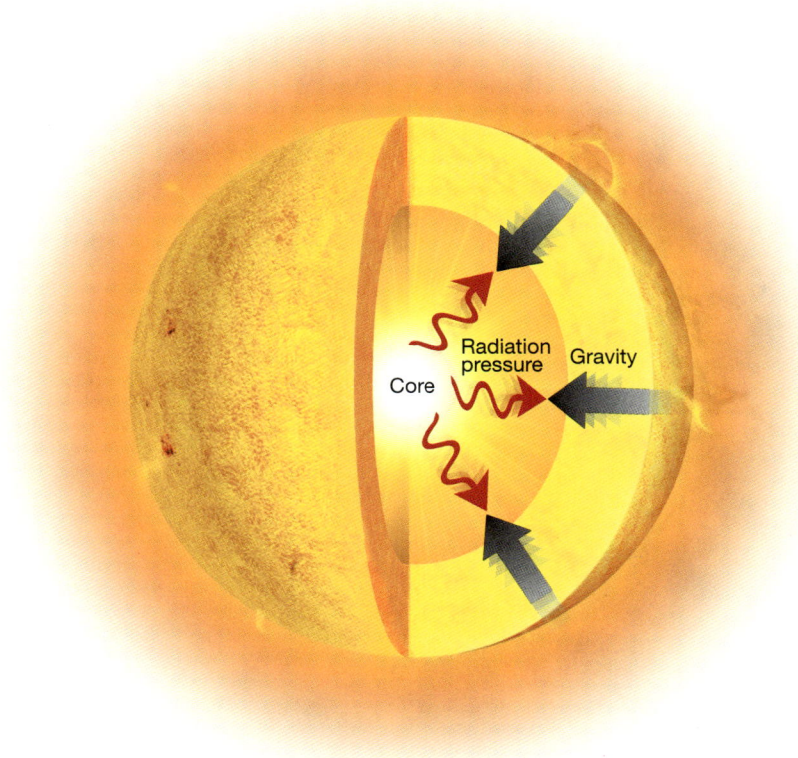

21.2 Balance of gravity and radiation pressure in a star.
Courtesy of Roen Kelly, *Astronomy* magazine

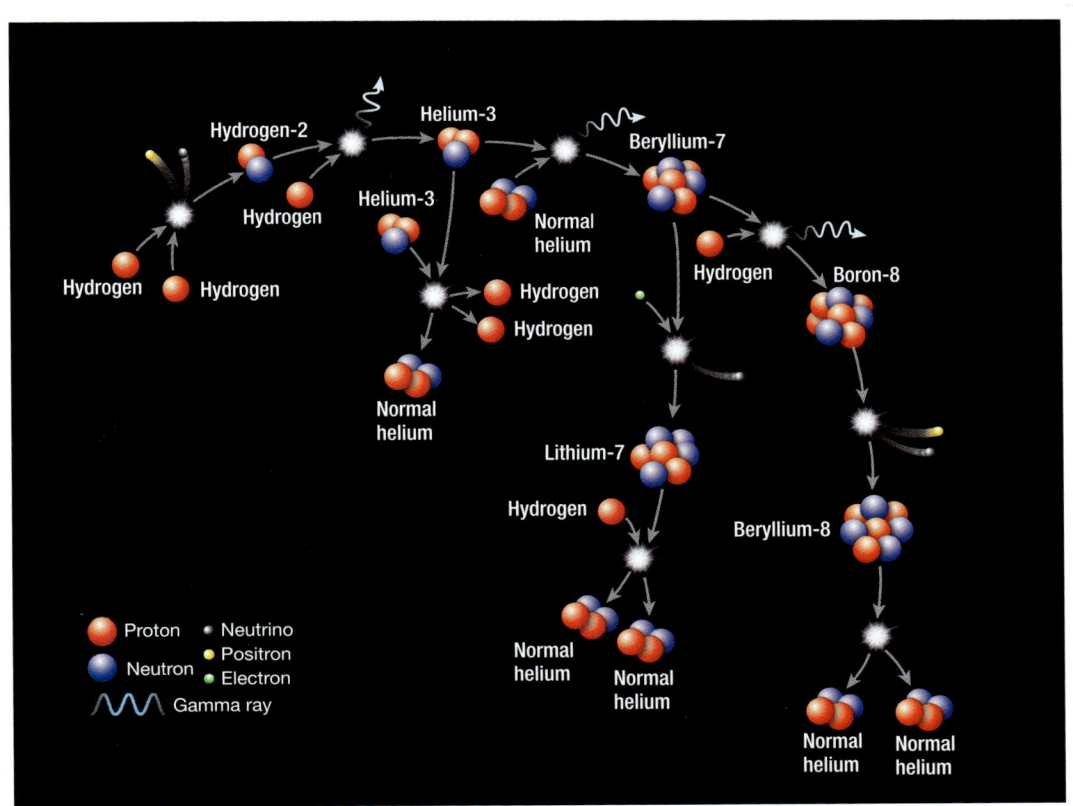

21.3 Nuclear fusion.
Courtesy of Roen Kelly, *Astronomy* magazine

The Hertzsprung-Russell Diagram   CHAPTER 21   409

The CNO Cycle

A different set of reactions becomes more important in main sequence stars with cores much hotter than our Sun's core. Four hydrogen nuclei still convert to a helium nucleus, but with the help of carbon, nitrogen, and oxygen isotopes (Fig. 21.4). This CNO cycle is the main reaction chain in stars greater than about 1.5 solar masses.

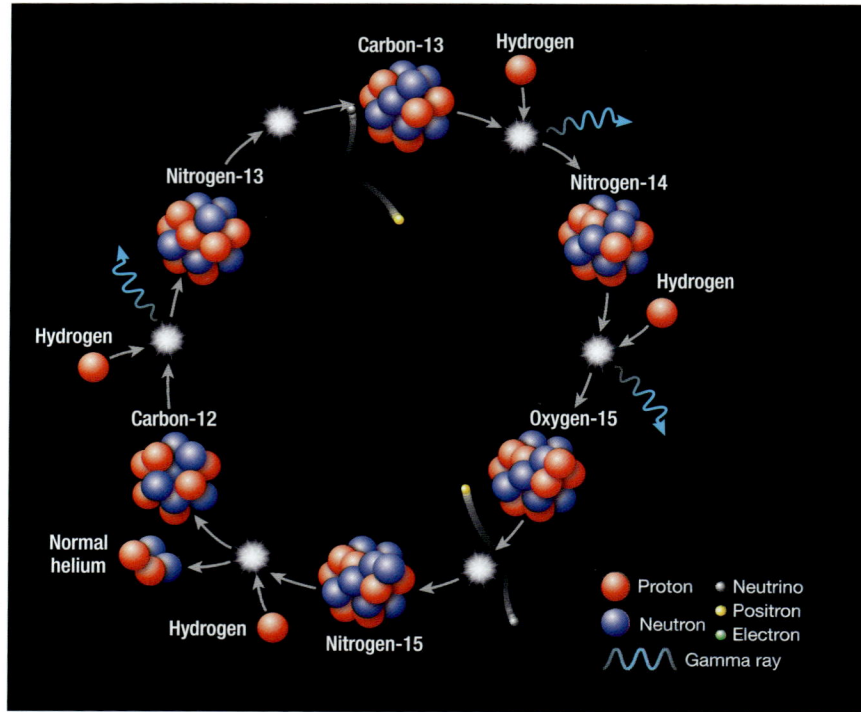

21.4 Mechanism of the CNO cycle.

Courtesy of Roen Kelly, Astronomy magazine

Check Your Understanding

2.1 If a star lies on the main sequence in the H-R diagram, how is it producing energy?

2.2 What two forces are in balance in a normal star?

2.3 Name the three elements besides hydrogen and helium that come into play in the CNO cycle.

Plotting Stars

Exercise 21.1

 Procedure

1. Using the graph on the data sheet, page 413, plot the stars from the data set given in Table 21.1 then answer the questions on pages 413–414 for the stars in the first data set.

2. On the same graph, plot the stars from the data set given in Table 21.2 then answer the questions on page 414 for the stars in the second data set.

3. Again using the same graph, plot the stars from the data set given in Table 21.3 then answer the questions on page 414 for the stars in the third data set.

4. Label the main sequence on your H-R diagram.
   What percent of your stars are on the main sequence? _____

5. Label "dwarfs" and "giants" on your H-R Diagram.

TABLE 21.1 Stars: First Data Set

		Visual Magnitude	Distance (Light-Years)	Temperature (Kelvin)	Luminosity (Sun = 1)
1	Sun	-26.7	0.00002	5,800	1
2	Alpha Centauri C	-0.01	4.24	5,800	1.5
3	Alpha Centauri A	1.4	4.36	4,200	0.33
4	Alpha Centauri B	11.0	4.36	2,800	0.0001
5	Wolf 359	13.66	7.7	2,700	0.00003
6	Lalande 21185	7.47	8.1	3,200	0.0055
7	Sirius A	-1.46	8.7	10,400	23
8	BL Ceti	12.5	8.7	2,700	0.00006
9	UV Ceti	12.9	8.7	2,700	0.00002
10	Ross 154	10.6	9.6	2,800	0.00041
11	Ross 248	12.24	10.3	2,700	0.00011
12	Epsilon Eridani	3.73	10.8	4,500	0.3
13	Ross 128	11.13	11	2,800	0.00054
14	61 Cygni A	5.19	11.1	4,200	0.084
15	61 Cygni B	6.02	11.1	3,900	0.039
16	Procyon A	0.38	11.3	6,500	7.3
17	Epsilon Indi	4.73	11.4	4,200	0.14
18	Altair	0.77	16.5	8,000	11
19	70 Ophiuchi A	4.3	17	5,100	0.6
20	Vega	0.04	26	10,700	55
21	Achernar	0.51	65	14,000	200
22	Delta Aquarii A	3.28	84	9,400	24
23	Tau Scorpii	2.82	233	25,000	2,500
24	Spica	0.91	260	21,000	2,800
25	Beta Centauri	0.63	300	21,000	5,000
26	Zeta Persei A	2.83	465	24,000	16,000
27	Delta Persei	3.03	590	17,000	1,300

TABLE **21.2** Stars: Second Data Set

		Visual Magnitude	Distance (Light-Years)	Temperature (Kelvin)	Luminosity (Sun = 1)
28	Arcturus	-0.06	36	4,500	110
29	Aldebaran	0.86	53	4,200	100
30	Antares	0.92	400	3,400	5,000
31	Betelgeuse	0.41	500	3,200	17,000
32	Delta Aquarii B	2.86	1030	6,000	4,300

TABLE **21.3** Stars: Third Data Set

		Visual Magnitude	Distance (Light-Years)	Temperature (Kelvin)	Luminosity (Sun = 1)
33	Barnard's Star	9.54	6	2,800	0.00045
34	Sirius B	8.5	8.7	10,700	0.0024
35	EZ Aquarii	12.58	11	2,700	0.00009
36	Procyon B	10.7	11.3	7,400	0.00055
37	Van Maanen's Star	12.36	14	7,500	0.00016
38	Fomalhaut	1.19	23	9,500	14
39	GJ 151	15.20	46	7,400	0.00021
40	Capella	0.05	47	5,900	170
41	GJ 742	13.19	49	9,800	0.0013
42	GJ 3306	14.10	63	6,300	0.00068
43	Canopus	-0.72	100	7,400	1,500
44	Alpha Crucis	1.39	400	21,000	4,000
45	Beta Crucis	1.28	500	22,000	6,000
46	Rigel	0.14	800	11,800	40,000
47	Deneb	1.26	1,400	9,900	60,000

Name _____   Section _____   Date _____

Graph for Plotting Stars

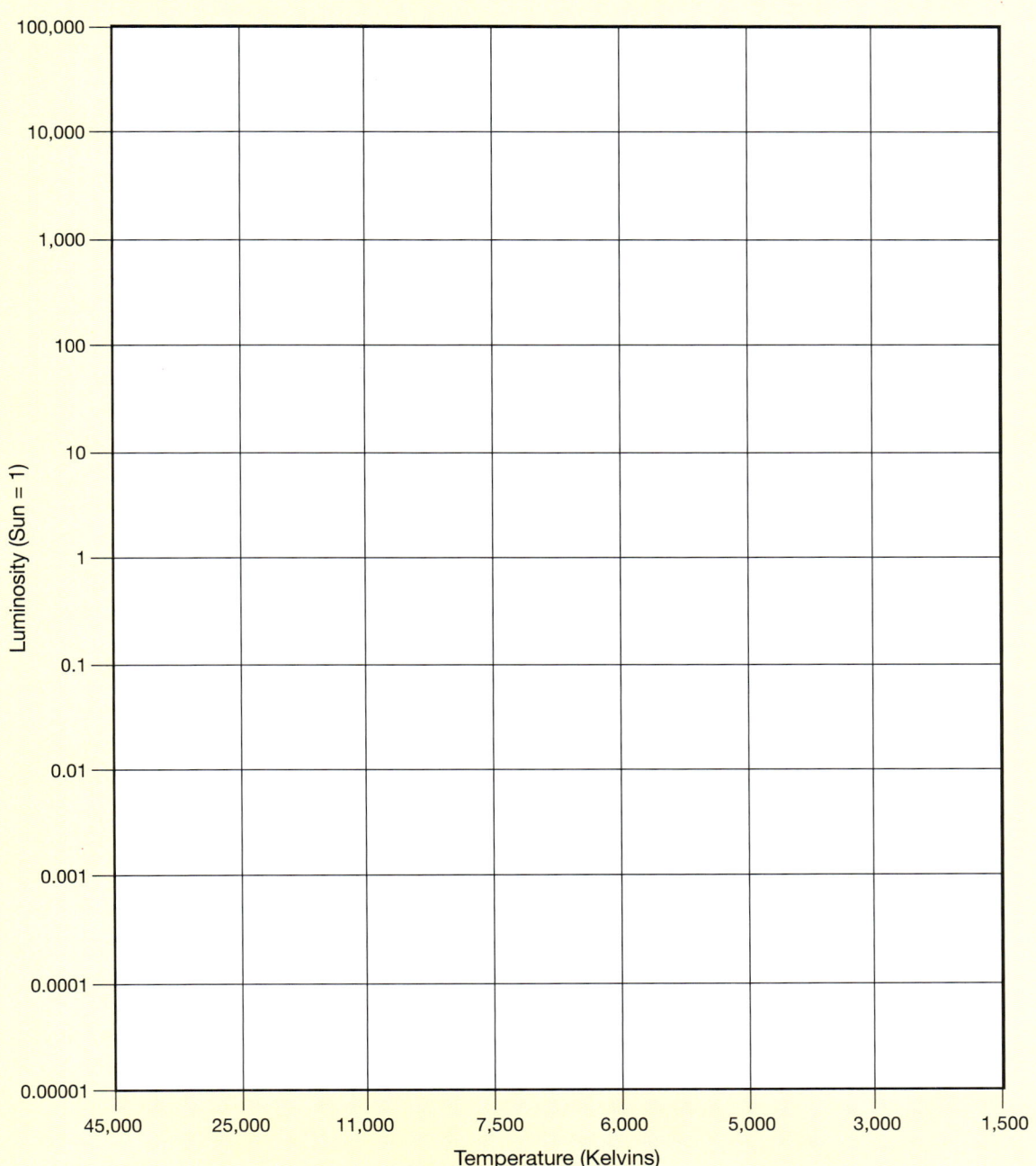

Questions for the stars in the first data set:

1. What would you tell someone who thinks that all stars are similar (discuss temperature and brightness)?

_____

_____

_____

_____

Exercise 21.1 Data Sheet

2 How does the Sun compare to other stars in brightness and temperature?

3 Is there a pattern to the stars that you plotted, or is it a random distribution? Explain your answer.

4 Would you expect hotter stars to be dim or bright? Does the graph agree with this answer?

Questions for the stars in the second data set:

5 Does the second set of stars follow the same pattern as the first set? Why or why not?

6 In a general sense, are the stars in this set bright or dim? _____

7 Is this set of stars hot or cool compared to the first set of stars? _____

8 Compare the two sets that you plotted. Does the relationship of brightness to temperature for the second set make sense? Explain.

Questions for the stars in the third data set:

9 Compare the areas of the graph where you plotted the second and third data sets. How are they different?

10 In a general sense, are the stars in this set bright or dim? _____

11 Are the stars in this data set hot or cool compared to the stars in the other two sets? _____

12 Does the relationship of brightness to temperature for the third set make sense? Explain.

Unit 6 Classifying Celestial Objects

Classifying Celestial Objects

Radioactivity

22

LEARNING OBJECTIVES
Upon completion of this chapter, you should be able to:

1. Describe the basic principles of radioactivity.
2. Differentiate between alpha particles, beta particles, and gamma rays.
3. Detail the basics about the Geiger counter.
4. Describe the concepts of radioactive decay and half-life.
5. Explain the applications of Avogadro's number, atomic mass, and mole.
6. Define isotopes.
7. Describe how scientists can determine the age of astronomical objects, such as Earth, the Moon, and meteorites.
8. Define all glossary terms.

Planetary nebula (NGC 5189)
Courtesy of NASA/ESA/the Hubble Heritage Team (STScI/AURA)

GLOSSARY TERM

Radioactivity decay of unstable radioactive atoms

Radioactive decay disintegration of unstable atoms

Alpha particles two protons and two neutrons bound together into a particle identical to the helium nucleus; emitted during radioactive decay; written as He^{2+} or $^4_2$He^{2+}; generally come from alpha decay

Beta particles electrons (e$^-$) or positrons (e$^+$) emitted during radioactive decay; emitted as an ionizing radiation or beta rays

Gamma rays high frequency electromagnetic radiation resulting in high energy photons; can be produced through decay of high energy states of atomic nuclei; emissions often accompany alpha and beta particle decay

Photon massless elementary particle that carries energy, momentum, and angular momentum created in reactions that involve electromagnetism; travels at the speed of light

Half-life (t$_{1/2}$) amount of time required for a quantity of material to fall to half of its value as determined at the beginning of a period of time

French physicist Antoine Henri Becquerel (1852–1908) discovered radioactivity in 1896 when, while working with photographic film, he noted that the film was exposed by being near a uranium salt source, even when he protected the film with an opaque covering. Further research showed that the activity that exposed the film traveled from the source along radial lines, thus the term **radioactivity**. This discovery led to later recognition through a Nobel Prize awarded to Becquerel, his doctoral student Marie Curie (1867–1934), and her husband Pierre Curie (1859–1906).

The discovery of radioactivity had to await the development of photography because there was no other easy method to detect it. Humans cannot directly detect radioactivity.

Radioactive Decay

Researchers soon discovered three types of invisible rays: alpha (α) rays, beta (β) rays, and gamma (γ) rays. The decay process produces these rays. This process, known as **radioactive decay,** or nuclear decay, occurs when the unstable atom loses energy by emitting alpha and beta particles. Any material that spontaneously emits these particles is radioactive. **Alpha particles** are the nuclei of helium atoms, **beta particles** are electrons, and **gamma rays** are energetic particles of light called **photons**.

Alpha, beta, and gamma rays are all harmful to life. When people encounter any of these three forms of radiation, it penetrates living cells and destroys DNA so that human cells cannot function properly.

Ernest Rutherford (1871–1937) and Thomas Royds (1884–1955) were the first to identify alpha particles as helium atoms in 1909. They were able to trap alpha particles in a glass tube that contained two metal electrodes. When they created an electric arc through the tube and examined the resulting spectrum, they identified the spectral lines of helium.

Ernest Rutherford, who also developed a model of the atom, first proposed the concept that as radioactive materials decay, they do so in a specified amount of time called the **half-life**. Figure 22.1 shows a diagram of the half-life. The half-life is 10 units of time (seconds, days, millions of years, etc.) long. As each 10 units of time passes, half of the radioactive material decays, and half is left. As radioactive materials emit particles and undergo radioactive decay, they can transform to atoms with a nucleus of a different state. The amount of time required for a quantity of material to fall to half of its value as determined at the beginning of a period of time is its half-life, represented as t$_{1/2}$. When he developed the idea of the half-life, Rutherford used it to study the ages of rocks by measuring the time it took for radium to decay into lead.

 22.1 Plot of the half-life of a radioactive material. Courtesy of Holley Y. Bakich

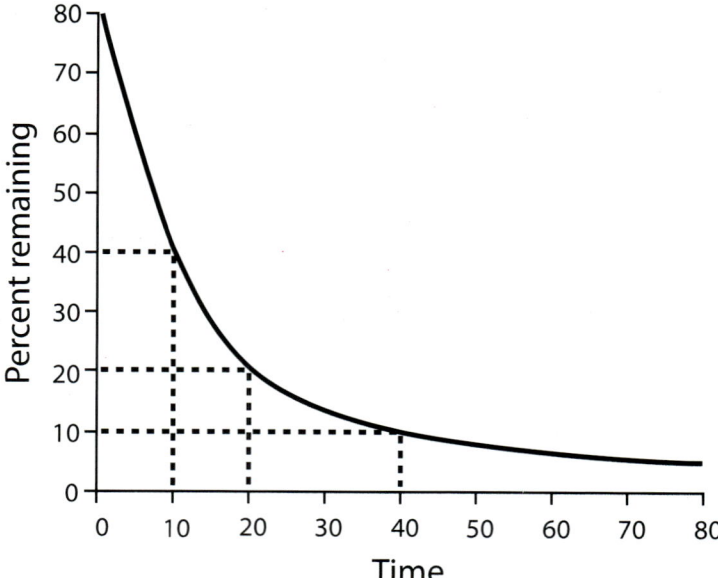

Radioactive elements, such as uranium, potassium, and thorium, have isotopes that decay to form lighter atoms. The energy these radioactive elements release during this decay process is in the form of fast-moving particles (alpha and beta particles) and high-energy waves (gamma rays).

Isotopes

Prior to the discovery of radioactivity, scientists believed that atoms were permanent and immutable. Radioactive **isotopes**, however, provided scores of examples of atoms that decayed naturally, producing these new atoms of other elements called **descendants**. Examples of uranium isotopes include uranium 234 (^{234}U), uranium 235 (^{235}U), and uranium 238 (^{238}U). What makes uranium uranium is that all its atoms have 92 protons. Yet each isotope has a different number of neutrons; uranium 234 (^{234}U) has 142 neutrons, uranium 235 (^{235}U) has 143 neutrons, and uranium 238 (^{238}U) has 146 neutrons. Early radioactivity research showed that the decay and half-life does not depend on the atom's environment (such as the temperature and pressure), the atom's location, or the state of the atom (whether it is solid, liquid, or gas).

Radioactive Dating

Radioactive dating is the process for determining the age of materials by analysis of radioactive isotopes and their descendants. Researchers have used this technique on numerous Earth and lunar rocks as well as meteorites. Radioactive dating is one of the methods astronomers and geologists used to verify the theory that Earth and the solar system are about 4.6 billion years old. Specific types of meteorites, such as carbonaceous chondrites, are key sources for determining this age. They represent some of the oldest objects that humankind has sampled.

An isotope's nucleus determines its radioactive properties, and they differ from those of all other isotopes. For example, the radioactive half-lives of uranium 238, uranium 235, and uranium 234 are 4.47 billion years, 0.704 billion years, and 0.246 million years, respectively.

Detecting Radiation

In 1908 German physicist Hans Geiger (1882–1945) invented a methodology to detect alpha and beta particles and gamma rays. In 1928 Geiger collaborated with one of his doctoral students, Walther Müller (1905–1979), to advance the techniques of radiation detection. Geiger and Müller developed a tube that could detect various forms of radiation.

The **Geiger counter** is a portable, easy-to-use, handheld particle detector. It works by ionization of a low-pressure gas in a metal tube called the **Geiger-Müller tube**, or G-M tube, which is the sensing element of the Geiger counter. An accessory speaker is plugged in, which lets the user hear clicks corresponding to particles detected (Fig. 22.2). The Geiger counter readout has two scales: the scale "mr/hr" (from 0 to 0.5) means "milliroentgens per hour," which expresses the strength of the radiation over time; the "C/M" scale, which means "counts per minute," expresses how strong the source is.

A Geiger counter can detect a single ionizing event. The limitations of the Geiger counter, however, are in determining high radiation rates and how much energy is in the radiation. The Geiger counter has uses not only in geology, experimental physics, and the nuclear industry, but also in the medical field. Applications include **dosimetry** (how much radiation someone receives) and health physics.

> **GLOSSARY TERM**
>
> **Isotope** variants of the same element that contain the same number of protons (which identify it as a specific element) but vary in the number of neutrons
>
> **Descendants** new elements produced by other elements that disintegrate by themselves
>
> **Radioactive dating** technique used to tell the age of materials, such as Earth and lunar rocks and meteorites; usually determined by a comparison between the observed, naturally occurring radioactive isotope abundance and its decay products, or descendants, using known decay rates; also known as radiometric dating
>
> **Geiger counter** type of particle detector that works by ionization of a low-pressure gas in the Geiger-Müller tube; a portable, handheld instrument that is easy to use and relatively inexpensive
>
> **Geiger-Müller tube (G-M tube)** sensing element of a Geiger counter, used to detect ionizing radiation
>
> **Dosimetry** measurement of the amount of radiation a substance (usually human tissue) receives

22.2 A 1950s-era Geiger counter in Geiger-Müller tube connected to a speaker.

Check Your Understanding

1.1 How was radioactivity first discovered?

1.2 What are the three types of high-energy radiation scientists identified?

1.3 What are the characteristics of alpha and beta particles?

1.4 Which of the three types of radiation only consists of high-energy electromagnetic waves?

1.5 What is half-life, and how is it determined?

1.6 What are isotopes? Give examples of the uranium isotopes.

1.7 How does the Geiger counter work?

1.8 What are the Geiger counter's limitations?

Radioactivity Calculations

Three definitions are useful in discussion of radioactive isotopes: **Avogadro's number**, **atomic mass**, and **mole**. A mole is a specified number of atoms of a substance, called Avogadro's number: 6.022×10^{23}. You also might see a reference to the Avogadro constant, which is the number of constituent particles, usually atoms or molecules, in one mole of a specific substance. For example, one mole of hydrogen atoms has a mass of 1 kilogram, while one mole of uranium has a mass of 238 kilograms. Ignoring typical errors of 1 percent, each atom has a mass in atomic mass units equal to its isotope number.

> **GLOSSARY TERM**
>
> **Avogadro's Number** 6.022×10^{23} atoms
>
> **Atomic mass** atom's mass; equal to its isotope number
>
> **Mole** specified number of atoms of a substance, called Avogadro's number: 6.022×10^{23} atoms
>
> **Exponential decay** way in which a quantity of a material decreases at a rate proportional to the material's current value

Determination of Mass

To determine the mass in grams of one atom of hydrogen, you will need the element's atomic mass number and Avogadro's number:

Atomic mass number/Avogadro's number = mass of 1 atom of the specific element

For hydrogen, its atomic mass number is 1.0 grams

$1.0 \text{ gram}/6.022 \times 10^{23} = 1.6605 \times 10^{-24}$ g

Probability of Decay

To determine the probability of decay per second for a naturally occurring isotope:

$0.693/(T)(N)$; where **T** is the half-life in a period of time and **N** is the number of seconds in a year

For Uranium 238:

$T = 4.47 \times 10^9$ years

$N = 3.16 \times 10^7$ seconds/year

$0.693/[(4.47 \times 10^9 \text{ years})(3.16 \times 10^7 \text{ seconds/year})] = 4.9 \times 10^{-18}$/seconds

Number of Disintegrations

To determine the number of disintegrations per second in a mole of uranium 238:

Disintegrations per second = (probability of decay/s for a naturally occurring isotope)(Avogadro's number)

$(4.9 \times 10^{-18}/s)(6.022 \times 10^{23}) = 2.951 \times 10^6$ disintegrations per second

Exponential Decay

Because the sample's activity is proportional to the number of radioactive atoms in the sample, the activity will decrease as the number of radioactive atoms decreases. This produces a type of change referred to as an **exponential decay**. Although the exponential decay curve never reaches zero mathematically, one mole of radioactive atoms will decay to a single atom in about 89 half-lives on average. So, thinking intuitively, a radioactive isotope will have completely transmuted to its descendants in less than 100 half-lives.

Consider an example of two isotopes of uranium found within a rock sample. As uranium 238 decays, it first emits an alpha particle (Fig. 22.3). The result is thorium 234. That substance then emits a beta particle and transforms into protoactinium 234. After completing 12 more steps (releasing seven alpha and five beta particles), uranium 238 becomes lead 206, which is stable and does not decay further. In this example, presume there are equal amounts of uranium 235 and uranium 238. From that starting point, you can determine the ratio of uranium 235 to uranium 238 after 4.47 billion years.

The number of uranium 238 atoms will drop by half in 4.47 billion years, while the number of uranium 235 atoms will drop in half in 704 million years, which is shown in Table 22.1. So, the number of uranium 235 atoms remaining after 4.47 billion years is:

$(0.50)(4.47/0.704) = 0.012$ or 1.2%

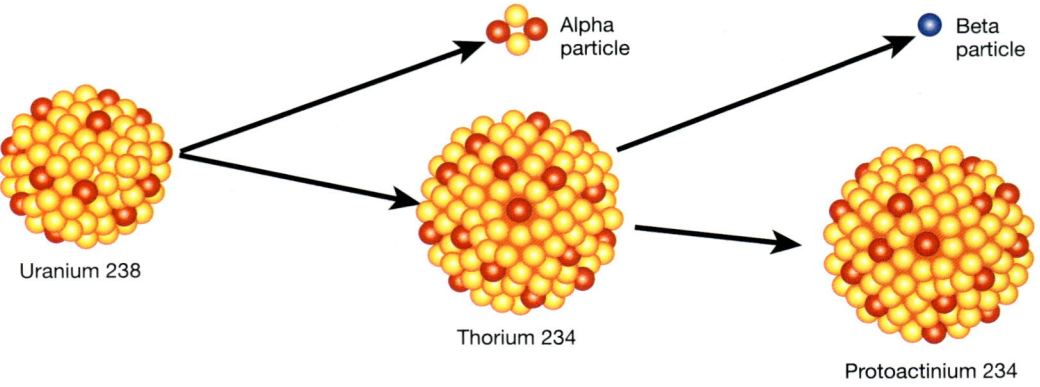

22.3 Partial decay of uranium 238.

Courtesy of Holley Y. Bakich

TABLE **22.1** Radioactive Half-Lives of Some Elements

Isotope	Half-life
Potassium 40	1.3 billion years
Rubidium 87	48 billion years
Samarium 147	106 billion years
Rhenium 187	41 billion years
Thorium 232	14.05 billion years
Uranium 234	0.246 million years
Uranium 235	0.704 billion years
Uranium 238	4.47 billion years

Radioactivity detection has come a long way since Becquerel's discovery in 1896. Inventors have created detectors for use in the laboratory and in the field. Understanding radioactivity and nuclear reactions has helped to explain many things that otherwise might not be logical. One example is how Earth's center can still be hot 4.6 billion years after our planet's formation. This heated core leads to convection, which is responsible for earthquakes, volcanism, continental drift, and Earth's magnetism. Calculations show that if it were not for the energy released by radioactivity, Earth's interior would have cooled long ago. Many nuclear reactions release a large amount of energy—a million times more energy per reacting atom than released in most energetic chemical reactions, such as combustion.

Check Your Understanding

2.1 What is Avogadro's number?

2.2 Define an element's atomic mass.

2.3 What is measured by a mole?

2.4 How do you determine the mass in grams of one atom of an element?

2.5 How do you determine the probability of decay per second for a naturally occurring isotope?

2.6 How do you determine the number of disintegrations per second in a mole of a specified element?

2.7 What is exponential decay?

2.8 Why do scientists believe that Earth is 4.6 billion years old?

2.9 Explain the method for dating materials.

2.10 How can different elements be used to date Earth, the Moon, or a meteorite?

Exercise 22.1   Measuring Radioactivity

 **Procedure**

Materials

- ❏ Geiger counter equipped with a Geiger-Müller tube
- ❏ Radioactive source (supplied by your laboratory instructor)
- ❏ Potential radiation blocking materials: paper, aluminum foil, plastic wrap, cardboard, lead foil, glass (supplied by your laboratory instructor)
- ❏ Graph paper

1 Obtain the Geiger counter and radioactive source from your instructor. Be certain to follow all directions for handling the radioactive source. Inspect the Geiger counter and radioactive source. Your instructor will provide specific directions for operating the Geiger counter you will use in this procedure. These radioactive sources are of low activity and are not harmful to you as used in this exercise.

CAUTION
Use radioactive source materials with caution and as directed by your laboratory instructor.

2 Using your radioactive source as directed, bring your Geiger counter near it. Note the clicks and/or the indicator; this is a count of ionization events by the radioactive source.

3 Collect counts for three 60-second periods at distances of 0, 2, 4, 6, 8, and 10 cm, and record your results in Table 22.2 on the data sheet, page 423. Record your data as counts per minute (cpm or c/m). Determine the average for the three counts.

4 Next, test the ability of various supplied materials to block radiation from the source. Place the material between the radioactive source and the Geiger counter, or as directed by your instructor. Again, collect counts for three 60-second periods at the distances of 0, 2, 4, 6, 8, and 10 cm, and record your data in Table 22.3 on the data sheet, page 424. As before, record your data as counts per minute (cpm or c/m). Determine the average for the three counts.

Name _____ Section _____ Date _____

Ionization Events Count

TABLE **22.2** Count of Ionization Events by Radioactive Source

Radioactive Source:				
Distance	Count 1	Count 2	Count 3	Average for the 3 Counts
0 cm				
2 cm				
4 cm				
6 cm				
8 cm				
10 cm				

1 Graph your results; average counts per minute (cpm) versus distance.

2 Based on your graph and data, what is the pattern that describes the change in counts per minute (ΔCPM) with distance? Consider: no change, linear decline, exponential decline, or some other pattern.

_____
_____
_____
_____

3 What would account for the pattern you graphed based on your data? Detail your explanation.

_____
_____
_____
_____
_____

Exercise **22.1** Data Sheet

Exercise 22.1 Data Sheet

TABLE **22.3** Count of Ionization Events with Various Blocking Materials

Material	Distance	Count 1	Count 2	Count 3	Average for the 3 Counts
	0 cm				
	2 cm				
	4 cm				
	6 cm				
	8 cm				
	10 cm				

4 Which material is the best at blocking radiation?

_____

5 Based on your original counts per minute, what percentage of the radiation is blocked?

_____

6 Which material is the worst at blocking radiation?

_____

7 Based on your original counts per minute, what percentage of the radiation is blocked?

_____

Classifying Celestial Objects

Hubble's Law

LEARNING OBJECTIVES
Upon completion of this chapter, you should be able to:

1. Describe Hubble's law, both a written description and also the equation.
2. Define Hubble's constant, H_0, and the importance of correctly determining the value of the constant.
3. Determine the differences in the age of the universe depending on different values for Hubble's constant.
4. Define all glossary terms.

M106. *Courtesy of NASA/ESA/the Hubble Heritage Team (STScI/AURA)/ R. Gendler (for the Hubble Heritage Team)*

The Relationship between Distance and Velocity

Hubble's law, named for American astronomer Edwin Hubble (1889–1953), describes the proportional relationship between an object's distance from us and its velocity. Astronomers interpret this relationship as a direct observation of the universe's expansion. For example, in Figure 23.1, six clusters of galaxies in a cube 100 million light-years across separate from one another as the cube doubles in size. There is no expansion center. From the perspective of an observer in any galaxy cluster, the other clusters all appear to be moving away. This creates the illusion that each observer lies at the center of the expansion.

Hubble's law was first described by Belgian astronomer Georges Lemaître (1894–1966). Lemaître published his work about two years prior to that of Hubble. Because of this, various individuals have suggested that Hubble's law should be called Lemaître's law.

Hubble's law states that an object's recessional velocity is proportional to the object's distance from the observer. The equation is:

$$v = H_0 d$$

Where:

- v = velocity of the object, in km/s
- d = the distance to the object, in megaparsecs, Mpc, where 1 Mpc = 1 million parsecs (1 parsec = 3.26 light-years)
- H_0 = the Hubble constant, a constant of proportionality between d and v; also known as the rate of expansion, in kilometers per second per megaparsec or simply $\dfrac{\frac{k}{ms}}{Mpc}$

Hubble's law is applicable for objects whose spectra are redshifted (receding) observed at distances of approximately 10 megaparsecs or greater. These redshifted objects are evidence of a spatial expansion of the observable universe, and thus are part of the evidence of the Big Bang.

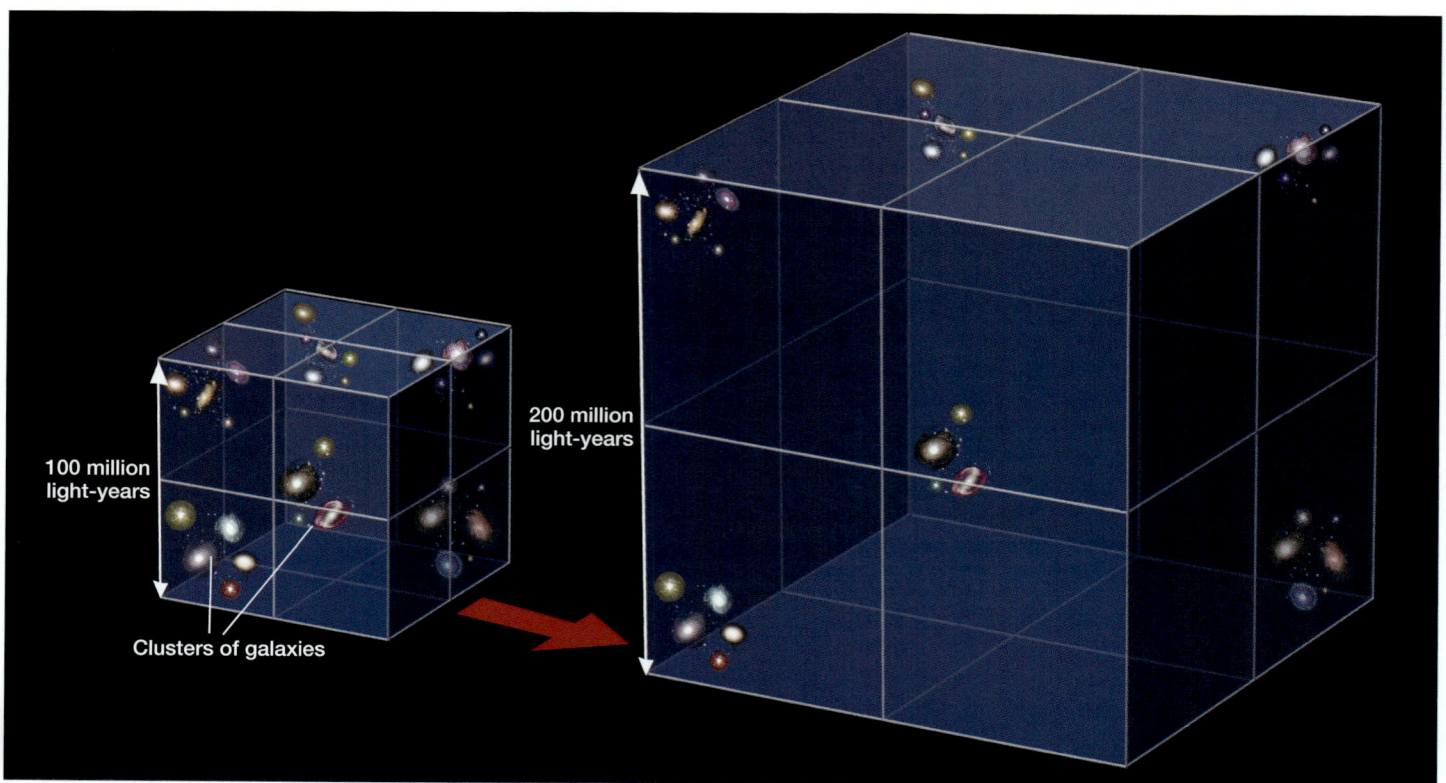

23.1 Expansion of six clusters of galaxies.

Courtesy of Roen Kelly, Astronomy magazine

Hubble's law infers that an object moving away from the observer twice as fast as another object is two times farther away, and an object that is moving away five times as fast is five times farther away.

Many often think that distant galaxies are moving away from each other. In reality, space is moving and the galaxies are being taken along for the ride.

Rearranging Hubble's law will provide you with the distances to deep space objects:

$$d = \frac{v}{H_0}$$

There has been, and still is, much debate over the value of H_0. Not only is the Hubble constant important to Hubble's law, it also gives the age of the universe. The age of the universe can be derived from Hubble's law through rearranging the equation:

$$v = H_0 d$$

$$\frac{v}{d} = H_0$$

Since $v = \frac{d}{t}$, substituting for v:

$$\frac{\frac{d}{t}}{\frac{d}{1}} = H_0$$

$$\frac{1}{t} = H_0$$

$$t = \frac{1}{H_0}$$

In recent years, variations in the Hubble constant derived by astronomers were in the range of 60 km/s/Mpc to 80 km/s/Mpc, which led to a range in the age of the universe of 12 billion years to 17 billion years.

Data collected by NASA's Wilkinson Microwave Anisotropy Probe allowed researchers to make fundamental measurements related to cosmology. They determined the Hubble constant was 69.32 ±0.80 km/s/Mpc. This also led cosmologists to determine the age of the observable universe to be 13.82 billion years. For an example: Assume Hubble's constant equals 72 km/s/Mpc

1 megaparsec = 30.86×10^{18} kilometers

H_0 = (72 km/sec)/(30.86×10^{18} km) = 2.33×10^{-18}/sec

t (the age of the universe) = $1/H_0$ = 4.29×10^{17} seconds

1 year = 31,557,000 (3.156×10^7) seconds

$4.29 \times 10^{17}/(3.156 \times 10^7)$ = 13.6×10^9 years or 13.6 billion years

Check Your Understanding

1.1 As stated, Hubble's law can be used to determine the age of the observable universe, knowing the value of Hubble's constant, H_0. Calculate the age of the universe using the following values of H_0.

a $50 \ \dfrac{\frac{k}{ms}}{Mpc}$

b $60 \ \dfrac{\frac{k}{ms}}{Mpc}$

c $75 \ \dfrac{\frac{k}{ms}}{Mpc}$

d $80 \ \dfrac{\frac{k}{ms}}{Mpc}$

e $100 \ \dfrac{\frac{k}{ms}}{Mpc}$

1.2 What is the range of values for the age of the universe you calculated?

1.3 Why could such a variation in range be an issue for astronomers, cosmologists, and scientists?

The Expanding Universe

Exercise 23.1

Procedure

Materials
- Images of galaxies (Figs. 23.2 and 23.3)

1. To simulate the expansion of part of the universe, you will measure distances between members of a group of galaxies during two times.

2. Measure the distances between G1 and the other 11 galaxies in Figure 23.2, which denotes Time 1. Then measure the distances between G1 and the other 11 galaxies in Figure 23.3, which denotes Time 2.

3. Note your measurements in Table 23.1.

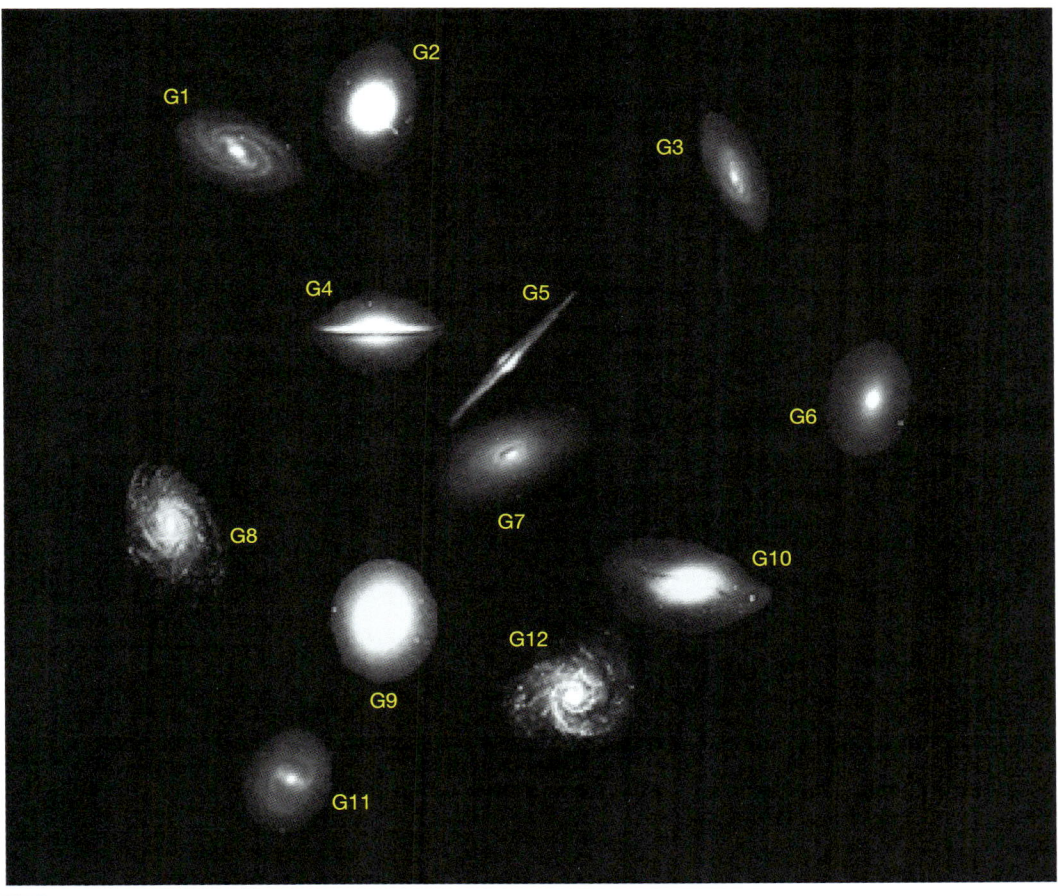

23.2 Twelve galaxies at Time 1.

Courtesy of Holley Y. Bakich

Hubble's Law CHAPTER 23 429

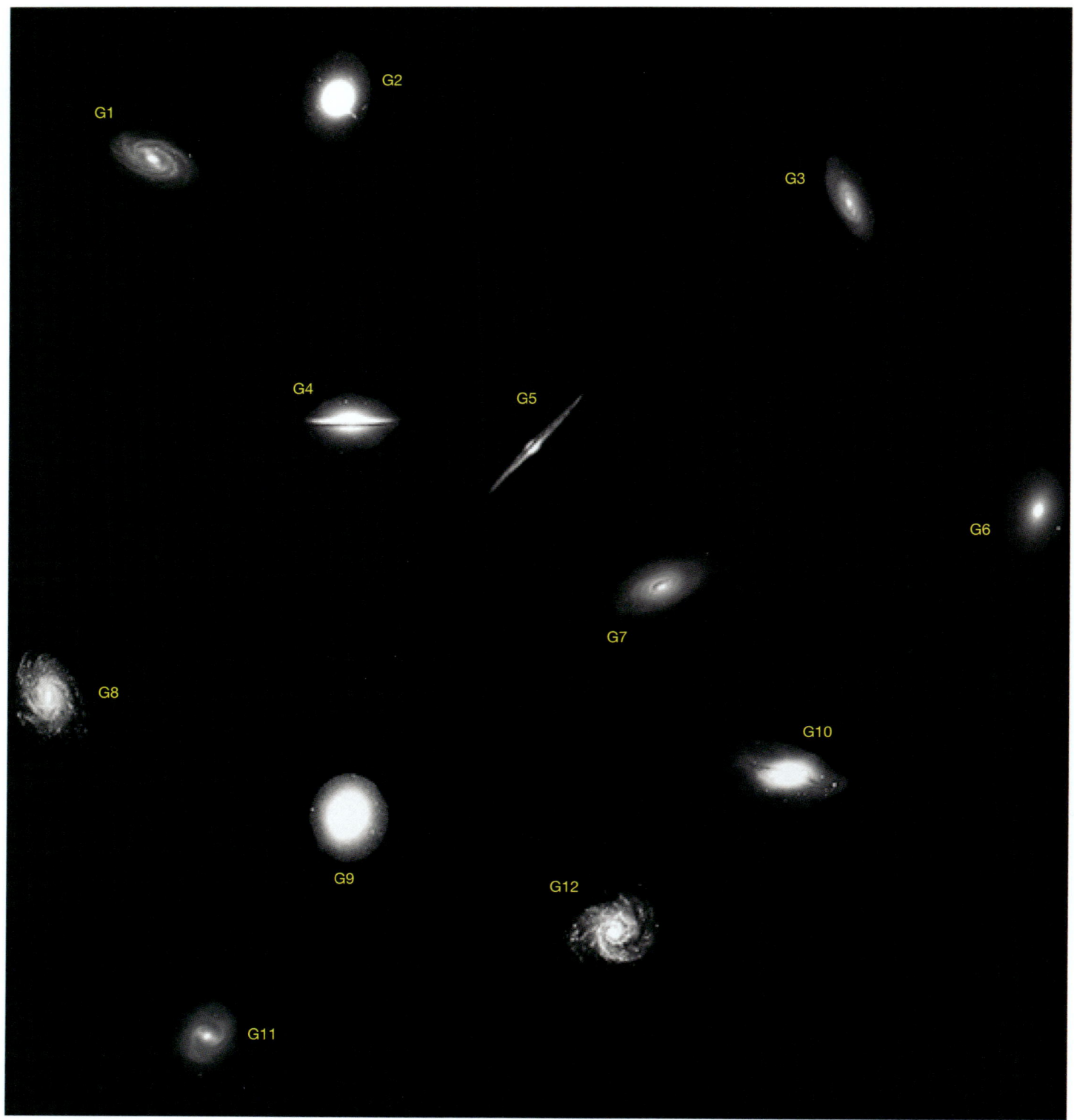

23.3 Twelve galaxies at Time 2.

Courtesy of Holley Y. Bakich

4 Measure the distances between G7 and the other 11 galaxies in Figure 23.2, which denotes Time 1. Then measure the distances between G7 and the other 11 galaxies in Figure 23.3, which denotes Time 2.

5 Note your measurements in Table 23.2.

Unit 6 Classifying Celestial Objects

TABLE **23.1** Constant Rate of the Expanding Universe from Galaxy G1's View

Time	Distance between G1 and galaxy _____, in mm					
	G1	G2	G3	G4	G5	G6
1	0					
2	0					
	G7	G8	G9	G10	G11	G12
1						
2						

TABLE **23.2** Constant Rate of the Expanding Universe from Galaxy G7's View

Time	Distance between G7 and galaxy _____, in mm					
	G1	G2	G3	G4	G5	G6
1						
2						
	G7	G8	G9	G10	G11	G12
1	0					
2	0					

6 Answer the following questions:

a These model galaxies show an expanding universe. Is this universe expanding at a constant rate? Why or why not?

b Hubble's law states the farther a galaxy is away from the observer, the faster it appears to be moving away from the observer. Do your measurements seem to confirm Hubble's law? Why or why not?

c Since all galaxies appear to be moving away from G1, does this infer that G1 is at the center of expansion? What about G7?

d Looking at the G1 versus G7 distances to other galaxies, do observers in each galaxy have the right to state that their galaxy is in the center of the expansion?

e Can any galaxy claim to be in the center of the expansion, and therefore the center of the universe?

Exercise 23.2 | Determining Galactic Distances

By using images of galaxies and their spectra, you can determine the distance and recessional velocity (how fast the galaxy is moving away from us).

In this exercise, you will use negative images of photographs taken with large telescopes. Astronomers often use negatives because details are easier to see.

In addition, we have included each galaxy's spectrum (the set of spiky lines). The strong hydrogen emission line called the hydrogen-alpha line is the one we are interested in. You will find that line at a rest wavelength (that is, the wavelength measured in a laboratory) of 6,563 Angstroms (Å). By examining the shift of this line to longer wavelengths (a redshift), you will be able to determine the object's recessional velocity.

Procedure

Materials
- ❏ Images of giant elliptical galaxies (Figs. 23.4–23.13)

1 Before beginning, answer the following questions:

  a If you were to see a galaxy's spectra shifted toward shorter wavelengths, what would this imply to you as the observer?

  b Why would you think a negative is better to examine than a positive?

  c The images of the galaxies you will study will be different sizes, even at the same magnification. What does that indicate? Why is this significant to Hubble's law?

2 You will use Figures 23.4–23.13 of 10 giant elliptical galaxies and their spectra to complete Table 23.3 in Part I, Table 23.4 in Part II, and Table 23.5 in Part III.

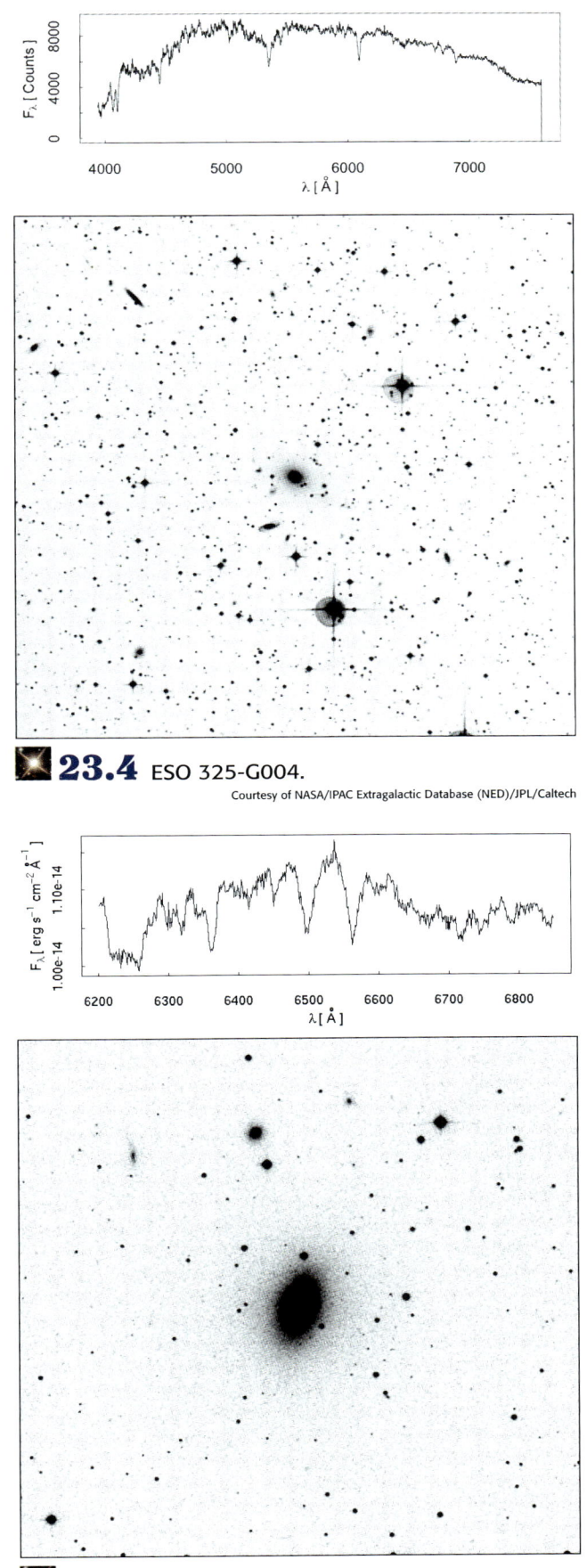

23.4 ESO 325-G004. Courtesy of NASA/IPAC Extragalactic Database (NED)/JPL/Caltech

23.6 M59. Courtesy of NASA/IPAC Extragalactic Database (NED)/JPL/Caltech

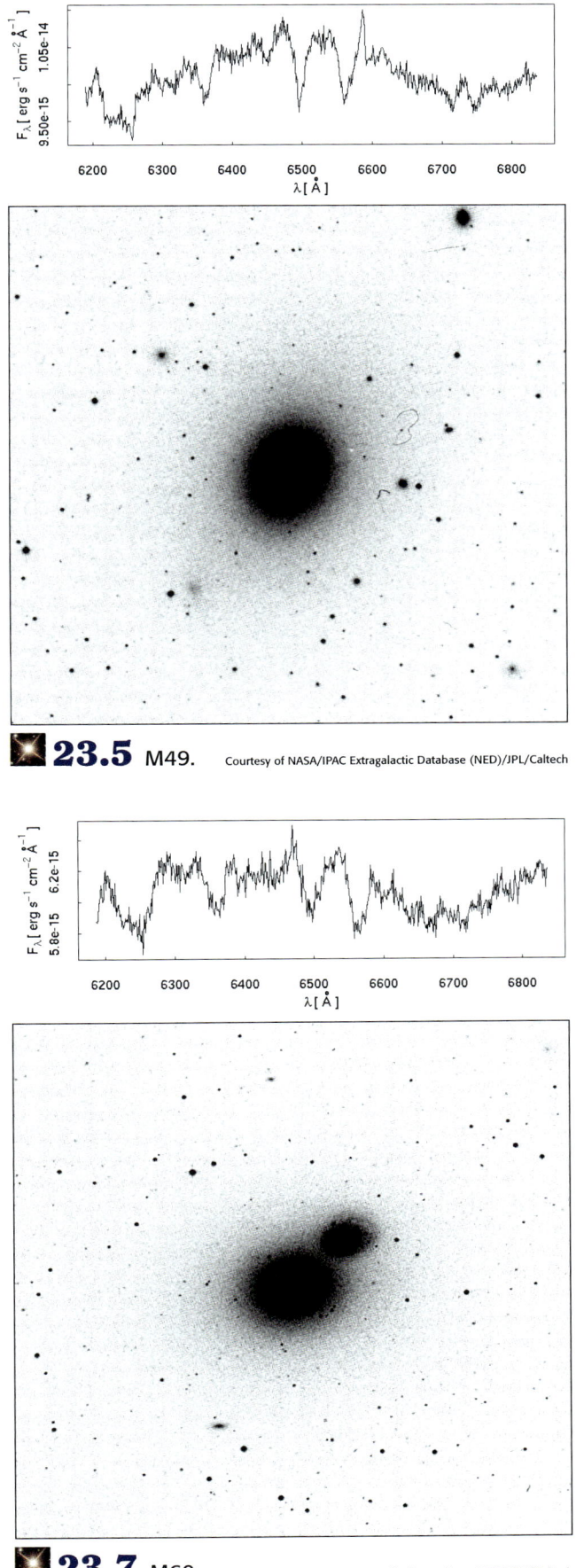

23.5 M49. Courtesy of NASA/IPAC Extragalactic Database (NED)/JPL/Caltech

23.7 M60. Courtesy of NASA/IPAC Extragalactic Database (NED)/JPL/Caltech

Hubble's Law **CHAPTER 23**   433

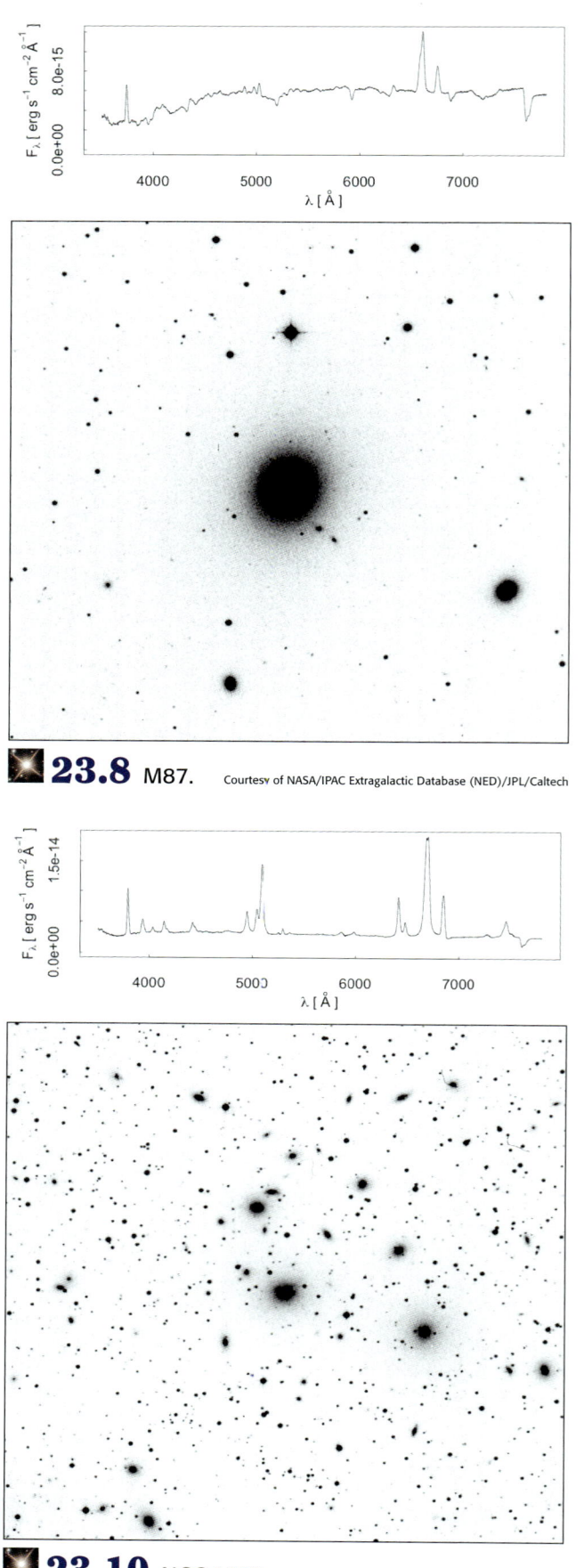

23.8 M87. Courtesy of NASA/IPAC Extragalactic Database (NED)/JPL/Caltech

23.10 NGC 1275. Courtesy of NASA/IPAC Extragalactic Database (NED)/JPL/Caltech

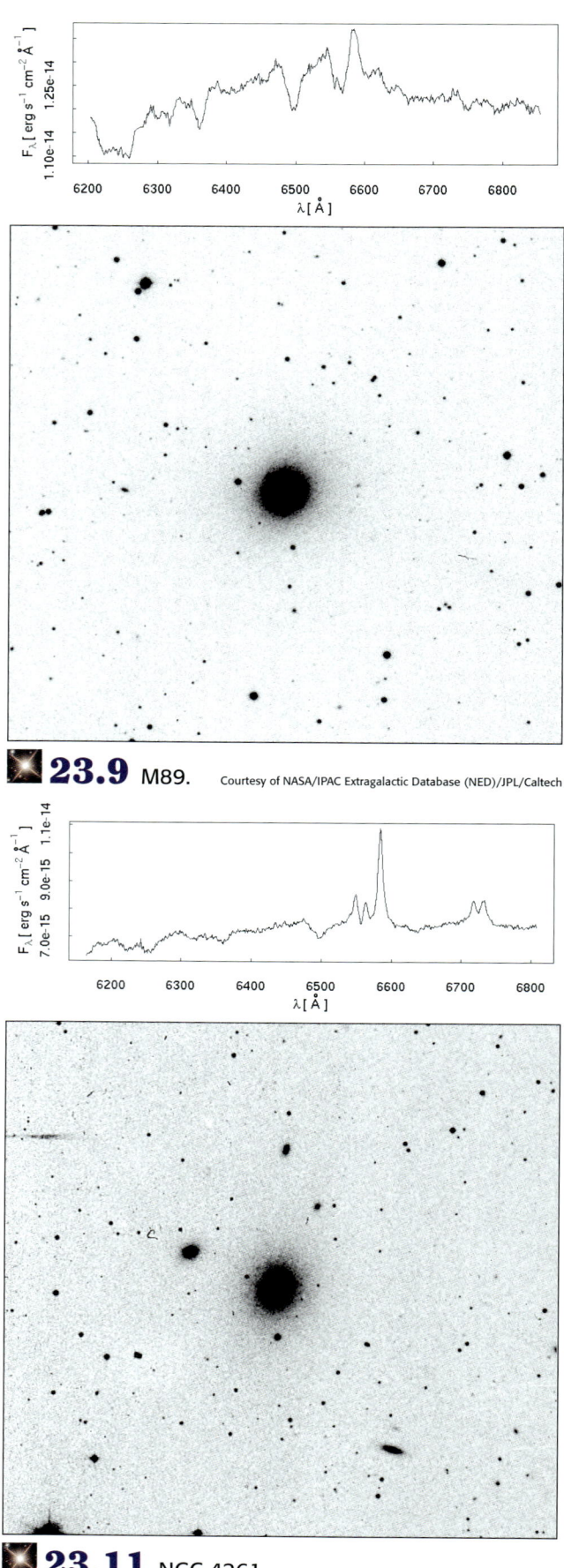

23.9 M89. Courtesy of NASA/IPAC Extragalactic Database (NED)/JPL/Caltech

23.11 NGC 4261. Courtesy of NASA/IPAC Extragalactic Database (NED)/JPL/Caltech

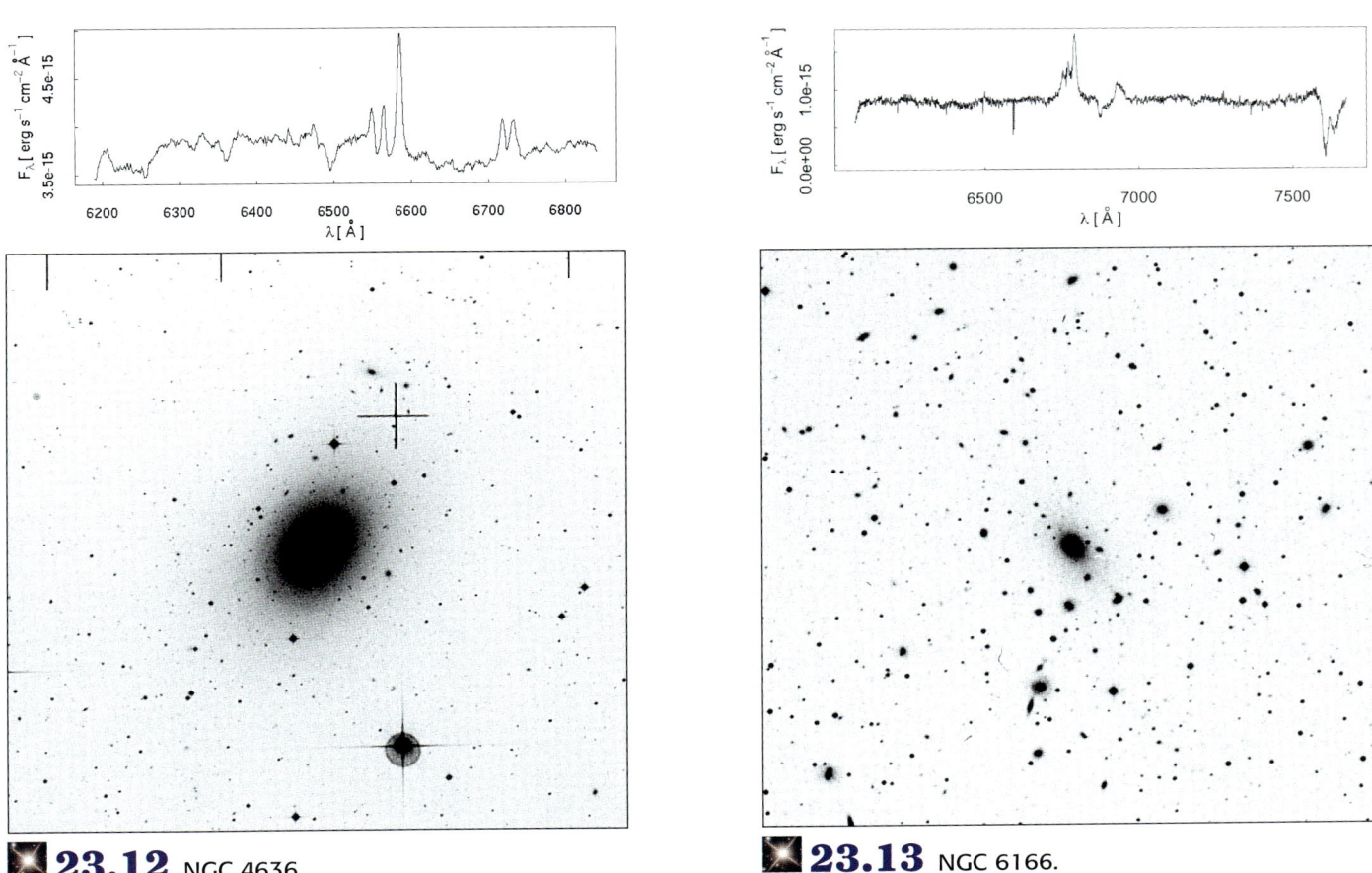

23.12 NGC 4636.
Courtesy of NASA/IPAC Extragalactic Database (NED)/JPL/Caltech

23.13 NGC 6166.
Courtesy of NASA/IPAC Extragalactic Database (NED)/JPL/Caltech

Part I: General Galactic Data

3 In column 1 of Table 23.3, identify the galaxy.

4 In column 2, note the diameter, in mm. To do this, carefully measure across the widest part of the galaxy in the image, looking for the edge of the galaxy's fuzzy disk. This is easier to see in negatives. Repeat your measurement twice, taking the average of the three measurements. Make your measurements accurate to at least 0.5 mm.

5 In column 3, note the hydrogen-alpha line wavelength from each spectrum. To do this, determine the position of the hydrogen-alpha line wavelength in Å. Repeat your measurement twice, taking the average of the three measurements.

6 In column 4, note the spectral shift in Å. This is the average of your measured hydrogen-alpha line minus 6563 Å.

TABLE 23.3 Spectral Shift Data

Galaxy	IMAGE: Diameter, in mm (measure three times)	SPECTRA: Hydrogen-alpha line wavelength (measure three times)	Spectral shift, in Å
	1. _____ mm 2. _____ mm 3. _____ mm Mean: _____ mm	1. _____ Å 2. _____ Å 3. _____ Å Mean: _____ Å	_____ Å
	1. _____ mm 2. _____ mm 3. _____ mm Mean: _____ mm	1. _____ Å 2. _____ Å 3. _____ Å Mean: _____ Å	_____ Å
	1. _____ mm 2. _____ mm 3. _____ mm Mean: _____ mm	1. _____ Å 2. _____ Å 3. _____ Å Mean: _____ Å	_____ Å
	1. _____ mm 2. _____ mm 3. _____ mm Mean: _____ mm	1. _____ Å 2. _____ Å 3. _____ Å Mean: _____ Å	_____ Å
	1. _____ mm 2. _____ mm 3. _____ mm Mean: _____ mm	1. _____ Å 2. _____ Å 3. _____ Å Mean: _____ Å	_____ Å
	1. _____ mm 2. _____ mm 3. _____ mm Mean: _____ mm	1. _____ Å 2. _____ Å 3. _____ Å Mean: _____ Å	_____ Å

(continues)

TABLE 23.3 Spectral Shift Data (continued)

Galaxy	IMAGE: Diameter, in mm (measure three times)	SPECTRA: Hydrogen-alpha line wavelength (measure three times)	Spectral shift, in Å
	1. _____ mm 2. _____ mm 3. _____ mm Mean: _____ mm	1. _____ Å 2. _____ Å 3. _____ Å Mean: _____ Å	_____ Å
	1. _____ mm 2. _____ mm 3. _____ mm Mean: _____ mm	1. _____ Å 2. _____ Å 3. _____ Å Mean: _____ Å	_____ Å
	1. _____ mm 2. _____ mm 3. _____ mm Mean: _____ mm	1. _____ Å 2. _____ Å 3. _____ Å Mean: _____ Å	_____ Å
	1. _____ mm 2. _____ mm 3. _____ mm Mean: _____ mm	1. _____ Å 2. _____ Å 3. _____ Å Mean: _____ Å	_____ Å

Part II: Determining the Galactic Distances, *d*

At first, it might seem relatively easy to determine the distances to galaxies. This is true for nearby galaxies, where astronomers observe objects within a galaxy, such as type I supernovae or Cepheid variable stars. Astronomers must use indirect methods for distant galaxies, however, and they also have to make some assumptions. In this lab, we will measure similar Hubble-classification galaxies, such as barred spirals (Sb) or giant ellipticals. These galaxies are of similar actual size regardless of their distance from the observer. For example we know that the size of giant elliptical galaxies is around 32 kiloparsecs (Kpc) or 0.032 Megaparsecs (Mpc). Astronomers call a presumption like this a standard ruler.

As an example, consider the Andromeda Galaxy (M31). We know its distance because of observations of Cepheids within it. We can then use these data to assume the distances to other, similar galaxies. Just remember, when you make such measurements from images, always use the same image scale, such as shown in Figure 23.14.

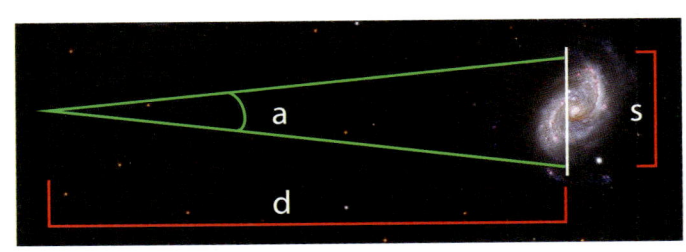

23.14 Angular size versus distance scale.

Hubble's Law **CHAPTER 23** 437

In this part, you will use a simple approximation to determine distances to the galaxies shown in Figures 23.4–23.13:

$$a = \frac{s}{d} \qquad d = \frac{s}{a}$$

Where:

- a = apparent angular size, measured in radians. 1 millirad = 0.057° = 206 arcseconds

  a in arcseconds = Diameter (in mm) times the scale of the image ($\frac{arcseconds}{mm}$)

- s = the galaxy's true size, or diameter
  For all galaxies here, $s = 206{,}265 \times 0.032 = 6{,}600.48$ (round off to 6,600)
- d = distance to the galaxy
  d in megaparsecs (Mpc) for giant elliptical galaxies = $(\frac{206{,}265 \times 0.032}{a})$

7 In column 1 of Table 23.4, identify the galaxy.

8 In column 2, copy your mean diameter data from Table 23.3.

9 In column 3, determine d by dividing your measured value of a into s.

10 In column 4, note your calculation of d in megaparsecs.

TABLE **23.4** Galaxy Distance Data

Galaxy	IMAGE Measured diameter in mm	$d = \frac{s}{a}$	Galaxy's distance, d, in megaparsecs
	Mean diameter: _____ mm	____/____	
	Mean diameter: _____ mm	____/____	
	Mean diameter: _____ mm	____/____	
	Mean diameter: _____ mm	____/____	
	Mean diameter: _____ mm	____/____	
	Mean diameter: _____ mm	____/____	
	Mean diameter: _____ mm	____/____	
	Mean diameter: _____ mm	____/____	
	Mean diameter: _____ mm	____/____	
	Mean diameter: _____ mm	____/____	

Part III: Determining Galactic Velocities, v

The receding velocity of each galaxy is simple to determine through use of the Doppler effect, using the equation:

$$\frac{(\lambda - \lambda_0)}{\lambda_0} = \frac{v}{c}$$

λ measured wavelength, in Å
λ_0 rest wavelength, in Å
v object's speed
c speed of light; $3 \times 10^5 \frac{km}{s}$
$\frac{(\lambda - \lambda_0)}{\lambda_0}$ is referred to as the redshift and is denoted by z

The equation can be rearranged to determine the object's velocity v:

$$v = cz$$
$$v = (3 \times 10^5 \frac{km}{s})z$$

Example: From a galaxy's spectra, you determine the hydrogen-alpha line is at 6625 Å. To determine the galaxy's velocity, *v*:

$$v = cz$$

$$z = \frac{(\lambda - \lambda_0)}{\lambda_0} = \frac{(6625 \text{ Å} - 6562.8 \text{ Å})}{6625 \text{ Å}} = 0.0093887$$

$$v = (3 \times 10^5 \frac{km}{s})(0.0093887) = 2861.6 \frac{km}{s} \text{ (receding)}$$

11 In column 1 of Table 23.5, identify the galaxy.
12 In column 2, copy your spectral shift data from Table 23.3.
13 In column 3, determine **z** by using the equation above.
14 In column 4, determine **v** by using the equation above.

TABLE 23.5 Galactic Velocity Data

Galaxy	Spectral shift, in Å	$z = (\lambda - \lambda_0)/\lambda_0$	$v = cz$
	Å		km/s
	Å		km/s
	Å		km/s
	Å		km/s
	Å		km/s
	Å		km/s
	Å		km/s
	Å		km/s
	Å		km/s
	Å		km/s

Appendix

Locations of Major Telescopes

Courtesy of Roen Kelly, *Astronomy* magazine

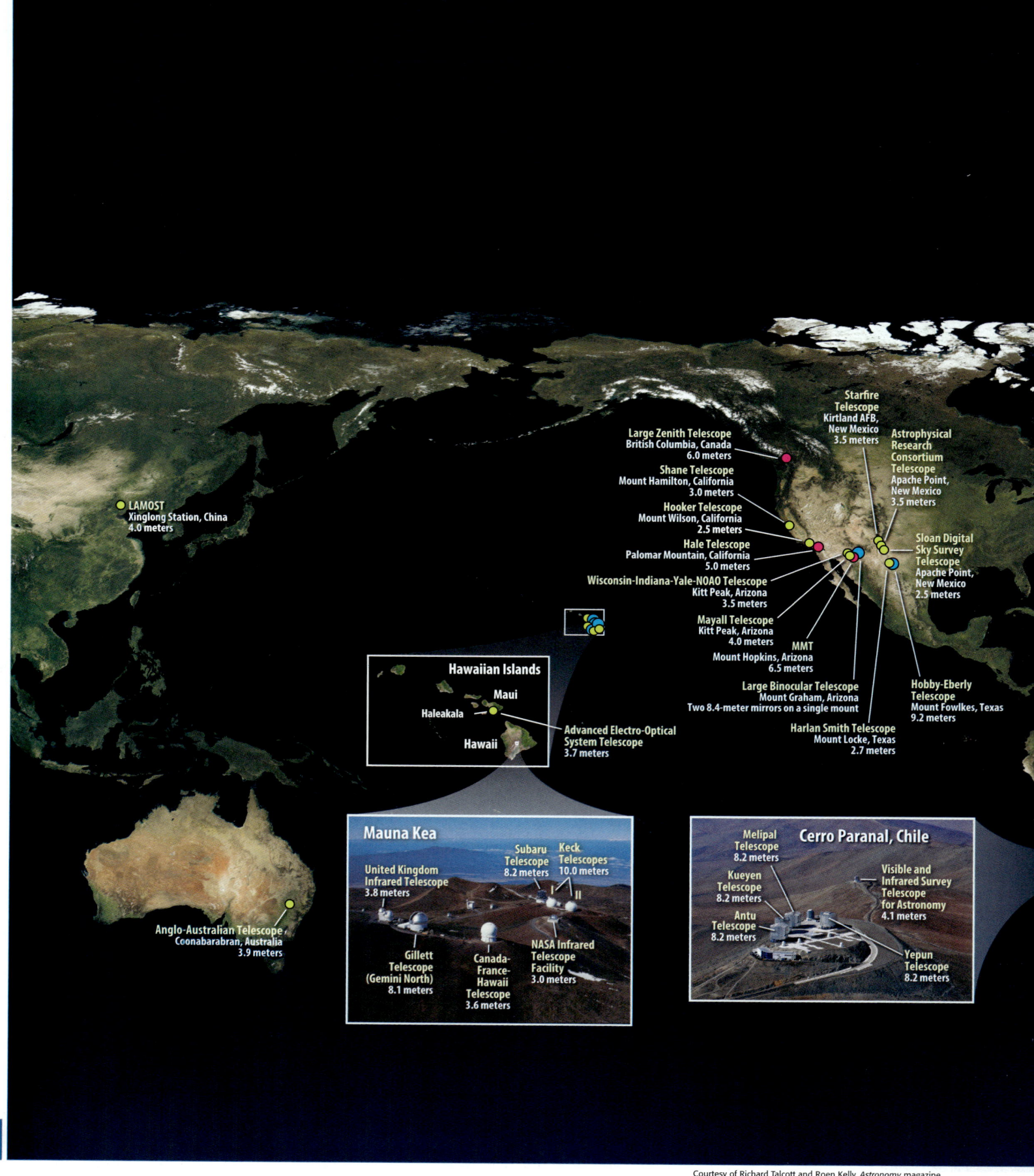

Appendix *Locations of Major Telescopes*

Courtesy of Richard Talcott and Roen Kelly, *Astronomy* magazine

Caldera de Taburiente

- Isaac Newton Telescope — 2.5 meters
- Nordic Optical Telescope — 2.5 meters
- William Herschel Telescope — 4.2 meters
- Telescopio Nazionale Galileo — 3.6 meters
- Gran Telescopio Canarias — 10.4 meters

Canary Islands

- La Palma

MPIA-CAHA Telescope — Calar Alto, Spain — 3.5 meters

Shajn Telescope — Nauchny, Ukraine — 2.6 meters

Bolshoi Azimuthal Telescope — Mount Pastukhov, Russia — 6.0 meters

Byurakan Astrophysical Observatory Telescope — Mount Aragatz, Armenia — 2.6 meters

Chile

- du Pont Telescope — 2.5 meters
- Walter Baade Telescope (Magellan I) — 6.5 meters
- Landon Clay Telescope (Magellan II) — 6.5 meters
- **Cerro Manqui**
- New Technology Telescope — 3.6 meters
- ESO 3.6-meter Telescope — 3.6 meters
- **Cerro La Silla**
- Victor Blanco Telescope — 4.0 meters
- Southern Astrophysical Research Telescope — 4.1 meters
- Gemini South Telescope — 8.1 meters
- **Cerro Tololo**
- **Cerro Pachón**

South African Large Telescope — Sutherland, South Africa — 11 meters

Locations of Major Telescopes **Appendix** **A** 443

Photo Credits

Photos are courtesy of Mike D. Reynolds or Michael E. Bakich unless noted here.

Abrams Planetarium, Michigan State University: Fig. **5.9**, p. 72; Fig. **5.10**, p. 73

Anthony Ayiomamitis: Fig. **7.2A, B**, p. 118; Fig. **7.4**, p. 122; Fig. **7.13**; p. 138; Fig. **7.15**, p. 140; Fig. **7.17**, p. 144; Fig. **7.18**, p. 145; Fig. **10.2**, p. 184; Fig. **10.31**, p. 191; Fig. **10.34**, p. 192; Fig. **11.1**, p. 212; Fig. **17.10**, p. 330; Fig. **17.12**, p. 332; Fig. **17.19**, p. 336

Association of Lunar and Planetary Observers: Table 11.4, p. 227–228

Astronomy magazine/Kalmbach Publishing: Fig. **5.11**, p. 78–79; Fig. **5.12**, p. 80–81; Fig. **5.13**, p. 82–83; Fig. **11.7**, p. 215; Fig. **11.9**, p. 216

Astronomy magazine: Roen Kelly: Fig. **2.5**–**2.6**, p. 20; Fig. **4.3**, p. 45; Fig. **4.9**, p. 49; Fig. **4.13**, p. 51; Fig. **4.14**, p. 52; Fig. **4.16A**, p. 52; Fig. **4.17**, p. 53; Fig. **4.20**, p. 55; Fig. **9.6A–C**, p. 175; Fig. **9.7**, p. 176; Fig. **9.8**, p. 179; Fig. **11.2**, p. 213; Fig. **14.17A, B**, p. 273; Fig. **15.3**, p. 286; Fig. **15.5A, B**, p. 287; Fig. **15.11**, p. 292; Fig. **15.12**–**15.13**, p. 293; Fig. **18.13B**, p. 350; Fig. **18.16**, p. 354; Fig. **19.12**, p. 376; Fig. **19.18**, p. 381; Fig. **20.3**, p. 391; Fig. **20.4**, p. 393; Fig. **20.7**, p. 395; Fig. **21.1**, p. 407; Fig. **21.2**–**21.3**, p. 409; Fig. **21.4**, p. 410; Fig. **23.1**, p. 426; **Appendix**, p. 441

Astronomy magazine: Roen Kelly after Stephen G. Cullen: Fig **15.16**–**15.17**, p. 295

Astronomy magazine: Roen Kelly and **Alison Mackey:** Fig. **20.5**, p. 393; Fig. **20.11**, p. 398; Fig. **20.12A–C**, p. 398

Astronomy magazine: Richard Talcott and **Roen Kelly:** Fig. **5.6**, p. 68; Fig. **5.8**, p. 71; Fig. **p. 93**; Fig. **p. 99**; Fig. **p. 105**; Fig. **p. 111**; Fig. **7.1**, p. 116; Fig. **7.3**, p. 121; Fig **7.5**, p. 125; Fig. **7.7**, p. 127; Fig. **7.9**, p. 131; Fig. **7.12**, p. 137; Fig. **7.14**, p. 139; Fig. **7.16**, p. 143; Fig. **18.7B**, p. 347; Fig. **18.10B**, p. 349; Fig. **18.13B**, p. 350; Fig. **18.14B**, p. 351; Fig. **18.15B**, p. 351; Fig. **18.17**–**18.18**, p. 355; Fig. **18.19**, p. 356; Fig. **18.20**–**18.21**, p. 357; Fig. **18.22**–**18.23**, p. 358; **Appendix**, p. 442–443

Holley Y. Bakich: Fig. **1.5**, p. 6; Fig. **2.1**, p. 16; Fig. **2.2**, p. 17; Fig. **2.3A, B**, p. 17; Fig. **2.4**, p. 19; Fig. **2.7**, p. 24; Fig. **3.1**, p. 32; Fig. **4.4**–**4.5**, p. 46; Fig. **4.7**, p. 47; Fig. **4.15**, p. 52; Fig. **4.18**, p. 53; Fig. **5.1**–**5.2**, p. 65; Fig. **5.3**–**5.4**, p. 66; Fig. **5.5**, p. 67; Fig. **10.1**, p. 184; Fig. **11.10A, B**, p. 218; Fig. **11.11**, p. 221; Fig. **12.1**, p. 232; Fig. **12.2**, p. 233; Fig. **13.3**, p. 256; Fig. **13.4**, p. 257; Fig. **13.5**, p. 258; Fig. **13.6**, p. 260; Fig. **14.3A, B**, p. 265; Fig. **14.6**, p. 267; Fig. **14.8**, p. 268; Fig. **16.8**, p. 322; Fig. **17.5**, p. 326; Fig. **17.13**, p. 332; Fig. **17.20**, p. 338; Fig. **18.1**, p. 344; Fig. **19.1**, p. 372; Fig. **22.1**, p. 416; Fig. **22.3**, p. 420; Fig. **23.2**, p. 429; Fig. **23.3**, p. 230

Big Bear Solar Observatory/New Jersey Institute of Technology: Fig. **11.5B**, p. 215

Adam Block/Mount Lemmon SkyCenter/University of Arizona: Fig. **7.6**, p. 126; Fig **7.8**, p. 128; Chapter **19 Opener**, p. 371; Fig. **19.3**–**19.6**, p. 373; Fig. **19.7**–**19.8**, p. 374; Fig. **19.9**–**19.11**, p. 375; **Image**, p. 377; Fig. **19.13**–**19.14**, p. 379; Fig. **19.15**–**19.16**, p. 380; Fig. **19.17**, p. 381; Fig. **19.19**–**19.24**, p. 383; Fig. **19.25**–**19.33**, p. 384; Fig. **19.34**–**19.39**, p. 387; Fig. **19.40**–**19.48**, p. 388; Chapter **21 Opener**, p. 405

Celestron: Fig. **3.2**–**3.3**, p.33; Fig. **3.4**, p. 34; Fig. **3.6A**, p. 38; Fig. **4.2**, p. 44; Fig. **4.8**, p. 48; Fig. **4.12**, p. 51; Fig. **4.16B**, p. 52; Fig. **4.19**, p. 55; Fig. **4.21**–**4.23**, p. 56; Fig. **4.24**, p. 57

John Chumack: Chapter **10 Opener**, p. 183; Fig. **10.3**–**10.6**, p. 185; Fig. **10.39**, p. 196; Fig. **17.3**, p. 325; Fig. **17.11**, p. 330; Fig. **17.15**, p. 334

John Chumack and **Holley Y. Bakich:** Fig. **17.7**, p. 327

Allan Cook/Adam Block/NOAO/AURA/NSF: Fig. **7.11**; p. 133

John Crawley: Chapter **1 Opener**, p. 1

David J. Eicher: Fig. **8.4**–**8.5**, p. 153; Fig. **8.6**–**8.7**, p. 154

David J. Eicher/Astronomy magazine: Fig. p. 254

©ESA, MPAe, Lindau: Fig. **15.2, p. 286**

ESA 2010 MPS for OSIRIS Team/MPS/UPD/LAM/IAA/RSSD/INTA/UPM/DASP/IDA: Fig. **15.15**, p. 294

ESO: Fig. **4.11C**, p. 50

Bill and **Sally Fletcher:** Fig. **5.7**, p. 70; Fig. **9.2**, p. 170

Giant Magellan Telescope—GMTO Corporation: Fig. **4.11A**, p. 50

Good-Lite Co.: Fig. **1.11A–C**, p. 13

Vikki Granger, Insight 360 Media: Chapter **11 Opener**, p. 211

445

P. James (University of Toledo), T. Clancy (Space Science Institute), S. Lee (University of Colorado), NASA: Fig. **14.11**, p. 269

Phillip Jones: Fig. **17.1**, p. 324

Larry Landolfi/Science Source: Chapter **5 Opener**, p. 63

Laurent Laveder/Science Source: Chapter **9 Opener**, p. 167

Steve Lee (University of Colorado), Jim Bell (Cornell University), Mike Wolff (Space Science Institute), NASA: Fig. **14.10**, p. 269

Lowell Observatory: Fig. **15.18**, p. 295

Craig Mayhew and **Robert Simmon, NASA GSFC:** Fig. **9.4**, p. 173

NASA: Chapter **4 Opener**, p. 43; Chapter **12 Opener**, p. 231; Fig. **12.3–12.8**, p. 239; Fig. **12.9–12.15**, p. 240; Fig. **12.16–12.17**, p. 242; Fig. **12.18–12.19**, p. 243; Fig. **12.20–12.27**, p. 244; Fig. **12.28–12.29**, p. 245; Fig. **12.30–12.34**, p. 246; Chapter **14 Opener**, p. 263; Fig. **14.2**, p. 265; Fig. **14.4**, p. 266; Fig. **14.7**, p. 268; Fig. **14.13**, p. 270; Fig. **14.16**, p. 273; Fig. **15.19–15.23**, p. 301; Fig. **17.4**, p. 326

NASA/ESA/Hubble Heritage Team (STScI/AURA): Chapter **3 Opener**, p. 31; Chapter **22 Opener**, p. 415

NASA/ESA/Hubble Heritage Team (STScI/AURA)–ESA/Hubble Collaboration: Chapter **8 Opener**, p. 149; Fig. **19.2**, p. 372

NASA/ESA/Hubble Heritage Team (STScI/AURA)/ R. Gendler (for the Hubble Heritage Team): Chapter **23 Opener**, p. 425

NASA/ESA/J. Hester/A. Loll: Chapter **7 Opener**, p. 115

NASA/ESA/R. O'Connell (University of Virginia)/ F. Paresce (National Institute for Astrophysics, Bologna, Italy)/E. Young (Universities Space Research Association/Ames Research Center)/the WFC3 Science Oversight Committee, and the Hubble Heritage Team (STScI/AURA): Chapter **18 Opener**, p. 343

NASA/ESA/SAO/CXC/JPL-Caltech/STScI: Chapter **6 Opener**, p. 89

NASA/IPAC Extragalactic Database (NED)/JPL/Caltech: Fig. **23.4–23.7**, p. 433; Fig. **23.8–23.11**, p. 434; Fig. **23.12–23.13**, p. 435

NASA/JPL: Fig. **14.14 A–D**, p. 271; Fig. **14.19–14.20**, p. 275

NASA/JPL-Caltech: Fig. p. 403, **top left**

NASA/JPL-Caltech/O. Krause (Steward Observatory): Chapter **13 Opener**, p. 253; Fig. **13.2**, p. 255

NASA/JPL-Caltech, N. Madhusudhan (Princeton University): Fig. **20.8**, p. 396

NASA, JPL-Caltech, UMD: Fig. **15.4A, B**, p. 287

NASA/SOHO: Fig. **17.2**, p. 325

NASA/Solar and Heliospheric Observatory: Fig. **11.4**, p. 214

NASA/Solar Dynamics Observatory: Chapter **20 Opener**, p. 389

NASA/STScI: Fig. **1.9A, B**, p. 11

NEAR Project, NLR, NASA/JHUAPL, Goddard SVS: Fig. **15.10**, p. 292

NOAO/AURA/NSF: Fig. **20.6**, p. 394

Palomar Observatory/California Institute of Technology: Fig. **4.10**, p. 49

Damian Peach: Fig. **10.9–10.13**, p. 187; Fig. **10.15–10.18**, p. 188; Fig. **10.21–10.22**, p. 189; Fig. **10.23–10.28**, p. 190; Fig. **10.30**, p. 191; Fig. **10.32–33**, p. 192; Fig. **14.9**, p. 268; Fig. **14.12**, p. 270; Fig. **14.15**, p. 271; Fig. **14.18**, p. 274; Fig. **15.1**, p. 286; Fig. **15.8C**, p. 289; Fig. **15.9**, p. 289; **Image** p. 290; Fig. **15.14**, p. 294

Damian Peach and **David Tyler:** Fig. **17.6**, p. 326

Jeremy Perez: Fig. **18.2**, p. 345; Fig. **18.3–18.5**, p. 346; Fig. **18.6**, p. 347; Fig. **18.7A**, p. 347; Fig. **18.8–18.9**, p. 348; Fig. **18.10A**, p. 349; Fig. **18.11–18.12**, p. 350; Fig. **18.13A**, p. 350; Fig. **18.14A**, p. 351; Fig. **18.15A**, p. 351

C. Pilchowski, M. Corbin, NOAO/AURA/NSF: Fig. **20.9**, p. 396

Rapho Agence/Science Source: Chapter **2 Opener**, p. 15

Gerard Rhemann: Fig. **15.7B**, p. 288

Erika Rix: Fig. **10.37–10.38**, p. 195

Chris Schur: Chapter **15 Opener**, p. 285; Fig. **15.6**, p. 287; Fig. **15.7A**, p. 288; Fig. **15.8A, B**, p. 289; Fig. **17.14A, B**, p. 333

George Seitz/Adam Block/NOAO/AURA/NSF: Fig. **7.10**, p. 132

N.A. Sharp/NOAO/NSO/Kitt Peak FTS/AURA/NSF: Fig. **20.2**, p. 390

Solar and HelioSpheric Observatory: Fig. **11.4**, p. 212

Babak Tafreshi/Science Source: Chapter **16 Opener**, p. 305

Craig and **Tammy Temple:** Fig. **11.3**, p. 213

TMT Observatory Corporation: Fig. **4.11B**, p. 50

David Tyler: Fig. **10.20**, p. 189; Fig. **11.12–11.19**, p. 229; Fig. **11.20–11.27**, p. 230

©UC Regents/Lick Observatory: Fig. **10.40 A–C**, p. 203–204

Wikimedia Commons: Fig. **13.1**, p. 254; Fig. **20.1**, p. 390

Yerkes Observatory, Richard Dreiser, Photographer: Fig. **4.6**, p. 46

The following images, which were previously published by Cambridge University Press, are reprinted with permission: Fig. **2.4**, p. 19; Fig. **3.1**, p. 32; **Image**, p. 37; Fig. **3.6A**, p. 38; Fig. **3.6B**, p. 38; Fig. **4.4**, p. 46; Fig. **4.5**, p. 46; Fig. **4.18**, p. 53; Fig. **5.1**, p. 65; Fig. **5.4**, p. 66; Fig. **8.1**, p. 150; Fig. **8.2**, p. 152; Fig. **9.3**, p. 173; Fig. **9.5**, p. 174; Fig. **14.1**, p. 264; Fig. **14.3A**, p. 265; Fig. **14.3B**, p. 265; Fig. **14.6**, p. 267; Fig. **17.5**, p. 326

Index

Abrams Planetarium *Sky Calendar* and star chart, 71, 72, 73, 74
absolute magnitude, 168-169, 406
afterimages, 4
albedo (Moon, Pluto, Mars), 236, 268
American Association of Variable Star Observers (AAVSO), 353
Angstrom, 216
aperture
    binocular, 33
    lens or mirror, 16
    Newtonian, 48
Apollonius, astronomer, 233
arcminutes, 66
arcseconds, 66, 169
Aristotle, quote, 353
association of young stars, 117
asterisms, 90, 122, 156–158
asteroid(s), 235, 240, 307
    Apollo and Aten, 293
    Ceres (Giuseppe Piazzi, discoverer), 292
    designations, 294
    Eros, 292
    Herschel (Caroline and Sir John), naming, 144, 292
    Kirkwood gap, 292
    Lagrangian points (Joseph Louis Lagrange), 293
    Main Belt, 292
    Minor Planet Center (MPC), 294
    Near-Earth (NEAs or NEOs), 292–293
    surveys, 295–296
    Trojan, 293
    types by chemical composition, 293
astigmatism/astigmatism chart, 11–12
astrology/astrological signs, 67
Astronomical League, 37, 175, 193
astronomical unit, Earth to Sun, 259
Astronomy with an Opera Glass (Garrett P. Serviss), 144
Atlas of the Stars, *Astronomy* magazine, 77
atlases. *See* star atlases
atmospheric extinction, 170–171
autumn sky, 91–92
    Andromeda Galaxy, 91
    Great Square, 91
    Milky Way, 91
    star chart, 93
    Summer Triangle, 91
averted vision, 5, 128
axis. *See* orbit

Barlow lens, 56
Big Dipper, 103–104, 125–128
binary star system/Capella, 121
binocular highlights
    January-February, 116–118
    July-August, 131–133
    March-April, 121–122
    May-June, 125–128
    September-October, 137–140
    November-December, 143–145
Binocular Messier Club certificate, 37
binoculars
    aperture, 33
    for comet viewing, 288, 289
    exit pupil, 35
    eyepieces, 34, 35
    field of view, 34–35
    focusers, 7, 34
    giant/Celestron, 33, 34, 38
    lenses, 33
    magnification, 33–34
    mounts, 37–38
    objective, 33
    selecting equipment, 37
Brahe, Tycho, 233, 254, 255
bright stars/bright planets, 116, 168, 178, 334. *See also* constellations
Bright Star Atlas (Wil Tirion), 74
brightness, 168–169. *See also* magnitude
    Algol, 369, 370
    light curve, 354
    Moon, 169
    night sky, 169
    variable star, 353

camera
    CCD, 353
    single-lens reflex, 3, 353
Carolina Biological® depth perception tester, 14
Cassini, Giovanni Domenico, Venus measurements, 266
Cassiopeia, 143–145
Catalina Sky Survey, 295, 296
celestial
    coordinates, 66–67
    equator, 65, 66, 67
    geometry (eclipses), 332
    luminosity, 406
    mechanics (Kepler), 254
    orbit, 254
    poles, 65, 327
    sphere, 64–65
Cerro Tololo Inter-American Observatory (CTIO), 169
chromatic aberration, 45, 46
circumpolar stars, 66
Citizen Science projects, 403
Clark, Alvin, telescope maker, 46
color blindness, 12–13
comet(s), 235, 240, 286–287, 307
    bright, 287–288
    coma, 286, 287, 289, 290
    degree of condensation (DC), 288, 289
    elliptical orbit, 286
    faint, 289–290
    fluorescence, 287
    Giotto spacecraft, 286
    Halley's comet, 286
    Hartley, 287, 289
    Holmes (Edwin), 287, 289
    magnitude, estimating, 289
    nucleus, 286
    outgassing, 290
    photographing, 290
    sketching, 153
    solar wind, 287
    tails, dust and ion, 287
condensed, definition, 122
constellations, 64, 67, 90, 116
    in autumn sky, 91–93
    Camelopardalis, 126
    Cassiopeia, 143–145
    Cigar Galaxy, 126
    Cygnus the Swan, 255
    Leo the Lion, 319, 322
    Lyra, 137, 153
    naming designations, 353
    Orion the Hunter, 78–79, 143, 179
    Perseus, 117, 118, 289, 381
    planetary nebulae, 380, 381
    Scorpius, 67, 70, 71, 109, 131, 132
    Scutum, 139, 140
    sketching those with light pollution, 180
    in spring sky, 103–105
    in summer sky, 109–110
    Ursa Major, 126
    in winter sky, 97–99
Copernicus, heliocentric model, 233
craters, Moon, 184, 186, 195, 203
    Eratosthenes, 195
    northeastern quadrant, 188

northwestern quadrant, 186–187
    ray, 186
    sketching, 194, 195
    southeastern quadrant, 191
    southwestern quadrant, 189–190
crystallography (Kepler), 254

de Vaucouleurs galaxy classification, 375–376, 377
declination, 66–67
deep-sky objects, 71, 116, 153, 379. See also double stars
depth perception, 14
double stars, 64, 117
    Albireo star chart, 350
    Algol, 369, 370
    autumn, 345–346
    Gamma star chart, 351
    Haas, Sissy, observer, 347
    Izar star chart, 347
    nomenclature (primary, companion, secondary, separation, and position angle), 344
    Omicron[1] (o^1) Cygni star chart, 351
    Pulcherrima (Wilhelm Struve), 347
    Smyth, William, obsrever, 348
    spring, 347–348
    star charts, 347, 349, 350, 351
    Steele, J. Dorman (author), 244
    summer, 348–352
    winter, 346–347

Earth, 239
    age determined by radioactive dating, 417
    in geocentric model, 232
    planetary motion (Kepler), 254–255
    radioactivity, 420
    rotational motion and revolution around the Sun, 212, 236
    transit of Sun disk, 327
    umbral shadow, 329, 330
earthshine, 185
Earth-Sun system, Lagrangian points, 293
eccentricity, orbit, 256
eclipses, 324. See also lunar eclipses; solar eclipses
    satellite, 271
    stellar, 369
ecliptic, 67, 265
    Moon, 328
    Venus' orbit, 325
electromagnetic spectrum, 390, 391
electromagnetic wave, 16
ellipses and eccentricity, 255, 256
Endeavour space shuttle, 11
equator, 267
exoplanets, 232
eye, human
    age of lenses/elasticity, 7
    anatomy, 2, 3
    astigmatism, 11–12
    blind spot, 5–6
    color vision/color blindness, 12–13
    cone, 2
    cornea, 2, 11
    depth perception, 14
    dominance, 8

focus/focal point, 7
iris, 2
lens, 2, 3, 11
peripheral vision/peripheral vision disk, 8, 9
photoreceptors, 2
pupil, 2
retina, 2, 4, 11
rod, 2, 8
in seeing and measuring the sky, 90
visual acuity tests, 10–11
eyeglasses/spectacles, origin, 16
eyepieces
    binocular, 35
    filters, 264
    telescope, 55

fall sky. See autumn sky
field of view, 34–35, 53, 254
filters
    color, 264, 265, 267
    solar, 215–216, 227
        hydrogen-alpha, 216, 327, 333, 335
finder scope, 56
focal length and equation, 19, 55
focal point/focus of lenses and mirrors, 18–19
focal ratio of mirror or lens, 45

galaxy, 64, 232, 430. See also Milky Way; solar system
    Andromeda, 437
    Bode's, 125, 153
    Catalogue of Galactic Planetry Nebulae (Lubos Perek and Lubos Kohoutek), 379
    definition, 125
    Cigar, 126
    clusters, and Hubble's law, 426–427
    Coma-Virgo cluster, 80-81
    de Vaucouleurs system, 375–377
    elliptical, 375
    Fireworks, 375
    Hubble's system, 375–377, 383–384
    northern sky, 377
    quasars, 396
    sketching, 152
    Spiral, 126, 375
    Whirlpool, 128
Galilean satellites (Jupiter), 271
Galileo, Galilei, 44, 168, 233, 254, 266
Geiger counter (Hans Geiger) and Geiger-Müller (G-M) tube, 417
geocentric model of solar system, 232
    Appollonius, 233
    crystalline sphere, 233
    epicycles, 233
    Hipparchus, 233
    Ptolemaic, 232–233
    retrograde motion, 232
geometry of similar triangles, 6
giant star, definition, 117
Giotto spacecraft, 286
globular cluster, 64, 131, 132, 153, 154
Great Bear. See Ursa Major
Gregory, James, telescope description, 48

Hale telescope at Palomar Observatory (George Ellery Hale), 49

Hall, Chester Moore, lens inventor, 46
Halley, Sir Edmond, and Halley's comet, 286, 327, 338
heliocentric model of solar system
    Brahe, Tycho, 233
    Copernicus, 233
    Galileo, Galilei, 233
    Kepler, Johannes, 233, 255
Herschel, Caroline (discovered cluster), 144
Herschel, Sir John (named asteroid), 292
Hertzsprung-Russell [H-R] diagram (Ejnar-Hertzsprung and Henry Norris Russell), 406. See also stars
    absolute magnitude, 406
    giant, supergiant, white dwarf, 406
    luminosity and stellar mass relationship, 406
    main sequence/CNO cycle, 406, 410
    spectral class, 406
    temperature/Kelvin temperature scale, 406, 407
Hevelius, Johannes, astronomer, 45
Hipparchus, Greek astronomer, 67, 70–71, 168, 233
Holmgren Yarn or Wool Test (Alarik F. Holmgren), 12
Hooke, Robert, astronomer, 270
Hubble, Edwin
    galaxy classification, 372, 375–377, 383–384
    tuning fork diagram, 372
Hubble Space Telescope (HST), 11, 269, 379, 390
Hubble's Law (Edwin Hubble and Georges Lamaître), 426–427

International Astronomical Union (IAU, 235
International Meteorite Collectors Association (IMCA), 311
International System of Units, 391
Ishihara Color Test (Shinobu Ishihara), 12–13

January-February binocular highlights, 116
    Andromeda and Cetus myth, 116–117
    Double Cluster, 117
    Perseus, 117
July-August binocular highlights, 131
    Antares, 131, 132
    Ptolemy's Cluster, 133
    Scorpius, 131, 132, 133
Jupiter, 168, 232, 233, 236, 240, 242, 264
    eclipses, occultations, and transits, 271–272
    Galilean satellites, 271
    Great Red Spot (GRS) anticyclone, 270
    Lagrangian points, 293
    North and South Equatorial Belts, 270
    occultations, 272
    poles, 270
    zones and belts, 270

Kelvin temperature scale, 406, 407
Kepler, Johannes, 233, 254
    first planet in a star's habitable zone, 255
    laws of planetary motion
        first—elliptical orbits, 256–257
        second—law of equal areas, 257–258
        third—period of revolution, 255, 258–259
nova, 255

Kuiper Belt/Kuiper Belt objects (KBOs), 235, 286, 293

latitude, 66
lenses
　achromatic (Chester Moore Hall), 45–46
　apochromatic, 45
　Aristophanes (*The Clouds* play), 16
　Barlow, 56
　binocular, 33
　converging, 18
　convex and concave, 16, 17, 44
　Egyptian hieroglyphs (simple lenses), 16
　focal ratio, 45
　magnification, calculating, 19–20
　Nimrud, 16
　objective, 33
　Pliny, 16
　in refracting telescope (Alvin Clark), 44, 45
　types, 16
Liebknecht, Johann, naming Sidus Ludoviciana, 127
light, wave characteristics
　absolute magnitude, 399
　blackbody curves, 398
　diffraction grating, 397
　interference, 397
　luminosity, 399
　spectrometer, 397
　Stefan-Boltzmann law, 398
　Wien's law, 398
light pollution, 173, 175–177
　Astronomical League, 175, 193
　clutter, 174
　of comets, 289
　dark adaption, 175
　full-cutoff light fixtures, 176
　glare, 174
　International Dark-Sky Association (IDA), 176
　LED lights, 176
　light trespass, 176
　nighttime lighting, 175
　rays, 16, 18
　sketching constellations, 180
　skyglow, 174
light-gathering power (LGP), 20
limiting magnitude, estimating, 170, 179–180
Lincoln Near Earth Asteroid Research Program (LINEAR), 289, 295, 296
Lipperhey, Hans, telescope inventor, 44
Little Dipper and Big Dipper, 66
logMAR eye chart, 10
longitude, 66, 325
Lowell Observatory Near-Earth Object Survey (LONEOS), 295, 296
luminosity, 399, 406
lunar calculations, 197. *See also* lunar eclipses; Moon
lunar eclipses, 328
　Danjon scale (Andre Danjon), 330
　future total (projected), 331
　observing, 330
　partial, 328
　penumbral, 328, 329
　photography, 331
　total, 328–329, 330, 331

magnification, 33–34. *See also* binoculars
　eyepiece calculation, 55
　lens calculation, 18–20
magnitude, 70–71. *See also* brightness
　absolute, 169, 406
　apparent vs. absolute, 168–169
　comet, estimating, 289
　limiting, 170
　magnitude difference and brightness ratio, 168
　minor planets, 295
　parsecs, 168–169
　Pogson, Norman R., research, 168
　of solar eclipse, 333
March-April binocular highlights, 121
　binary star, 121
　Capella, 121
　Cigar Galaxy star chart, 125, 126
　Hodierna, Giovanni Battista, 121
　Salt and Pepper Cluster star chart, 121, 122
　zenith, 121
Mars, 232, 236, 239, 254, 264
　albedo Syrtis Major, 268, 269
　clouds, 268-269
　oppositions, 268
　rotation, 268
May-June binocular highlights, 125
　Big Dipper, 125–128
　Bode's galaxy, 125
　Cigar Galaxy, 126
　Mizar's companion, 127
　Ursa Major star charts, 125–127
　Whirlpool Galaxy star chart, 127, 128
Mercury, 232, 236, 239, 248, 255, 258
cusp, 266
　"evening" and "morning" stars, 265
　greatest elongation, 265
　inferior conjunction, 325
　points of contact, 326
　transits/future transits, 325, 327
meridian, 221, 327
Messier (Charles) catalog of interstellar objects, 64, 71, 90, 122, 126, 143, 310, 379
meteor (shower, radiant, trail, fireball), 306, 319
meteorite(s)
　age determination, 417
　buying, 311
　collecting, 309–310
　definition, 306
　display methods, 310
　iron, 308, 311
　stone, 307
　stony-iron, 308
　Widmanstätten pattern, 310
Milky Way, 64, 90, 91, 97, 109, 131, 144
　galaxies in northern sky, 377–378
　luminosity, 406
　planetary nebulae in, 378
　star charts, 82-83, 139, 143
minor bodies/planets, 295. *See also* asteroid(s); comet(s); planet(s)
mirrors
　aluminum coated, 49
　coma, in parabolic mirrors, 48
　concave, convex, and flat, 17, 18, 19

first-surface and second-surface, 17
　focal ratio, 45
　in Newtonian telescope (primary and secondary), 44, 48, 49
　Nimrud lens, 16
　speculum-mirror reflectors, 49
Moon, 232, 239, 244. *See also* craters; lunar eclipses
　aperture mask to observe, 193
　brightness, 168, 169
　crescent, 185
　feature size calculation, 196
　first and last quarters, 185
　full Moon, 184, 185, 192, 289, 328
　in geocentric model, 233
　gibbous, waning and waxing, 185
　lunar month, 184
　lunation, 184
　mare/maria, 186
　new Moon, 184, 332
　observing with naked eye and with telescope, 205, 207
　orbital period, 185
　path of totality, 332
　phases, 184–185
　rays, 186
　rille/rimae, 186
　sketching, 152, 194–195
　synodic period, 185
　terminator, 192, 194
　wrinkle ridge, 186
Moon quadrants
　northeastern, 188
　northwestern, 186–187
　southeastern, 191
　southwestern, 189–190
mounts. *See* telescope mounts

NASA, Wilkinson Microwave Anisotropy Probe, 427
Near-Earth Asteroids (NEAs or NEOs), 292–293
nebula, 64, 153. *See also* planetary nebulae
Neptune, 240, 243, 275
　Transneptunian objects (TNOs), 293
　Triton moon, 275
New General Catalogue (NGC) (John Louis Emil Dreyer), 64, 71, 90, 117
Newton, Isaac, 390
　gravitation work, 254
　Newtonian telescope, 48, 52, 56
　solar system research, 233
　spectrum definition, 390
NGC objects, *New General Catalogue*, 117, 388
night sky, 116, 169, 406. *See also* binocular highlights
non-stellar objects. *See* galaxy; planetary nebulae
North celestial pole (NCP), 65, 66, 67
Norton's Star Atlas and Reference Handbook, 74
nova/novae, 255, 353
November-December binocular highlights, 143–145
nuclear fusion/reactions, 408, 409

Omega Centauri, 154. *See also* globular cluster
open cluster, 64, 117
Ophiuchus Sun sign, 67
optical bench/optical table, 16, 24
optical components
   lenses, 16–17
   mirrors, 17
optics calculations
   focal point, 18
   light-gathering power, 20
   magnification, 19
orbit/orbital period/orbit eccentricities, 254, 255, 256
   aphelion, 257
   apogee, 257
   perigee, 257
   perihelion, 256
Orion and star chart, 78–79, 143, 179

Palomar Observatory (Hale Telescope), 49
parallax, 255
penumbra, 214, 215, 227, 332, 328
perception, definition, 14
peripheral vision/peripheral vision disk, 8, 9
photosphere, 212, 216
Pisgah Astronomical Research Institute (PARI), 403
planet(s). *See also* asteroid(s)
   albedo (Moon, Pluto, Mars) 236, 268
   angular size, 268
   anticyclone, 270
   characteristics, 235
   cloud features, 264
   dwarf, 235
   ellipses, 255
   gas (Jovian) giants, 235, 236, 248, 250
   in geocentric model/epicycles, 233
   Greek *planete*, 232
   habitable zone, 255
   IAU criteria, 235
   inner, 264
   longitude, 325
   minor, magnitude and discoveries by year, 295
   opposition, 232
   orbits, 254–255
   and planetary moons, 235
   rocky, 235, 236, 249
   rotation, 236
   sketching, 152, 250–252
   viewing through telescope, 236
planetary atmospheres, 248–250
planetary nebulae, 378, 381
   Bruce Balick research on shapes, 380
   classifications, 387–388
   de Vaucouleurs, 375–376
   Dumbbell, 379, 380, 381
   fusion, 378
   history (Charles Meissier, William Herschel, William Huggins, Lubos Perek, Lubos Kohoutek), 379
   Hubble, 376, 377
   Hubble Space Telescope, 379
   NGC, 388
   red giant, 378

Ring Nebula, 380
   shapes/classes, 379–380, 381
   stars as, 378
Plato's five geometrical solids, 254
Pluto, 235, 240
Polaris/polar alignment, 53, 65, 92, 103, 170
prism, 390
Pseudoisochromatic Plate Ishihara Compatible (PIPIC) Color Vision Test, 13
Ptolemaic solar system model (Ptolemy), 233

quasars, 396

radiation
   detecting (Geiger counter and Geiger-Müller (G-M) tube, 417
   dosimetry, 417
radioactive (nuclear) elements and decay, 416
   alpha, beta, and gamma rays, 416
   alpha particles, 416 (Ernest Rutherford and Thomas Royds), 416
   dating, 417
   descendants, 417
   DNA, 416
   half-life, 416, 417, 420
   isotopes/uranium, 417, 419
   photon, 416
radioactivity (Antoine Henri Becquerel, Marie Curie, Pierre Curie), 416, 420
   atomic mass, 419
   Avogadro's number, 419
   exponential decay, 419–420
   measuring, 422
   mole, 419
reflecting telescopes, 48
   advantages and disadvantages, 48
   aluminized mirrors, 49
   coma, 48
   European Extremely Large Telescope, 49, 50
   first, construction of (Justus von Leibig), 49
   Giant Magellan Telescope, 49, 50
   Hale Telescope (George Ellery Hale), Palomar Observatory, 49
   Herschel, William, discovery of Uranus, 49
   light pathway, 48, 49
   mirrors (William Parsons) and coating (John Donovan Strong), 48, 49
   Newtonian (Isaac Newton), 44, 48
   reflective surfaces, 48
   Thirty Meter Telescope (TMT), 49, 50
   University of Arizona Steward Observatory Mirror Lab, 49
refracting telescopes, 44
   achromatic lenses, 45–46
   advantages and disadvantages, 45
   apochromatic lenses, 47
   chromatic aberration, 45
   early (Chester Moore Hall, Alvin Clark, Christiaan Huygens, Johannes Hevelius), 45–46
right ascension, 66, 67
Royal Astronomical Society, and Isaac Newton, 48

Sagittarius, 67, 82-83
satellite eclipse, occultation, and transit, 271

satellites, 235
   Callisto, 244, 271
   Deimos, 244
   Dione, 244
   Enceladus, 246
   Europa, 246, 271
   Galilean, 271
   Ganymede, 244, 271
   Io, 246, 271
   Mimas, 244
   Miranda, 246
   Moon, 244
   Phobos, 244
   Rhea, 244
   Titan, 246
   Titania, 245
   Triton, 245
Saturn, 232, 236, 240, 242
   belts and zones, 273
   Cassini Division, 273
   rings and divisions, 273–274
Schmidt-Cassegrain telescope (SCT) (Bernhard Schmidt), 51
Scientific Revolution, 254
SCOPE (Stellar Classification Online Public Exploration), 403
Scorpius constellation and star chart, 67, 70, 71, 82-83, 103, 133
seeing, astronomical definition, 90, 116
September-October binocular highlights, 137
   Altair, 139
   Lyra constellation, 137-138
   Vega, 137
   Wild Duck Cluster/Scutum Star Cloud, 140, 141
Serviss, Garrett P. (*Astronomy with an Opera Glass*), 144
Sirus magnitude, 168
sketching
   asterisms, 156–158
   circles, 150
   color-in, 154
   consellations with light pollution, 180–181
   craters, 195
   dark adaptation, 150
   deep-sky objects (galaxy, nebula, comet), 153–154
   Eberhard Faber Ebony art pencil, 150
   features, 152
   negative image, 155
   Omega Centauri, 154
   red flashlights, 150
   reference point, 153
   Ring Nebula, 153
   smudge/smudging, 153, 155
   solar system objects (Sun, Moon, planets), 152
   techniques, 151–152
   variable stars, 354
Sky Atlas 2000.0 (Wil Tirion), 74
sky brightness. *See also* brightness
   at various Moon phases, 169
Smyth, William Henry, naming Wild Duck Cluster, 110
Snellen eye chart (Herman Snellen), 10
solar cycle, 214

solar eclipses. *See also* Sun; total solar eclipse
   annular/*annulus*, 332, 333
   chromosphere, 212, 213, 216
   faculae, 212
   filaments, 214
   granulation, 212
   hybrid, 333
   limb darkening, 212
   magnitude, 333
   new Moon, 332
   partial, 332, 333
   penumbra and umbra, 332
   photosphere, 212
   plages, 214
   prominence, 214, 216
solar features, 213
solar filters/telescopes, 215–216, 227, 327
solar flares, 214, 216
solar spectrum. *See* spectra
solar system, 232
   geocentric model (Ptolemy), 232–233
   heliocentric model (Copernicus), 233
   Kuiper Belt objects (KBOs), 235, 286, 293
   major and minor bodies, 235
   Oort Cloud, 235, 286
   planetary types, 235–236
   radioactive dating, 417
   retrograde motion, 232
   Transneptunian object, 293
solar telescopes and filters, 215, 216
South celestial pole (SCP), 65, 66, 67
Spacewatch, 295, 296
spectra/spectrum
   absorption (Léon Foucault), 391, 393
   continuous, 391
   electromagnetic, 391 (James Clerk Maxwell)
   emission (Charles Wheatstone), 390, 391
   Fraunhofer lines (Joseph von Fraunhofer), 390
   nanometer to define wavelength, 391
   Newton's definition, 390. *See also* solar spectrum
   quasars (Maarten Schmidt), 396
   redshift, 396
   stellar, 403
   tube, 402
   visible, 395–396
   Wollaston (William Hyde) discovery of lines, 390
spectroscopy/spectroscope
   Kirchhoff's laws, 391
   SCOPE classification, 403
   spectral analysis (Gustav Kirchhoff and Robert Bunsen), 390, 391
spectrograph (Anders Jonas Ångström), 391
spectrometers (Christiaan Huygens), 397, 400
spiral galaxy M100, 11
spring sky, 103–104
   Big Dipper, 103, 104
   Cancer the Crab and Beehive Cluster, 104
   color differences between stars, 104
   star chart, 105
star atlases, 70–71
   Abrams Planetarium *Sky Calendar*, 71, 72, 73, 74
   Atlas of the Stars, 77

Bright Star Atlas (Wil Tirion), 74
Great Atlas of the Sky, 74
Messier objects, 71
New General Catalogue (NGC), 71
Norton's Star Atlas and Reference Handbook, 74
Sky Atlas (Wil Tirion), 74
Uranometria 2000.0, 74
star brightness
   absolute magnitude, 399
   comparison of various stars, 178
star chart binocular highlights. *See* binocular highlights; star charts
star charts, 64, 65, 90, 168
   Abrams Planetarium, 73
   apps, 75
   Atlas of the Stars, using, 77
   autumn sky, 93
   Cassiopeia, 143
   Coma-Virgo cluster, 80–81
   constellation Orion the Hunter, 78–79
   constellation Scorpius, 71
   Double Cluster, 116, 117
   double star, 347, 349, 350, 351
   January, 68
   Leo the Lion, 319
   Lyra (harp) constellation, 137, 138
   Milky Way, 82-83, 139, 143
   Omicron^1 (o^1) Cygni, 351
   Orion, 78–79, 179
   planisphere, 75
   Polaris area magnitudes, 170
   Ptolemy's Cluster, 131, 133
   scale, 65
   Scorpius, 71, 82–83
   software, 75
   spring sky, 105
   summer sky, 111
   Summer Triangle, 139
   Ursa Major, 125, 126, 127
   variable stars, 355–358
   Vega, 137
   Virgo the Maiden, 80–81
   Whirlpool Galaxy, 127, 128
   Wild Duck Cluster, 139, 140
   winter sky, 99
star clusters, sketching, 153
star color differences, 104
star maps. *See* star charts
stars, 212, 232, 406. *See also* double stars; Sun; variable stars
   absolute magnitude, 399
   association, 117
   circumpolar, 327
   in geocentric model, 233
   GNO cycle, 410
   giant and supergiant, 406
   gravity and radiation, 409
   Hertzsprung-Russell diagram, 406
   life cycle, 406
   main sequence, 406, 410
   nova, 255
   nuclear fusion, 408, 409
   open cluster, 117
   photosphere, 406
   as planetary nebulae, 378

   planets in habitable zone, 255
   spectral class, 406
   temperature, 406
stellar system and evolution, 232, 254, 378. *See also* spectra
Strong, John Donovan, physicist, 49
summer sky, 109–110
   Arcturus, 109
   Big Dipper, 109
   Hercules Cluster, 110
   Milky Way, 109
   Northern Cross, 109
   star chart, 111
   Summer Triangle, 109
Sun, 213, 232, 235, 239. *See also* eclipses; solar eclipses; solar features; total solar eclipse
   analemma, 212
   brightness, 168
   corona during total eclipse, 332, 333, 335
   diurnal changes, 218
   eclipse, 332
   ecliptic, 67, 265
   in geocentric model, 233
   greatest elongation of Mercury or Venus, 265
   in heliocentric model, 233
   maximum altitude/transit, 221
   Moon rotation around, 185
   path from Earth, 67
   photosphere, 212
   rotation days, 139
   sketching, 152
   spectrum, 391
sunspots, 212, 215
   cycle, 214
   McIntosh classification system, 227
   penumbra, 214, 215, 227
   Schwabe, Heinrich, astronomer, 214
   umbra, 214, 215, 227

telescopes. *See also* Hubble; Kepler; refracting telescopes; reflecting telescopes
   for asteroid observation, 294
   astronomical (Kepler), 254
   catadioptric/compound, 44, 51
   with CCD cameras, 353–354
   Celestron, 53
   color filters, 264
   coma, 48
   for comet viewing, 289
   first, by Hans Lipperhey, 44
   focusers, 7
   Galileo, 44, 168, 254
   with GPS (global positioning system) electronics, 54
   large/giant, and atlases for, 49–50, 74
   lens, 3, 7, 45, 46
   mirrors, 48
   Newtonian, 48, 52, 56
   polar alignment, 53
   Schmidt-Cassegrain (SCT), 51
   solar filters, 215, 216
telescope accessories
   Barlow lenses, 56
   eyepieces, 55

finder scopes, 56
focusers, 57
star diagonals, 56
storage cases, 57
telescope mounts
   alt-azimuth, 52
   Dobsonian (John Dobson), 52
   driven alt-azimuth, 53
   equatorial, manual and motorized, 53–54
   go-to, 53–54, 265
total solar eclipses, 332, 336. *See also* solar eclipses
   Bailey's beads (Francis Baily), 333, 335
   bright stars, 334
   chromosphere, 333, 335
   corona, 333
   diamond ring, 333
   future projected, 334
   partial phases, 333
   path of totality, 332
   planets, 333, 223
   shadow bands, 333, 334
transit(s), 221, 324, 327
   calculating distances (Sir Edmund Halley), 327, 338
   Jupiter, 271
   longitude, 325
   Mercury, 324–325
   nodes/longitudes, 325, 328, 332
   points of contact, 326
   Venus, 325, 338
   viewing, 326
Transneptunian objects (TNOs), 293
transparency measurement of sky, 90, 116
tripod mounts, 33, 37-38
twinkling (scintillation), 117

umbra, 214, 215, 227, 328, 330, 332
unresolved, definition, 117
uranium, 417, 419–420
Uranometria atlas (Johann Bayer), 64
Uranus, 236, 240, 243, 275
Ursa Major (Great Bear), 109, 126, 127

variable stars, 71, 353
   discoveries by David Fabricus, Geminiano Montanari, John Goodricke, Benjamin Apthorp Gould, 353
   intrinsic and extrinsic, 354
   light curve, 354, 369
   Mira, first discovery, 353
   naming conventions (Friedrich Wilhelm August Argelander), 353
   star charts, 355–358
Vega star, 137
Venus, 168, 232, 236, 239, 255, 258, 264
   atmosphere discovery (Mikhail Vasilyevich Lomonosov), 325–326
   daytime observations, 266
   Galileo and Cassini observations, 266
   inferior conjunction, 325
   nodes/longitudes, 325
   orbit around sun, 267
   points of contact, 326
   transits/future transits, 325, 327
vernal equinox, 67
Virgo the Maiden star chart, 80–81
virtual image, 18
visual acuity, 10–11
von Leibig, Justus, chemist, 49

wavelength
   characteristics of light, 397–389
   measuring, 390, 391, 395
winter sky, 97–98
   Orion, 97–98
   Pleiades, 98
   Sirius, 98
   star chart, 99
   Winter Triangle, 98

Yerkes Observatory telescope (by Alvin Clark), 46

zenith, 121, 170, 221
zodiac in astrology, 67